A CULTURAL HISTORY
OF THE HUMAN BODY

VOLUME 3

A Cultural History of the Human Body

General Editors: Linda Kalof and William Bynum

Volume 1

A Cultural History of the Human Body in Antiquity
Edited by Daniel H. Garrison

Volume 2

A Cultural History of the Human Body in the Medieval Age
Edited by Linda Kalof

Volume 3

A Cultural History of the Human Body in the Renaissance
Edited by Linda Kalof and William Bynum

Volume 4

A Cultural History of the Human Body in the Age of Enlightenment
Edited by Carole Reeves

Volume 5

A Cultural History of the Human Body in the Age of Empire
Edited by Michael Sappol and Stephen P. Rice

Volume 6

A Cultural History of the Human Body in the Modern Age
Edited by Ivan Crozier

A CULTURAL HISTORY
OF THE HUMAN BODY

IN THE
RENAISSANCE

Edited by Linda Kalof and William Bynum

Bloomsbury Academic
An imprint of Bloomsbury Publishing Plc

B L O O M S B U R Y
LONDON · OXFORD · NEW YORK · NEW DELHI · SYDNEY

Bloomsbury Academic

An imprint of Bloomsbury Publishing Plc

50 Bedford Square
London
WC1B 3DP
UK

1385 Broadway
New York
NY 10018
USA

www.bloomsbury.com

Hardback edition first published in 2010 by Berg Publishers, an imprint of
Bloomsbury Academic
Paperback edition first published by Bloomsbury Academic 2014
Reprinted by Bloomsbury Academic 2015

British Library Cataloguing-in-Publication Data
A catalogue record for this book is available from the British Library.

ISBN: HB: 978-1-84788-790-0
PB: 978-1-4725-5464-2
HB Set: 978-1-84520-495-2
PB Set: 978-1-4725-5468-0

Series: The Cultural Histories Series

Library of Congress Cataloging-in-Publication Data
A catalog record for this book is available from the Library of Congress.

Typeset by Apex CoVantage, LLC, Madison, WI, USA

CONTENTS

ILLUSTRATIONS

CHAPTER 6

CHAPTER 7

CHAPTER 8

CHAPTER 9

CHAPTER 10

SERIES PREFACE

A Cultural History of the Human Body is a six-volume series reviewing the changing cultural construction of the human body throughout history. Each volume follows the same basic structure and begins with an outline account of the human body in the period under consideration. Next, specialists examine major aspects of the human body under seven key headings: birth/death, health/disease, sex, medical knowledge/technology, popular beliefs, beauty/concepts of the ideal, marked bodies of gender/race/class, marked bodies of the bestial/divine, cultural representations and self and society. Thus, readers can choose a synchronic or a diachronic approach to the material—a single volume can be read to obtain a thorough knowledge of the body in a given period, or one of the seven themes can be followed through time by reading the relevant chapters of all six volumes, thus providing a thematic understanding of changes and developments over the long term. The six volumes divide the history of the body as follows:

Volume 1: A Cultural History of the Human Body in Antiquity (750 B.C.E.–1000 C.E.)

Volume 2: A Cultural History of the Human Body in the Medieval Age (500–1500)

Volume 3: A Cultural History of the Human Body in the Renaissance (1400–1650)

Volume 4: A Cultural History of the Human Body in the Age of Enlightenment (1650–1800)

Volume 5: A Cultural History of the Human Body in the Age of Empire (1800–1920)

Volume 6: A Cultural History of the Human Body in the Modern Age (1920–21st Century)

General Editors, Linda Kalof and William Bynum

Introduction

WILLIAM BYNUM

Historical periodizations have real significance, even if they are sometimes fuzzy around the edges. The words used to describe them can vary according to the academic discipline that is being practiced. The "Renaissance" in the title of this volume spans roughly two and a half centuries, from 1400 to 1650. Thus, it deals with the period traditionally known as the Renaissance, but also with what is generally called the "Early Modern Period." It roughly coincides with the epoch historians of science would dub the scientific revolution. It includes what historians of religion call the Reformation and what historians of geography would think of as the Age of Exploration. In addition, some of the chapters in this volume reach back, or look forward, to tie up topics the contributors seek to develop.

These temporal shifts are almost inevitable, as time is seamless and historical periodizations are generally post hoc constructions. People have always been aware that they are living in a particular time, but no "pre-Socratic" philosopher knew Socrates was just around the corner. "Enlightenment" was a term that was contemporary with the ideals that shaped it, but the Enlightenment occurred at different times in different places, and with varying meanings. The Bulgarian Enlightenment, for instance, happened a century after it was experienced in Western Europe. All historical labels are culture bound: English-speaking scholars know what is meant by "Victorian," but to use that word to describe nineteenth-century values in Russia or Italy strains meaning. In any case, the world is much larger than we often imagine, and even the ubiquitous word "Renaissance" is geographically bound, and has meaning to a Japanese or Tibetan scholar only if he or she is thinking about Europe. There was an Enlightenment in North America, and there were even Victorian values in that region of the world, but there was no Renaissance. The ordinary historical terms we routinely use still leave most of the world untouched.

The chapters in this volume thus deal with themes circumscribed both chronologically and geographically: they center on Europeans and the perceptions of Europeans. They have been written by scholars from a variety of disciplines and specializations, which is entirely appropriate for a volume of cultural history. They all naturally touch on "bodies" (all history that deals with people does, willy-nilly: our bodies are all we have, in the end); but they also all reflect in varying degrees five important developments that mark out the Renaissance as a particular epoch in Western history. At the risk of stating the obvious, we shall deal with these general themes, before teasing out some of the important insights of each individual chapter. These generalities may be grouped as follows: print culture, religious turmoil, the expanding world, the classical inheritance, and the beginnings of modern science and medicine.

The first general innovation that characterizes the Renaissance was the rise of print culture. Long ago, Francis Bacon (1561–1626) identified the printing press (along with the compass and gunpowder) as one of the technological developments that had made his own world what it was. The history of the book is now an academic subject in its own right, but the impact of printing is a subject no historian of the Renaissance and early modern period can neglect, regardless of her or his primary focus. Print culture made literacy more vital, and it helped create the common scholarly culture that can be discerned to emerge during our period. It changed the nature of academic debate and disputation, and although it was of course one of those top-down developments, print came to change the lives of ordinary people as well.

The classic modern study of the role of the printing press in social change is that of Elizabeth Eisenstein: her monumental volumes, published in 1979 and partially inspired by the media studies of Marshall McLuhan, examined print as an instrument of social change. Her arguments about the communication of knowledge, the growth of literacy, and the commerce of books reminded historians how fundamental the printing press was in the formation of European culture. More recently, the equally important work of Adrian Johns has challenged Eisenstein's basic framework. Johns argues that print culture was not simply a given after Gutenberg's innovation of the 1430s. Rather, everything that had to do with printing, publishing, selling, and trusting books had to be *made*, in a much more gradual and piecemeal manner than Eisenstein's model presupposed. Johns offers rich case studies of the care scholars took to ensure that their books contained what they really wanted to say. Often, they would travel long distances, and stay for months, to supervise the printing and production of their treasured works. He also points out that the absence of copyright meant readers had to beware that what they purchased was what the stated author had actually written. Print culture did not simply spring onto the world in 1439; it was gradually made over a century or more.[1]

As both Eisenstein and Johns recognized, print culture was not simply about books; it also took in posters, pamphlets, tracts, and many other forms of communication. It allowed authors to write about bodies, and it also allowed them to illustrate them. Woodcuts and engraving techniques were used in Europe even before Gutenberg's moveable type introduced printing, and it was an obvious step to combine the two. Anatomy books routinely did just this. Vesalius's great work of 1543 used illustrations to great effect, and it is probably the first book in which the illustrations are actually more important than the text. The history of Renaissance anatomy is inseparable from the history of Renaissance publishing, as anatomists sought to present their discoveries in ever more striking ways. Monographs by Sawday, Roberts, and others explore this strand to science, medicine, and publishing during the period.[2]

Renaissance artists also contributed to the field, and although artistic anatomy concentrated on the musculoskeletal system and what was later called the anatomy of expression, Renaissance depictions of the human body, male and female, meant that artists kept abreast of, and sometimes contributed to, discoveries in anatomy. Leonardo da Vinci is probably the best-known example of this fusion of art and science, although his immediate impact in the area was limited, because much of his private work remained unknown until much later. Had he used the printing press to disseminate the insights he recorded in his notebooks, both the history of art and the history of anatomy, science, and technology might have been different.

The novelty of print culture thus runs through the period and permeates the chapters in this volume. So does a second theme, the importance of religion, and especially, the Protestant Reformation. The Reformation, more powerful in northern Europe than in the Mediterranean lands, catalyzed the cultural fragmentation of Europe. The nature of specific religious confessions began to influence what was acceptable, and what was not, and in a roundabout way to reinforce the nationalism that was beginning to emerge in the period.

Protestantism put a premium on literacy, because it emphasized each person's relationship with God.[3] Being able to read the Bible was seen as a duty for all good Christians. This in turn created a market for religious books, which was also reflected in more general printing and publishing. At the same time, the polarization of the various confessions created tensions that often resulted in persecutions and wars. Most of the European wars of the period were ostensibly fought in the name of religion: we might wish to analyze them in political terms, but the rhetoric of the time suggests that religion was frequently the language in which politics expressed itself. Henry VIII of England may have used religious observance as the excuse to do what he wished to do, but both Henry and his chief advisers were acutely aware of the power religion played in both international and national affairs. The Reformation may have started as an internal issue within the Catholic Church, but it soon became much more. Whatever the confession, however, Europe remained overwhelmingly Christian during this period, and religion profoundly influenced the way most people perceived the world.

Religion also influenced what people could and could not say or publish. The Counter-Reformation had its own power, and the Catholic Church reacted to the threats to its autonomy in ways that were often firm, even brutal, but hardly surprising. The Index—the list of books Catholics were forbidden to read—was first created in 1529, and a number of books, some even by devout Catholics, appeared on the Index. Most famously, Galileo's *Three Discourses*, which presented the balance of probability in favor of the Copernican worldview, earned its author the condemnation of the Church authorities, and Galileo spent his last years under house arrest.[4] Others were not so lucky, and heretics were routinely burned at the stake or otherwise disposed of. Giordano Bruno was burned in 1600 for his philosophical and astronomical heresies.[5] The Roman Catholic Church had no monopoly on such acts of brutality, of course: Michael Servetus was burned at the stake in Geneva by John Calvin for his doubts about the Trinity, and Henry VIII dispatched more than one of his enemies, including Sir Thomas More, one of the most gifted men in England.

The religion of the period was thus often extreme and violent. At a crude level, one could argue that souls were placed at a much higher premium than bodies: if your soul was tainted, I could do what I liked with your body. Bodies mattered too, and one of the major divergences between Catholicism and Protestantism concerned the most sacred body of all: that of Jesus. At the Council of Trent, first convened on 1545 and lasting almost two decades, the Roman Catholic Church set out its stall in the face of the rising tide of Protestantism. One key affirmation was that, during the Eucharist, the host literally becomes the body of Christ. Protestants interpreted the ritual symbolically, a stance that was easier to square with the scientific innovations of the period.

A third persistent influence on mentality during the Renaissance was the expanding geographical world. The discovery of the New World in the late fifteenth century was the most striking example, but Europeans also became much more familiar with parts of Africa and the Far East. The exploitation of these newfound lands began almost immediately; colonization was more selective but is of course a feature of the early European presence in both North and South America. The disease toll of sub-Saharan Africa was a formidable obstacle to early European penetration, but trading posts in the Malay Archipelago and other areas in the Far East were established, and India, like China, fascinated Europeans. India felt the thrust of European ambitions earlier than China, but both countries presented the remarkable realty of ancient, literate cultures. They were of course not newly known to Europeans, but they became part of the heady mix of the expanding world that so dominates some of the art and literature of the period.[6]

The impact of all this on the consciousness of Europeans was aided by print and illustration, and the confrontation with new physiognomies and exotic cultural practices was influential for thoughtful Europeans. Michel Montaigne, whose reflections on his world are invoked in several chapters in this volume, is a particularly striking instance of someone whose perceptions of the world were challenged by his realization that there was a world elsewhere. John Donne's poems also articulated the liberating, as well as disturbing, implications of a world (and universe) expanded by the compass, sail, and telescope. The world, it now seemed, was filled with a vast variety of bodies of differing shapes, sizes, and colors. Equally significantly, Europeans discovered that customs, conventions, morals, and beliefs also came in many shades. The anthropology of the period, based on travelers' tales and reports from sea captains, explorers, and merchants, contained a mix of wonder, awe, revulsion, and fantasy; but it widened perceptions of what bodies might be and do.[7]

Exploration forced Europeans of the period to look outward; but for all that was new, people in the Renaissance were also uncovering and revering the old. The classical tradition inevitably looms large in any work on the Renaissance, but much that was new in art, architecture, design, philosophy, and literature was inspired by ancient practitioners of these and other domains. If Aristotle had been the overarching presence in the Middle Ages, he was at least partially dethroned by his own teacher, Plato, during the Renaissance. Neoplatonism became one of the most powerful intellectual forces during the early Renaissance, pervasive from the Italy of Marsilio Ficino (1433–1499) to the England of Ralph Cudworth (1617–1688). A range of themes, such as the Great Chain of Being, the correspondence between the macrocosm and the microcosm, and the reality of astral forces, were elaborated within the context of ancient wisdom, of which Plato was only one authority. The Hermetic tradition, first explored in detail by Francis Yates, provided an undercurrent throughout the period in philosophy, science, and even religion.[8]

More soberly, perhaps, the steady discovery and editing of yet more ancient texts provided meat and drink for scholars and was a major force in the humanistic movement. People in the Renaissance realized how rich was the classical tradition, and our own appreciation of it rests on foundations established in the period. Many of the chapters in this volume quite appropriately look backward, as major figures sought to express their own modernity in ways that were deeply and explicitly colored by their readings of their favorite classical authors.

At the same time, the centuries covered in this volume also coincide with what historians of science and medicine call the "scientific revolution," and the achievements of the period in fields as diverse as anatomy and astronomy, challenged people's attitudes both to the classical past, and to the world they found around themselves. What came to be called "the Battle of the Books"— about whether the ancients or the moderns know more—was a debate about science and medicine, not other areas of human endeavor. By the early seventeenth century, it had been answered pretty conclusively in favor of the moderns, even if there were still a few individuals around who still believed Galen knew best about medicine, or Aristotle and Ptolemy about physics and astronomy. The tide was rapidly turning, however, and the progressive nature of scientific knowledge was being accepted as a comfortable fact of modernity.[9]

The science most relevant to perceptions of the body was medicine. Its practice was and largely still is as much a matter of art as of science, even if we hope that art is infused with scientific insights. There are some paradoxes within medicine, however. First, although Galen was still a force within academic medicine, even in 1650, toward the end of the period covered in this volume, Hippocrates (who now merely stands for a group of writers) began to be the most revered medical authority of antiquity. Thomas Sydenham (1624–1689) is not known as the "English Hippocrates" for nothing. The humors, elaborated by the Hippocratics but put into their transmitted form by Galen, continued to dominate explanations of health and disease at the bedside. But Galen began to be seen increasingly as representing the dogmatic within medicine and Hippocratics as the open, empirical element within the science and practice of medicine. The rise of Hippocratism, in contrast to Galenism, still drew from ancient inspirations, but it was seen as a progressive, modernizing trend within medicine.[10]

Second, new theories and new experiments began to change the way doctors understood both the structure and the functions of the body. Anatomy remained the queen of the medical sciences, and although Vesalius hardly arose out of a vacuum, his anatomical research inspired others to explore not only the surface but also the interior of the body. The organs became better known, and new ones were discovered. As always, those seeking to explain what the organs did reached for the known, familiar items in their world. Humoral physiology continued to be a favorite framework (Vesalius's explanations of function were still largely Galenic, despite his attacks on Galen's anatomy), but other models also surfaced. Followers of Paracelsus favored notions of the body as an elaborate chemical alembic, generally with a variety of occult forces thrown in. With the rise of mechanical philosophy, and the invention of elaborate clocks with automata to strike the hours, mechanical explanations gained purchase. The culmination within our period was Descartes' hypothesis that animals are simply complicated machines, and humans are complicated machines with added souls. The one physiological discovery of the period that stands out in hindsight was William Harvey's revelation of the circulation of the blood, published in 1628. Descartes reinterpreted Harvey for his own mechanical ends, and the notion that the heart is simply a pump, designed to take venous blood and send it to the lungs, and then take the blood it received from the lungs and send that to the rest of the body, was easy to place within the context of the mechanical philosophy. That was one of its consequences, but it misreads Harvey's aims and intentions. Of all the major figures in the scientific revolution, Harvey was probably the most conservative, deeply influenced by Aristotle and innovative only because his experiments led him to his conclusions.[11]

The impact of these developments on ideas of health and disease, on medical practice, and on popular notions of how to preserve health and avoid disease, are explored in several of the chapters to follow. As the chapters explicate various aspects of the cultural history of bodies during this period, these background themes surface in many ways. Each of the broader topics covered here could easily be the subject of a separate monograph, so rich are the possibilities. Nevertheless, the chapters' sharper foci open up its specific theme, and the full notes guide readers to the extensive literature in the field. The Renaissance in this volume is primarily European; England, France, and Italy are most densely analyzed, although northern Europe and the wider perceptions of Europeans are also included. We are born, live for a time and die. Or, as T. S. Eliot put it,

> Birth, and copulation, and death,
> That's all the facts when you come to brass tacks.[12]

This is a stark way of summarizing human (and animal) life, and few people, either in the Renaissance or today, would be quite so brutally blunt. Nevertheless, birth and death are fundamental events in all human cultures, and Lianne McTavish's decision to view both primarily through the birthing chamber produces a powerful impression. Both maternal and infant mortality were relatively high during the period, but a number of legal, gender, social, and medical issues were invoked in the birth bed, especially one in which the child was feared dead, or was born dead or very sickly. The baptism of the child, so important in the Roman Catholic Church, colored attitudes toward the dead or dying infant and could influence the way difficult deliveries were conducted. Protestant confessions tended to be more unambiguous about saving the life of the mother, even if it meant the death of her infant. In practice, in many of these difficult deliveries, both mother and child perished. Physicians or surgeons for the most part were called in only in cases of urgency; as a result, the medical literature probably exaggerated the dangers of childbirth. Midwives or neighbors generally tended uncomplicated deliveries, and these would pass unnoticed in the literature. Toward the end of our period, a few male doctors began to handle routine deliveries, although the male midwife did not become common until the eighteenth century. Even then, however, parturition was still overwhelmingly women's business.[13]

McTavish considers several other problematic aspects of childbirth, including postmortem birth, the much-debated topic of Cesarean section, and supposedly miraculous births after death. She also examines the different attitudes toward death in Catholic and several Protestant confessions, as well as highlighting the perennial fear of many people that they would be buried before they were actually dead. The possibility of awakening in a coffin seems to have terrified many individuals, before and after the Renaissance. Given the nature of medical care and deathbeds, it probably occurred more frequently in earlier periods than it does today, but its rarity has never been correlated with the anxiety it has created.

Anxiety of another sort was central to concern about who fell ill, and why. Margaret Healy contrasts the more public persona of bubonic plague and the private suffering caused by the stone, to dissect the moral dimensions of health and disease in our period. Plague has attracted much historical attention, partly because of its profound economic, social, and societal impact, and more recently, because certain characteristics of the epidemics are hard to explain on the assumption that the plague epidemics that swept Europe between the 1340s and the mid-seventeenth century were caused by the bacterium of modern bubonic plague, *Yersinia pestis*. That speculative and sometimes heated debate is happily beyond the remit of this volume, as people then experienced what they called the plague or pest. The historical consequences of these periodic epidemics can be analyzed without worrying what the disease really was.[14] People at the time gave these scourges both moral and medical meanings, and they saw no conflict in interpreting them as God's punishment for wicked humanity, but at the same time seeking medical help, or submitting (though not always passively) to the stringent series of control measures imposed by European rulers. As the old advice to soldiers put it, "Praise the Lord but keep your powder dry."

Epidemics like the plague invited interpretations that embodied collective sin, guilt, laziness, or social failings, but these collective meanings still left unanswered the altogether trickier question of why some but not all were affected during such communal events. Individual afflictions, such as the all-too-frequent stone, raised other issues. Renal and bladder stones were much more common in earlier periods than they are now. Even today, they are exquisitely painful and frightening. In the Renaissance, when bladder and kidney infections were poorly understood and inadequately treated, and when dietary consumables were largely unregulated, with the consequence that metals and other impurities could be innocently consumed, stones were a real risk for anyone. Healy looks at several sensitive and articulate sufferers from the stone, including Erasmus, Samuel Pepys, and Montaigne. Unsurprisingly, all three men appear in other chapters: their values and the insights they offer through correspondence, diaries, essays, and books have helped to define how we understand their historical epochs.

Just as there were pamphlets, treatises, printed regulations, and other literature dealing with plague, so sufferers of the stone could look to an extensive medical literature offering advice about how to cope with this affliction. Both conditions could wrack the body in pain, and both could be fatal. Samuel Pepys forever celebrated the day when he was successfully operated on for his bladder stone, in much the same way we celebrate our birthdays or Christmas. (St. Valentine's Day was another of Pepys' red-letter days, given his roving eye.) All of Healy's examples survived their episodes of stone, and each had a reasonable life span, but each of them knew firsthand about the pain of their affliction and regulated their lives according to the dictates of medical opinion in the period. Healy's choice of plague and stone reminds us that bodies in the past suffered much pain, and had far less effective treatments for it. Pain has been described as the greatest of evils, and psychological tolerance for it has diminished as its potential for control has increased.[15]

Erasmus also makes an appearance in Katherine Crawford's exploration of sexuality during the period. Her chapter ranges widely, reminding us of the continued shadow of classical authors such as Plato and Ovid in Renaissance writings about sexuality, but also of the power of the Church, both Catholic and Protestant, in laying down rules supposed to govern the sexual conduct of men and women. Like most writers on sexuality nowadays, Crawford finds Michel Foucault's analysis inadequate, and she concentrates instead on the rediscovery, during the Renaissance, of classical authors, even if the readings were modified by the dominant Christian ideology of the period. The value that Catholics place on celibacy for monks and nuns was reevaluated within the Protestant tradition, although most confessions continued to place a premium on faithfulness within marriage. Most writers followed Aristotle in viewing the sexual anatomy of men and women as complementary, with men the norm, of course. Humoral explanations of menstruation, lactation, and ejaculation continued to provide the main paradigm, and Renaissance fascination with "marvels" meant hermaphrodites received a good deal of attention.

Both doctors and theologians wrote about so-called unnatural sexual practices, such as homosexuality and bestiality, and many commentators felt the need to assimilate news of the exotic sexual mores, physiques, and genitalia of the Aboriginal peoples sailors and explorers brought back from the East and the West. The natives of North and South America figured prominently in many accounts of humans and their nature. At the same time, the nature of love and sexuality was central to many aspects of the strong Neoplatonic tradition that flourished in Italy but had many disciples elsewhere. The courtly tradition of love found expression in many major works of literature and reinforced the centrality of European values in trying to make sense of, and to regulate, passion and sexuality.

The role of print looms large in Susan Broomhall's evaluation of medical knowledge of the body during the Renaissance. Print culture changed the nature of medical education, and it also facilitated the dissemination of knowledge of newly discovered diseases, such as syphilis. Even while the writings of doctors like Paracelsus, who challenged ancient medical knowledge and theories, were being published, translations of newly discovered texts from Galen and other classical authors offered traditional physicians a greater insight into their heritage. Vernacular works devoted to popular health and disease prevention increased access to medical ideas among the literate. At the same time, as recent scholarship has demonstrated, the printing press was never completely controlled, so academic medicine had to compete in the marketplace with the ideas and practices of healers of all stripes. It created what Broomhall calls the "leaky corpus," in which access to medical knowledge was no longer the preserve of only a few.

Karen Raber examines some of the consequences of this new state of affairs in her analysis of popular beliefs about the body during the Renaissance. Using a variety of printed sources, she argues convincingly that the period was one of competing models of how the body works, as traditional classical (largely humoral) explanations were integrated (rather than simply replaced) by a series of alternative models. Many of these were influenced by the new science, but their realization in popular culture was varied. Raber considers the five senses, and how new theories of sight, hearing, taste, touch, and smell were translated into notions of how the body works and why it can fall ill. She draws on a wide range of literary sources and relates these to medical writings that discussed the nature of the human senses.

The senses are also satiated with food and drink, and in "Belly Lore," Raber shows how important Galenic ideas about digestion and coction were in the Renaissance reflections on the dinner table. Gluttony was, as always, easily attacked by moralists, and temperance has ever been a medical virtue. It still is today, as is our appreciation of the evolutionary basis of what earlier generations called "healing by nature," or the *vis medicatrix naturae*. A modern monograph tracing this notion through history would be extremely illuminating; Raber provides some suggestive hints about the cultural context of this idea during our period. She also looks at another persistent theme in popular culture, the anxiety provoked by dead bodies, and the framework within which these emotions worked themselves out in burial practices.

Weddings, not funerals, are more apposite to Mary Rogers's fine analysis of theories of beauty in Renaissance art and culture. Like so many other chapters in this volume, hers looks back to classical authorities. Aristotle, Cicero, and Vitruvius, as well as the achievements of (mostly anonymous) ancient architects, artists, and sculptors, were scrutinized by those in the Renaissance seeking to understand and create beauty. There were, by the Renaissance, many styles of expression, and most commentators recognized that beauty was actually a matter of taste. Nevertheless, certain standards needed to be maintained, and certain conventions found widespread adoption. The basic principles of physiognomy, for instance, that the face was the window of the soul, influenced the way artists and writers depicted human beings. Proportion was likewise fundamental, in representing both male and female figures; although dress was of course a gendered affair, it was also discussed for both males and females. Differences between masculinity and femininity were also significant, and reflected in paintings, in aesthetic theorizing, and in etiquette books, which had a great vogue during the period.

Rogers concentrates in the Italian Renaissance and on the brilliant high art it produced. Patrizia Bettella stays in Renaissance Italy, but concentrates on the other end of the spectrum of humanity: on "the other." Prostitutes and courtesans, peasants, black Africans, Muslims, and American Indians all join the parade of her characters. So do the aged, the odd one out, as it happens to everyone who lives long enough, and the literature on the old, and the pleasures and pains of old age, probably deserves separate treatment. Bettella's chapter makes instructive reading alongside Rogers's, for both draw on the same historical (and in many ways, timeless within Western culture) modes of representation. Physiognomy features in both chapters, and so do dress codes. These two contributions address a single problem: How does one portray beauty, on the one hand, and ugliness and grotesquery on the other? White and black: us and them. These are still powerful sentiments, not entirely eradicated by modern liberalism. Bettella makes excellent use of a wide range of historical genres, from travel narratives to imaginative literature. But the fact that the underlying conventions in each chapter are consonant should not surprise; in both cases, the speakers are the intelligent, literate male speaking about the two ends of the spectrum. Of course, their values are carried through.

The marked bodies of Diane Purkiss's chapter remind us that the Renaissance was a period of genuinely creative renewal, but also the age of the massive witch hunts that also define the epoch. Witches were of course defined not only by the putative actions, but also by signs on their bodies: moles, extra teats, or areas of the body lacking ordinary feeling, and therefore insensitive to pinpricks. She uses Shakespeare's most famous witches—those in *Macbeth*—to make the powerful juxtaposition of the world of witches with that of saints' relics. Modern scholarship has alerted us to the strong undercurrents of magical beliefs in many areas in Renaissance thought and activity, including science and medicine. Many such occult beliefs still persist today.

Purkiss's linking of the witches of *Macbeth* with the religious literature and practice of relics provides a powerful insight into the underlying commonalities of the sacred and profane. Just as the Black Mass used the sacred to blaspheme, so many of the actions and powers attributed to witches drew

on the wider beliefs of the veneration and power of the relics that were treasured in many of the churches of Europe. Drawing on both anthropological
perspectives and literary sources, such as Thomas Malory, John Donne (who
wrote a poem entitled "The Relic"), and Shakespeare, Purkiss shows how
widespread were beliefs in the potency of body parts. The ultimate power
of body parts was embodied in the Eucharist, of course, the interpretation
of which divided Catholics and Protestants, and the latter cast aspersions on
Catholic veneration of saints' relics. At the same time, both Catholics and
Protestants attributed to witches the capacity to effect dark magic through
their acquisition of a part or possession of their victim. This belief structure
surfaces in Shakespeare's creation of his witches in *Macbeth*, but suggestions
that he was a crypto-Catholic are not convincing.

On a more mundane level, but with equally sharp social consequences, body
parts featured in the stereotyping of other marginal groups during the period.
Thus, Jews were depicted with hooked noses and old women with sagging
breasts: these and many other bodily characteristics were fused in Renaissance
literary conventions as well in their realization in social attitudes, legislation,
religious doctrine, and outright persecution.

Margaret Healy's second chapter, "Fashioning Civil Bodies," revisits a number of individuals and themes that earlier chapters have explored. She draws
on a wonderful array of authors and genres to dissect how manners and civility, as so brilliantly explicated by Norbert Elias, were expostulated through
the many popular volumes on etiquette and deportment, poems and books
describing rules of health maintenance, and imaginative literature, in which
characters exhibiting praiseworthy characteristics were presented in a good
light. The reverse was also true, of course, and as ever, the perceived good was
contrasted with the perceived less good: proper Christian, European behavior,
for example, thrown into relief by the barbarism of American natives, or the
dark attributes of evil people at home.

Fashionable bodies also sometimes contracted loathsome diseases, most tellingly in our period, syphilis. The new disease (historians still debate whether
it was entirely new in late fifteenth-century Europe) became very widespread
among all classes and social groups. Its visible manifestations and insalubri-
ous mode of spread made it an easy disease to moralize about. The traditional
joke about its naming—the French Pox, the Neapolitan Disease, the Polish
Pox, etc.—hides a deeper cultural fear about the affliction. Ironically, given the
social and moral opprobrium attached to the disease, it was given its modern
name and assigned a dignified poetical and classical pedigree by one of the
major medical theorists of the period, Girolamo Fracastoro (1478–1553).[16]

We are all, of course, embodied in our bodies. What that "we" (or, more
precisely, "I," as in Descartes' "ego") is has been given many interpretations, and is probably less settled in our postmodern period than it was in the

Renaissance. But it has always been significant if one were female or male, young or old, rich or poor, powerful or powerless. Margaret King looks at four contrasting pairs of bodies with radically different forms, social standing, behaviors, and expectations: young and old; prostitutes and Jews; mothers and saints; and warriors and courtiers. Her examples are well chosen, for they allow her to explore a wide variety of meditative, imaginative, legal, and medical writings. She begins with a salutary reminder for our age of obsession about pedophilia: that Renaissance paintings frequently depict young children nude, and that in the Madonna and Child genre, the infant Jesus is often not only shown naked, but with his genitalia rather prominently displayed.

King's eloquent evocation of the bodies of her four pairings makes wonderfully clear how life trajectories can differ so much. Her geographical center of gravity is Italy, but she has salient things to say about matters in northern Europe as well. She ends with an examination of Baldassare Castiglione's *Book of the Courtier*, published in 1528, but composed more than a decade earlier, and describing four evenings on the hilltop palace at Urbino. There a group of men and women discussed the ideal attributes of the courtier as the ultimate gentleman, and how he relates to women. The roles of class, gender, and social status that were there elaborated reverberated for more than three centuries within European culture.

That culture has now passed away, but, as the chapters in this volume subtly show, men and women in the Renaissance were grappling with what it means to be human. They lived in their bodies as we live in ours, although the meanings they assigned to this mysterious phenomenon were different from ours. That should not surprise us: truth is ever born anew in its own time, and in its own country.

CHAPTER ONE

Birth and Death in Early Modern Europe

LIANNE McTAVISH

Figure 1.1, a German woodcut from 1475, portrays a woman lying prone on a bed within a compressed domestic space. Her belly has been crudely slashed open, and a midwife lifts a fully formed child from it. Demonic creatures flank the scene. On the left, a grimacing monster removes the recumbent woman's soul—depicted as a tiny human figure—from her mouth. On the right, a goat-like animal with swollen teats stands ready to receive and possibly feed the newborn. This violent image represents the birth of the Antichrist, an evil figure whose life story was narrated in opposition to that of Christ during the early modern period.[1] In the New Testament, Christ is born without injuring his mother. She remains a virgin; her body eternally closed. In contrast, the Antichrist brutally ruptures the body of his mother, killing her in the process. The woodcut highlights this splitting of the maternal body while also perverting the visual conventions employed to depict the lying-in chamber. Many early modern images of childbirth feature the newly delivered woman resting comfortably on a bed while nurses care for her infant. In the *Birth of Saint Edmund*, 1433, reproduced in chapter 10 of this volume (Figure 10.7), a nurse warms the swaddled child by the fire, her back turned to the midwife and female helpers who serve Saint Edmund's mother. Birth is shown as a productive event fostered by the bonds of female sociality. In the slightly later woodcut, however, birth and death coincide in a frightening way that disrupts human relationships.

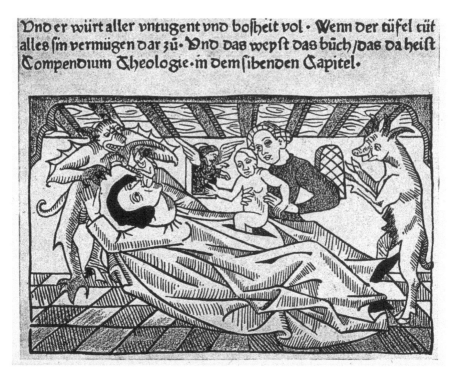

FIGURE 1.1: *Birth of the Antichrist*, 1475. Woodcut, German. Wellcome V0014917.

The German woodcut provides a useful introduction to this chapter because it presents birth and death together. The two categories are often placed
in opposition to one another, with birth marking the beginning and death the
end of a life. Scholars who research the early modern period tend to analyze
birth and death separately, specializing in one or the other. They explore the
shifting political, cultural, and medical significance of childbirth by drawing
on a range of sources, including medical case books, women's diaries, legal
records, and published texts. Among other things, they examine how female midwives were educated, perceptions of the act of giving birth, and the
ways in which the lying-in chamber could be rife with discontent and female
sociability.[2] Researchers who focus on death likewise approach it as a historically and culturally specific practice, not simply a biological fact. They
utilize an equally diverse array of primary sources to discover the fears, religious ceremonies, and burial practices associated with death.[3] When birth
and death are analyzed together, scholars position them at different ends of
the early modern life cycle, as rites of passage that were experienced according to a person's gender, social status, and age.[4]

This chapter is informed by such scholarship but avoids discussing birth
and death as spatially and temporally discrete categories of knowledge and
experience. My analysis centers on those early modern situations when birth

and death overlapped or were not readily distinguishable. Sometimes death could even precede birth, or else coincide with it, as in the woodcut depicting the birth of the Antichrist. By emphasizing these cases, I examine the ways in which birth and death were continually defined and redefined in relation to one another. My approach positions birth and death as flexible categories, reflecting on why their uncertainty both fascinated and terrified early modern people. Furthermore, the cases I have selected reveal the shifting boundaries of early modern bodies. They are literally unthinkable within a modern epistemology that regards bodies as simultaneously predictable machines, sites of pleasure, and something in need of control through diet and exercise. In the discussion that follows, early modern bodies appear as fluid entities that were unpredictable, fantastical, and ultimately unknowable.

The first case study considers the efforts made by medical men, women, and laypeople to determine whether or not the contents of a woman's womb were alive or dead. Male medical practitioners were usually called to assist in the early modern birthing room after a woman had labored unfruitfully for several days and was nearing the point of death. These men would then decide if a living child could be born or if it was dead and should be removed with surgical instruments. The second case focuses on the Cesarean operation, performed almost exclusively after the death of an undelivered woman during the early modern period. In these situations, the pregnant woman's friends, relatives, and medical practitioners had to judge whether the woman was dead and if her unborn child might still be living and thus worthy of the sacrament of baptism.[5] As indicated in the German woodcut, the Cesarean operation was linked with unnatural birth during the early modern period and could easily produce a grisly scene instead of a living child. The importance of baptism returns in the third case study, which involves the birth of stillborn children. The birth of a dead child was particularly upsetting if the child had died without being baptized, a Christian ritual that allowed its entry into heaven. Early modern parents prayed for the miraculous revival of these children, so that they might receive the sacrament before returning to a state of death. The fourth and fifth cases concern anxieties about the dead bodies of adults. Though autopsies designed to reveal the cause of death had been practiced since at least the Middle Ages, anatomical dissections were undertaken with greater frequency during the early modern period. Various fears accompanied the methodical dismemberment and display of the human body, among them the possibility of opening a living body. Stories about dissected bodies suddenly showing signs of life circulated during the early modern period, suggesting that this situation may actually have occurred. Early modern tales of bodies buried alive were potentially even more horrifying. The fifth case study contemplates the worries about what could happen when bodies that merely appeared to be dead returned to life after burial.

All of these cases include the possibility of mistaken identity, when living bodies could be declared dead and dead bodies considered alive. In every example, birth (or rebirth) and death occur in close proximity, producing both miraculous and terrifying results. The cases are additionally bound together by their emphasis on visual analysis. Many of the signs of birth, death, and life were observed by early modern participants, and their looking took place within realms infused with meaningful visual exchanges, such as the birthing room or anatomical theater. These visual elements are accented in the discussion that follows, in keeping with my primary training in art history. At the same time, my analysis engages with a number of scholarly disciplines, including literary studies, anthropology, women's studies, and the history of medicine. Much of my primary evidence stems from a selection of medical publications that address both popular and scholarly beliefs, two strata that can scarcely be distinguished during the early modern period. Given my own research on the visual culture of early modern France, many of these primary documents are French, but I include references to other parts of Europe.

DEATH IN THE BIRTHING ROOM

The rate of maternal mortality during the early modern period has long been of interest to historians. Determining how many women died in childbed has proven difficult because few reliable records survive before the eighteenth century. Some scholars speculate that maternal mortality was higher than fifteen percent of the total births because laboring women were assisted by untrained female midwives working in unhygienic conditions.[6] Yet archival research by such historians as Doreen Evenden, Hilary Marland, and Nina Rattner Gelbart, among others, shows that female midwives in early modern Europe were often literate and capable of handling emergencies.[7] Quantitative research by Irvine Loudon and Roger Schofield indicates that women in Great Britain, continental Europe, and Scandinavia died in childbed at relatively low rates, though many of their records date from the late eighteenth and nineteenth centuries.[8] Adrian Wilson's study of man-midwifery in seventeenth- and eighteenth-century England confirms, however, that only one or two percent of labors were complicated, corresponding with a similarly low rate of maternal mortality.[9] Other scholars argue that the number of deaths increased during the eighteenth century, when more women were delivered in lying-in hospitals, served by male midwives who spread infection with their forceps. Lisa Forman Cody's careful research counters such claims, indicating that for the most part the death rates in London hospitals did not surpass those of women giving birth elsewhere, and that skilled female midwives, not men, performed hospital deliveries.[10] Despite lacking clear evidence for the fifteenth and sixteenth centuries, it seems that the number of women who died in childbed was low throughout the early modern period. All the same, surviving diaries and letters reveal that some early modern women feared suffering injury or death during childbirth and made plans to safeguard their families' welfare.[11]

Scholars may have overestimated maternal mortality rates because childbirth is often portrayed as dangerous in the obstetrical treatises written by male physicians and surgeons. According to Jacques Gélis, hundreds of these treatises were published and exchanged between countries in multiple editions and translations in early modern Europe.[12] Typically covering all aspects of childbirth, from theories of conception to the diagnosis of pregnancy, labor and postpartum ailments, the books were primarily designed to display the knowledge of their respective authors.[13] Before the middle of the eighteenth century, obstetrical treatises featured difficult births, accompanied by detailed case studies in which authors portrayed as heroic their efforts to save the lives of laboring women.[14] In his obstetrical treatise of 1695, French surgeon man-midwife François Mauriceau described being called on February 25, 1671, to assist a thirty-five-year-old woman whose water had broken eight days before she went into labor.[15] Convinced that the large child lodged in her vagina was dead because it had not moved for two days, this woman and her relatives begged Mauriceau to remove it using any means necessary. The surgeon man-midwife refused to apply his instruments, hoping that the woman would give birth naturally to a living child. His cautious approach was vindicated when the child was born two hours later, an outcome drastically different from a case occurring on November 17, 1670, when Mauriceau used his *crochet* [hook] to pull a dead child from a sixteen-year-old woman.[16] Mauriceau and other male authors narrated many instances in which children died while their mothers barely survived, arguing that skilled male midwives should be immediately summoned to the birthing room because female midwives were incapable of remedying these situations.

Mauriceau explained that lack of movement was not a sure indication of an unborn child's death. In an earlier obstetrical treatise published in 1668, he outlined the means by which to distinguish a dead child from a living one when it remained in the womb and could not be seen.[17] Echoing most of the signs of death provided by the German physician Eucharius Rösslin in 1513, Mauriceau argued that a dead child could produce fetid discharges from the woman's body, causing her pain.[18] It could fall like a heavy ball from side to side when she moved. If a surgeon inserted his hand into the womb, the child would feel cold and lack a pulse. In contrast, a living child would be lively, feel warm, and have a strong pulse. Other signs were visible rather than tactile. A pregnant woman harboring a dead child would have flaccid breasts, a gray complexion, and a languid expression. Mauriceau affirmed, however, that even when all of this evidence was present, a medical practitioner should

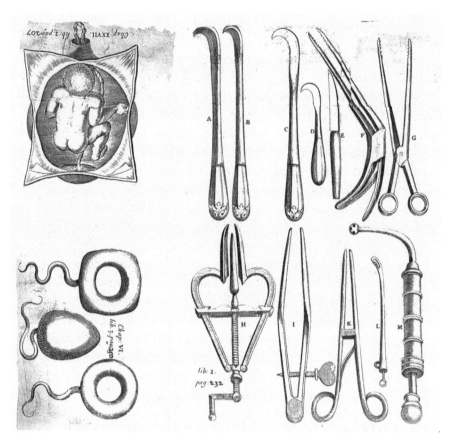

FIGURE 1.2: Pessaries, fetus, obstetrical instruments. *Crochets* (top middle) and baptismal syringe (bottom right), engraving from François Mauriceau's *The Diseases of Women with Child, and in Child-bed*, English trans. (London: A. Bell, 1697). Wellcome L0014388.

err on the side of timidity rather than temerity, treating the child as if it were alive.

Mauriceau's circumspection was encouraged by the ritual space in which the act of birth took place. When a woman went into labor during the early modern period, her husband summoned her female midwife, friends, and relatives to a warm room, known as a lying-in chamber, especially prepared for the birth.[19] Men were usually excluded from this space, but husbands and male relatives could be present, especially if the delivery did not progress as expected.[20] When male midwives entered the lying-in room, they worked in front of an audience that evaluated their interventions and could blame them if the laboring woman died.[21] Male midwives might even cast aspersions on each other. Mauriceau accused rival surgeon man-midwife Philippe Peu of being a dangerous murderer who thoughtlessly used *crochets* on living children. Peu insisted that the unborn children described in his obstetrical treatise of 1694 were already dead, and that it was Mauriceau who in fact killed children with his *tire-tête*, a head-puller invented to remove impacted fetuses from the womb.[22] This heated dispute indicates how crucial it was for male midwives to determine the status of an unborn child. Displaying a mutilated living child to the spectators within the lying-in room would ruin a male midwife's reputation, and producing a child that was whole rather than in pieces could enhance it.

Mauriceau and other medical practitioners defined the signs of death and life in opposition to each other. Death was linked with coldness, heaviness, and grayness, and life was associated with heat, movement, and a colorful hue. The two states remained difficult to distinguish within the lying-in chamber, and efforts to discern between them were linked with far more than maternal mortality during the early modern period. Because male midwives were striving to increase their activity in the potentially lucrative realm of childbirth, they needed to be associated with live births rather than exclusively with death. According to Adrian Wilson, once English men-midwives began using forceps to remove living rather than dead children from suffering women, the men not only received praise but were also subsequently called more quickly to assist in the birthing room.[23] Early modern concerns with the differences between life and death in the lying-in room were informed by the desire to save lives, but also by the goals of enhancing social identity and attaining economic reward.

POSTMORTEM BIRTH

In his obstetrical treatise of 1596, Italian physician Scipione Girolamo Mercurio defied mainstream belief by arguing that women could survive a Cesarean operation.[24] Several woodcuts in his book illustrate the operation in progress. In Figure 1.3, a weak, partially nude woman with downcast eyes is supported by male assistants. A well-dressed physician shown brandishing a scalpel approaches her from the left, apparently to realize the long, vertical incision marked upon her abdomen. Mercurio's conviction that opening the female body could produce living children without killing the laboring woman was not based on firsthand knowledge, for he had had little if any experience in the birthing room. In contrast to many other parts of Europe, male medical practitioners never established their presence in Italian lying-in rooms during the early modern period, a difference Nadia Maria Filippini attributes to the dominance of the Catholic Church in Italy.[25] Mercurio was primarily influenced by the arguments made in 1581 by François Rousset, a French physician claiming to have witnessed the successful performance of Cesarean sections.[26]

Before the eighteenth century, most medical practitioners affirmed that the Cesarean operation was unnatural and horrifying, in keeping with the image of the birth of the Antichrist introduced at the beginning of the chapter. In 1609, the French surgeon Jacques Guillemeau maintained that his disgust with the

FIGURE 1.3: Cesarean operation. Woodcut from Scipione Girolamo Mercurio's *La Commare o riccoglitrice* (Venice: C. Cioti, 1601). Wellcome L0006090.

procedure was based on having attempted it and seen it attempted by others in a few desperate cases, all without success.[27] Guillemeau's warnings against the procedure were echoed in the treatises written by French surgeon men-midwives, including Mauriceau in 1668, Cosme Viardel in 1671, Peu in 1694, and Pierre Dionis in 1714, as well as by English writers such as Percivall Willughby in the seventeenth century.[28] According to the vast majority of authors, a Cesarean section should not be contemplated until after the pregnant woman had taken her last breath, in the hopes of baptizing a living child.

The delivery of an infant via an incision in its mother's abdomen became known as a Cesarean birth because of the Roman *Lex Caesarea*, a law specifying that a fetus had to be removed and buried separately from the body of a dead pregnant woman.[29] In early modern Europe, the Christian practice focused more on baptizing an unborn child if it showed any signs of life. The first church official to advocate performing Cesareans on dead women was Odon de Sully, a Parisian archbishop who died in the early thirteenth century.[30] Renate Blumenfeld-Kosinski argues that male surgeons gained early entry to the birthing room primarily by practicing postmortem Cesareans. Other scholars disagree. According to Mireille Laget, few postmortem Cesareans were attempted because the men were summoned too late to contemplate removing a living child.[31] Male medical practitioners furthermore feared opening the body of a woman who was still alive, although she appeared to be dead, producing a hideous spectacle for the audience in the lying-in room, and receiving blame for her death.[32]

Many obstetrical treatises nevertheless include information about how to undertake a postmortem Cesarean. In his sixteenth-century German publication, Rösslin vaguely advised:

> Furthermore if it happens that the mother is dead which one can recognize through signs of a dead person / and if there is still a hope that the baby lives / Then you should hold the woman's mouth / uterus / and genitals open / so that the baby has air and breath / as women usually know. Then you should cut open the dead woman along the length of the left side with a razor / for the left side is freer and emptier than the right side/ this is because the liver lies in the right side.[33]

Rösslin did not elaborate on "the signs of a dead person," but later writers, including Pierre Franco, an itinerant surgeon working in France during the sixteenth century, provided more detail. Drawing on an earlier publication by Paré, Franco argued that practitioners should not begin a Cesarean section until the pregnant woman's movements had ceased entirely, and she had "taken her last breath."[34] This emphasis on the last breath is repeated in many early modern publications, indicating both the time-sensitive nature of the operation and that an absence of breathing was considered a crucial sign of death.

Rösslin likewise refers to breathing in his brief account of the Cesarean operation, though he recommends opening the mouth and genitals of the dead woman's body so that the unborn child can breathe. In later treatises, male midwives continued to offer this advice, without reinforcing the long-standing belief in a continuous link between a woman's mouth and uterus. As late as 1714, Dionis advised putting a gag in the dead woman's mouth to keep it open, but claimed that the infant did not breathe in its mother's belly because there was in fact no communication between the mouth and uterus. He explained that the female birthing assistants would cast blame on the surgeon if he forgot to undertake this ritual.[35] Advice about the Cesarean operation was informed not simply by medical or theological concerns but also by efforts to please the predominantly female audience gathered in the lying-in chamber.

By the middle of the eighteenth century, a greater number of medical writers insisted that Cesareans could be undertaken on living women, claiming that these women might survive despite the continuing dangers of the operation. In his publication of 1752, the famous British obstetrician William Smellie argued that when the laboring woman's pelvis was particularly narrow or distorted, a Cesarean could save the life of the child and possibly even that of the mother.[36] If the pregnant woman was strong and "of a good habit of body," Smellie encouraged the practitioner to serve her some sustaining broths or cordials before proceeding with the operation. If, however, she was weak, exhausted, or bleeding, he should wait until after her death before making a longitudinal incision along the left side of her body, where the liver did not extend as far. Smellie affirmed that the medical man working on a living body should divide the muscles and cut through the peritoneum and uterus, pushing the intestines back inside if they started to protrude. After manually removing the child and any coagulated blood from the uterus, the operator should suture the abdominal opening and ensure that the woman get plenty of rest.

This last example indicates that birth practices gradually changed during the early modern period, albeit at different rates throughout Europe. In the case of the postmortem Cesarean section, the birth of a child usually followed the death of its mother. Anxiety arose about how to ensure that the pregnant woman was in fact dead, without hesitating too long and endangering her unborn child's chance to be baptized. In keeping with cases involving the uncertain status of an unborn child, distinguishing between a living and dead maternal body involved making decisions in relation to particular circumstances. Factors to consider included the soul of the unborn child, the suitable performance of long-standing cultural practices within the birthing room, and the reactions of the parturient woman's family.

MIRACULOUS BIRTH AFTER DEATH

In contrast to the low rate of death in childbed, infant mortality remained high during the early modern period. The precise number differed according to economic and climactic conditions, but children regularly perished from malnutrition, accidents, and various diseases. Jean-Louis Flandrin calculates that as many as 300 per 1,000 children died in northern France during the seventeenth century.[37] According to Arthur Imhof, the frequency of death in childhood produced historically specific understandings of mortality. Using a demographic approach to study the distribution of deaths before 1875 in Dorotheenstadt, a parish in Berlin, he finds that half of all deaths occurred before the age of eight, and the remainder were dispersed evenly between nine and ninety years of age.[38] Imhof argues that if an early modern person had survived childhood, death was behind him, not looming in front as it is for Europeans today. Early modern adults continued to fear their eventual death, but for the most part they could prepare for it, having survived the risk of a sudden, early demise.

The high rate of infant death meant that many parents would experience the loss of a child, and women would give birth to more children than could be expected to survive. In 1960, Philippe Ariès claimed that early modern parents were not emotionally attached to their frail children; the romantic concept of childhood as a crucial phase of development was invented by the upper classes during the seventeenth and eighteenth centuries.[39] His important study focused attention on the family as a historically diverse entity, suggesting that such categories as "child," "mother," and "love" were culturally produced rather than part of an unchanging human nature. Even so, Ariès's thesis has been harshly critiqued for the past forty years as historians increasingly rely on in-depth studies of particular regions of Europe, rather than on sweeping demographic methods. Scholars now analyze letters, diaries, and children's books, arguing that parents experienced intense grief after the death of a child, and they attended carefully to the training of their living offspring.[40] The degree of continuity between past and present understandings of childhood nevertheless remains subject to debate.

Much surviving evidence indicates that parents suffered when a pregnancy and labor produced a dead or stillborn child. According to Gélis, early modern Europeans could resign themselves to the physical death of a child, but they were tormented by thoughts of its spiritual death; an unbaptized child was forbidden burial in consecrated ground and would remain forever in a state of limbo between heaven and hell.[41] Hoping for a miracle, relatives might take the child to a holy place and lay its body before a sacred image of the Virgin Mary while praying for its temporary resurrection and subsequent baptism. Gélis's study reveals that this practice was not uncommon in rural parts of northeastern France, and in Belgium, Austria, and Switzerland, from the sixteenth through the eighteenth centuries. Between 1569 and 1593 in Faverney in Haute-Saône there were, for example, 459 registered cases of children baptized after their brief return to life.[42] Though religious authorities were suspicious of the ritual, the Roman Catholic Church did not attempt to suppress it until 1729. After that the number of sanctuaries diminished, but some remained active into the twentieth century.

Accounts of these miraculous resurrections were recorded by the curés of various sanctuaries, providing a vivid picture of the ritual. Those caring for the child's body frequently traveled long distances by foot to the sanctuary, carrying corpses described as rotten, stinking, stiff, and black, details affirming their morbidity.[43] Once at the sanctuary, the parents or relatives watched the body carefully for any signs of life justifying its baptism. These signs were remarkably consistent: a rosy hue swept over the child's face, sweat appeared on parts of its body, traces of blood became visible at its nostrils, it opened and closed its mouth, or it moved its arms ever so slightly. In keeping with previous discussions, death was linked with stiffness and coldness, while life was associated with color, moisture, blood, and movement.

Despite displaying signs of life, the miraculously revived child was usually baptized conditionally, with the declaration "if you are alive, I baptize you in the name of the Father, the Son, and the Holy Spirit."[44] This cautionary phrase was meant to forestall the baptism of a dead body that merely appeared to be alive. Baptizing a body that was in the process of dying was, however, considered crucial. Within the Catholic tradition, removing the taint of original sin was so important that laypeople who had themselves been baptized could perform the ritual in emergency situations. In some parts of Europe, female midwives were authorized to baptize an endangered child, while male midwives might baptize children in utero, using the curved syringe pictured in the bottom right corner of an engraving in Mauriceau's obstetrical treatise (Figure 1.2). Mauriceau's rival, Philippe Peu, claimed that when members of the "so-called reformed religion" attempted to prevent him from undertaking such baptisms, he did so secretly, using a syringe he had hidden in his pocket.[45] Protestants did not reject baptism, but they held a range of opinions about its meaning. According to David Cressy, early members of the Church of England debated when, how, and by whom the ritual should be conducted, wondering if it was rightly administered to children rather than adults, or if it should be exclusively performed by clergy.[46] Baptism remained an important but ambivalent practice for Protestants throughout the early modern period.

ACCIDENTAL VIVISECTION

Whereas baptism was valid only when performed on living bodies, anatomical dissections were reserved for the dead. Though never banned by the church, the dismemberment of the human body conflicted with early modern burial practices, and people were not eager to subject themselves or their relatives to the anatomist's knife.[47] Those surgeons and physicians who wished to conduct anatomical dissections had difficulty acquiring a sufficient number of corpses. They occasionally received the bodies of criminals who had been executed and were condemned to suffer continuing humiliation by being publicly dissected.[48] Yet the number of criminal bodies was inadequate, and even the famous sixteenth-century Italian anatomist Andreas Vesalius reported disinterring bodies to use their bones as study material. Some early modern mourners kept watch over the graves of newly deceased relatives, waiting until their bodies were too decayed to be of use to anatomy students.[49]

Although Katharine Park has shown that in certain cases opening the human body was linked with respect rather than shame, early modern anatomists needed to distance themselves from the corporeal punishment meted out by executioners.[50] To raise their social status, an increasing number of anatomists published treatises replete with written descriptions of different parts of the body and vivid illustrations displaying human bodies undergoing dissection. Around 1521, the Italian physician Berengario da Carpi, for example, refashioned the fourteenth-century anatomical treatise of Mondino de'Liuzzi, adding numerous woodcuts to it. The woodcut shown in Figure 1.4 shows a male figure standing in a nondescript landscape while peeling back his own viscera to reveal the muscles underneath. According to Jonathan Sawday, the image depicts a living body that actively invites viewers to look inside its abdomen, forestalling their feelings of guilt at having transgressed a social taboo.[51] Rather than offering spectators a violent scene of dismemberment, this reassuring image of a compliant corpse could have made dissection seem more acceptable. At the same time, the woodcut's conflation of life and death provides an appropriate picture of anatomy, for dissections involved examining dead bodies in order to learn more about living ones.[52]

Confusing the boundaries between life and death could also produce an appalling spectacle, especially when a dissected body began to sigh or move. Living pigs, dogs, and other animals had been surgically opened since ancient times, often with the goal of examining their beating hearts.[53] Though the vivisection of the human body was forbidden, rumors of its practice circulated throughout the early modern period. In 1611, the French anatomist Jean Riolan claimed, for example, that King Louis XI had permitted the vivisection of a condemned soldier in 1474 so that doctors could learn about bladder stones.[54] Other stories portray bodies that were accidentally opened when they were still alive. The sixteenth-century surgeon Paré asserted that an unnamed famous anatomist (likely Vesalius) had delivered the second razor blow to the body of a woman declared dead from "the suffocation of the womb" when she began to move, much to the astonishment of witnesses.[55] The English antiquary John Stow similarly recorded that on February 20, 1587, a "strange

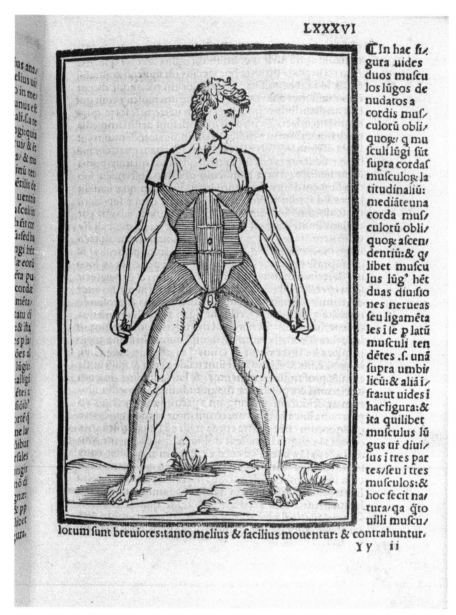

FIGURE 1.4: Male anatomical figure holding his trunk open. Woodcut from Berengario da Carpi's *Commentaria* (Bologna: Colophon, 1521). Wellcome L0033532.

thing happened" after the body of a hanged felon was cut down, stripped, and then transported to the Surgeon's Hall in London for dissection, "the chest being there opened, and the weather extreeme cold hee was founde to be alive, and lived till the three and twentie of Februarie, and then died."[56] Related stories of criminals who were revived after their execution were long-standing, dating from the Middle Ages. William Cragh was hung at Swansea in 1260 and deemed dead by at least nine eyewitnesses because his body was seen "lying like a dead man," without breath, its face black, and its tongue protruding.[57] The criminal's subsequent recovery was attributed to the miraculous intervention of the Bishop Thomas de Cantilupe, but it raised persistent questions about the means used to declare bodies truly dead.

The determination of death remained ambiguous throughout the early modern period. Various authors argued that even when a body lacked heat, movement, and respiration, and did not respond to stimuli, it might still be alive.[58] According to Christian tradition, death occurred when the immaterial soul left the material body, a process that was both immediate and invisible. Attempts to represent the moment of death portray the soul as a small human figure departing from the body as angels or devils battle to claim it, a convention informing the fourteenth-century image of the birth of the Antichrist discussed at the beginning of the chapter.[59] Popular opinion held, however, that a soul could linger near its former bodily abode for quite some time, particularly if an individual had met a sudden, violent end.[60] Recently dead bodies had a liminal status, existing between life and death.

This intermediary understanding of the dead body was encouraged by the Catholic doctrine of purgatory, which became entrenched during the twelfth century. Purgatory was an "other world" in which souls destined for heaven went to be purified by having their venial or minor sins "burned" out of them.[61] As a particular state between death and the Last Judgment, purgatory differed from limbo, which was occupied by unbaptized infants, and hell, where the damned suffered eternal punishment. Purgatory was often imagined as an actual place, in no small part because of Dante Alighieri's *Purgatorio*, a section of his early fourteenth-century *Divine Comedy* in which famous Italians do penance on an island-mountain near Jerusalem.[62] The Catholic Church advocated praying for the dead, holding masses for them, and donating money in the form of indulgences in order to lessen their time in purgatory. Relatives and friends of the dead were obligated to engage in these practices, especially immediately after a death; prayers were considered most effective when said in the presence of the corpse. Attempts to help relatives in purgatory could also continue for many years after their deaths, establishing a long-lasting bond between the living and the dead.[63]

Protestant reformers of the sixteenth century rejected the doctrine of purgatory, arguing that it was not based in scripture and encouraged the papal distribution of indulgences. Martin Luther refuted the doctrine of purgatory around 1530, and later claimed that the dead slept unconsciously.[64] Undermining belief in purgatory entailed severing the relationship between the living and the dead and lessening the need for elaborate funerals and displays of mourning.[65] In theory, the rejection of purgatory encouraged Protestants to view the corpse as soulless flesh detached from the previously living person, a view that could have made anatomical dissection seem more acceptable. In practice, beliefs were slow to change, and Protestants advocated remembering and praising the dead, allowing corpses to retain some human identity.[66]

LIVE BURIAL

Sometimes corpses maintained a hold on the living because they were apparently still alive while inside their coffins. According to the man-midwife Willughby, in 1650 Mrs. Emme Toplace died in labor and was immediately buried by her abusive husband. This haste alarmed several neighborhood women, as did the report that "as shee was carried to the grave, some thought, that they heard a rumbling in the coffin. A noise was heard like the breaking of a bladder, after which followed a noisome smell."[67] Suspecting that Mrs. Toplace might not be dead, one witness laid her ear to the grave, thinking she discerned the sighs of a dying person, while a nearby soldier affirmed that he additionally heard a crying child. The disinterred coffin revealed the disheveled body of Toplace and that of a newborn child, its placenta still attached. These contents were inspected by curious passersby, some of whom insisted that the bodies of women who died suddenly in childbirth should display signs of putrefaction before their burial.

This tale of premature burial and "coffin birth" provides a paradigmatic example of the indistinct boundaries between life and death during the early modern period. Fears about being buried alive increased during the eighteenth and nineteenth centuries, even as new techniques were devised to revive those who merely seemed dead. In 1788, the English physician Charles Kite proposed applying electric shocks to resuscitate the bodies of victims of drowning and other mishaps.[68] The wealthy increasingly took precautions to ensure that their bodies would be truly lifeless before burial. French society woman Suzanne Necker published *Des inhumations précipitées* [Premature burials] in 1790, and instructed her family to place her body in a lead coffin for three months before submerging it in a vat of alcohol.[69] Other devices were meant to allow an "undead" corpse to alert the living of its presence. As late as 1905 William Tebb published images of a "safety coffin" featuring a flag that would wave if a supposedly dead body began moving, but there is little indication that this or similar inventions were ever used.[70] For the most part, the designs provide evidence of a long-standing and serious concern about dying beneath rather than above the ground.

Premature burial was particularly unsettling because those subjected to it, such as Emme Toplace, were unable to achieve a "good death." Throughout the early modern period, instruction manuals provided advice about the *ars moriendi*, or art of dying. The fifteenth-century *Book of the Craft of Dying* focused on preparing for death, something denied to those who died suddenly or violently.[71] In this context, death was not portrayed as the opposite of life, as it was in many medical treatises, but rather as a potentially lengthy process that could be managed and performed properly.[72] Donald Duclow argues that death was a "public drama" in which a dying person was ideally in bed at home, surrounded by family, friends, physicians, and clergy, as well as by demons, angels, and the Trinity.[73] The dying figure was encouraged to avoid clinging to life or worldly property by accepting death willingly and even gladly.[74] The final moment of life before death was especially important, depicted in the *ars moriendi* as a battle in which repentance remained possible even for the worst sinners. For Protestants a person's mental state at death could not determine salvation, but it could provide information about whether or not he or she was among the chosen.[75]

The rules for dying well differed according to gender. In her analysis of seventeenth-century diaries, Lucinda Becker argues that the physical suffering of dying women was not highlighted in deathbed accounts. Instead, good women died quietly, passively, and with decorum, their eyes lifted toward heaven. At the moment of her death in the sixteenth century, Katherine Stubbs, for example, focused on resisting temptation and sin rather than on the husband and children she was leaving behind.[76] While women were not supposed to resist their final illnesses in a masculine fashion, men were encouraged to face death heroically, "with physical endurance in a noble fight against the temptation to sin."[77] Figure 1.5, a woodcut from 1493 shows a man who has just died well. He reclines on a bed after having received the last rites, indicated by the priest holding a taper. His soul—depicted in the act of praying while gazing toward heaven—is separated from his body and whisked away by an angel. The physicality of the man's lifeless body is emphasized by his exposed torso, yet the rest of him is discretely covered, and his arms are modestly crossed. The man's wife or another female relative prays at the side of the newly dead body, while a woman standing in a doorway covers her mouth and nose with a cloth. Her gesture may connote mourning, but it might also refer to the putrid smell of dead flesh, in keeping with images of the miraculous raising of Lazarus from the dead in which spectators cover their noses to muffle the smell.[78]

This female character invokes possible reactions to dead and dying bodies, a topic of interest to historians of the early modern period. In her study of the corpse in Paris from 1550 to 1670, Vanessa Harding explores what actually happened to dead bodies in the crowded city. She finds that not all bodies were treated alike; whereas the poor had little control over the disposition of their bodies, and often ended up in mass graves, the corpse of a wealthy person might be preserved by embalming, protected in a lead coffin, and buried in a sacred location specified before death.[79] During the early modern period, all bodies were usually interred within three days of their death, in close proximity to the living. These bodies were removed from cities to distant cemeteries during the eighteenth century, a displacement that scholars identify with the

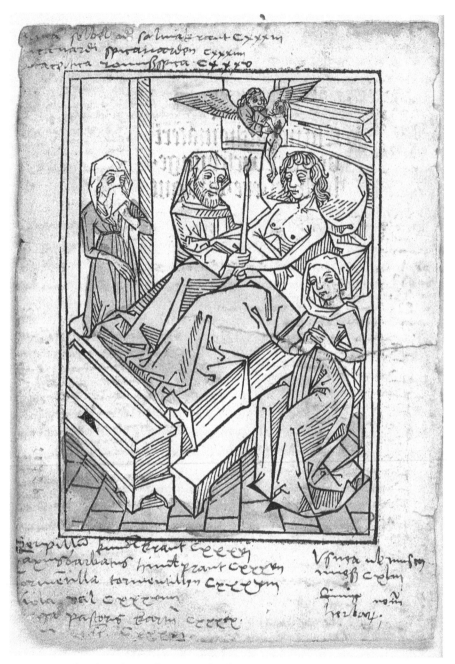

FIGURE 1.5: A priest gives a dying man his last rites, 1493. Woodcut with watercolor, German. Wellcome V0042055.

growing separation between the living and the dead, and the development of a culture that would rather deny the presence of death than meditate upon it.[80]

Elaborate tombs were sometimes constructed for the most wealthy and powerful members of society, including popes and kings. Pope Julius II began to plan his tomb long before his death in 1513, commissioning Michelangelo to sculpt some forty life-sized marble figures to surround a massive freestanding structure destined for St. Peter's Basilica. The much simplified but still grandiose result is now positioned against a wall in San Pietro in Vincoli in Rome, Julius II's family church.[81] When English and French kings died, the royal organs considered noble or even sacred were removed from their bodies, and dispersed in several significant tombs. In 1547 the will of the dead French king Francis I stipulated that his heart, brain, and entrails be taken to the abbey of Notre-Dame-des-Hautes-Bruyeres, while the rest of his body remain at the cathedral of Saint-Denis, where many other French monarchs were buried.[82] These examples reveal that the dead bodies of particular individuals could both retain their identity and be considered worth revering. The distribution of body parts could also be disquieting, however, for during the Last Judgment the soul was supposed to be reunited with its material body. The precise nature of this resurrected body was subject to theological debate throughout the Middle Ages and into the early modern period: would amputees, for example, have their limbs reattached or not?[83] The importance of the physical body diminished in Protestant burials, with the revised Edwardian prayer book of 1552 speaking of the corpse in the third rather than second person.[84] Yet tombs remained authoritative among the wealthy in England during the late sixteenth century, supplanting the decaying body with permanent sculptures that positioned the deceased within the aristocratic ranks of honor.[85]

CONCLUSION

All five cases discussed in this chapter feature the uncertainty of the body. It could be unclear whether a child in the womb was alive, the body of a pregnant woman was dead, a child born dead had begun to move, a corpse on display in the anatomy theatre had started to sigh, or bodies interred beneath the ground were crying out. Death, life, and birth could become entangled, defying the expected chronology of events. A child might be born from death, as in the case of a Cesarean operation; the corpse of a stillborn child might temporarily be endowed with life; and an executed criminal might be partially dissected and then taken to his bed in the hopes of recovery; a dead woman might give birth in her coffin. Even as birth and death were opposed to one another in medical treatises, they could overlap when the body hovered in between the two states. Death could also be considered a lengthy process begun at birth, rather than a unique event at the end of life. At the same time, the high mortality rate in childhood meant that early modern adults had overcome death and moved farther away from it. The locations and definitions of birth and death were continually shifting during the early modern period.

The human body was intensely scrutinized for signs of life and death. The pallor and rigidity of a body might indicate decay, whereas a rosy complexion, trace of respiration, or slight motion might represent life. Such visible evidence was not always trusted, but it provided a guideline for those spectators called to make or confirm a judgment. Audiences were present in each case outlined in the chapter. Friends, family, neighbors, and colleagues gathered in birthing rooms, at holy sanctuaries, at public executions and dissections, around death-beds, and at funerals. They used a range of other senses to determine the status of the dying or dead person, listening for sighs or breathing, touching for indi-cations of heat or pulse, and discerning the smell of rotting flesh. Though the body was subject to interrogation by those medical practitioners and laypeople attempting to interpret its condition, it might refuse to provide the desired in-formation. Sometimes waiting for visible signs of decomposition was the only way to be sure a body was dead, although badly decomposed stillborn children had been known to cry or move. In the end, miraculous interventions could override the materiality of the body.

Bodies in between birth and death were portrayed in visual images during the early modern period, and a few of them have been analyzed in this chapter. These representations shed light on each other. The woodcut showing the birth of the Antichrist is informed by conventional depictions of the lying-in cham-ber, but it also responds to scenes of death. Unlike the peaceful image of the man who died well, attended by a priest and his family, the Antichrist's mother performs death badly. With eyes wide open and arms flung upward in horror, she meets death with surprise rather than acceptance. Both the Antichrist's mother and the recently dead man are in recumbent poses that contrast with the standing posture of da Carpi's dissected man. His upright posture conveys the movement and energy of life. Distinct from the others he is heroically nude, resembling an antique statue. The midsection of the Antichrist's mother is ex-posed in a potentially shameful way, while the torso of the recently dead man conveys vulnerability rather than strength. Interestingly, the pose and dress of the woman undergoing Cesarean section in Mercurio's treatise adhere to none of these conventions. Although shown on a bed, she is propped up by assistants, indicating that her life depends on others. It is difficult to interpret her exposed breasts, for this reference is rare in early modern scenes of birth unless related to breast-feeding. Within the context of this particular woodcut, the breasts figure her as female and may also suggest the hurried nature of the procedure. The only independent human body in the images discussed in this chapter is that in da Carpi's anatomical tract. This self-dissecting man is not a public figure, existing in a domestic setting like the others; he stands apart from social identity as a corpse, albeit an engaging instead of a frightening one.

This figure might be the one most familiar to us today. Comparable im-ages of dead bodies that seem alive are featured in Bodyworlds, the successful anatomical exhibition still touring the Western world in 2007. In this case human corpses preserved by plastination, a process invented by the German anatomist Gunther von Hagens, do not show signs of decay.[86] Endowed with color, lively expressions, and active postures, they are uncanny, invoking both fascination and disgust from audiences. Some viewers understand the figures simply as flesh that lacks human identity, while others feel uncomfortable about the permanent display of human remains. Responses will depend on a person's religious beliefs, social status, and cultural background, in keeping with the factors affecting reactions in the past. Though we might be tempted to consider the uncertainty of the early modern body entirely foreign to our modern ways of thinking, and the cases discussed in this chapter strange or unbelievable, we clearly remain fascinated by the confusing terrain in between life and death.

CHAPTER TWO

Why Me?
Why Now? How?

The Body in Health and Disease

MARGARET HEALY

> There is a wisdom in this beyond the rules of physic: a man's own observation, what he finds good of, and what he finds hurt of, is the best physic to preserve health.[1]

Francis Bacon begins his essay "Of Regiment of Health" effectively counseling "Know thyself"—arguably the most important maxim of the Renaissance. Medical self-scrutiny involves examining one's "customs of diet, sleep, exercise, apparel, and the like" (termed by physicians the "non-naturals") in relation to the body's feeling of well-being or otherwise and, if anything "thou shalt judge hurtful, to discontinue it by little and little."[2] Such modest "regimes," linked to moderation in everything, including in Bacon's estimation the choice of one's physician (he should be of a "middle temper") are, according to all the writers on medical matters in this period, the best route to maintaining health. Compared with today's standards, "helth" as defined by a noted authority on these matters, Sir Thomas Elyot, is a relatively modest affair:

> The Conservation of the body of mankynde, within the limitation of helth, which (as Galene sayth) is the state of the body, wherin we be neyther greved with peyne, nor lette from doing our necessary busynesse.[3]

Indeed, carrying on one's business was, as we shall see, frequently accompanied by considerable pain and discomfort—chronic "dis-ease" may well have been more normative than "helth."

While Galenic, humoral medicine faced many challenges to its authority in this period, including Paracelsian, "chymical" medicine,[4] and Vesalian anatomy, which to some extent reconfigured bodily models, "physic" remained a hazardous activity for the patient often involving excruciating and invasive cures in the form of bleeding, purging, cauterization, sweating treatments, and poisoning (such as mercury for syphilis), which probably had very few positive consequences. Knowing one's own constitution and trying to prevent disease in the first place through attending to the non-naturals (Bacon notably left out an important non-natural from his list, "matters of Venus"—a euphemism for sexual intercourse) was a sensible alternative to employing a costly physician whose very cures might make one sicken or even die. Indeed, the treatments for early modern syphilis were particularly barbaric and probably caused significant harm; however, because the pox was understood as a sinful, venereal disease, the just reward of lechery, the punitive cures were undoubtedly construed as beneficial for the soul (Figure 2.1).[5]

Until very recently, in modern Western societies the discourse of medicine about disease, involving such tangible but low-affect culprits as microbes, aberrant physiology, and genes, was so loud that it readily droned out all the others. But weighty and costly matters, such as increasing obesity levels and alcohol consumption, have captured the attention of governments and economists, and are serving to foreground that disease is never simply an objective biological issue. In fact, the understanding of bodily dysfunction and the interpretation of the exterior signs of morbidity are shaped by powerful behavioral, social, political, and economic forces.[6] Diseases are, therefore, most appropriately understood as unstable constructs (this is not to downplay the importance of the biological component), which are historically and culturally determined; and, far from being rarefied, discrete forms of communication about the human body, medical discourses can be shown to constitute themselves through their intersection with other discourses.[7]

Early modern medicine may have been short of effective cures, but what it did usefully achieve was the important provision of plausible stories about bodily dysfunction, giving at least the illusion of knowledge and thus alleviating some of the anxiety associated with the unknown, with fears about bodily and community chaos. In the case of contagious illnesses such as bubonic plague, convincing explanations also enabled rudimentary containment measures to be implemented, which may, indeed, have had a positive outcome, impeding the spread of infection. Medical explanations can be exopathic (disease as external invasive force) or endogenous (disease as internal disorder), and there is always an interplay between the two as disease is constructed within a culture's wider

FIGURE 2.1: Treatments for syphilis by sweating, inhalation, and cautery, engraved title page, *Die belagert une entsetzte Venus*, by Stephen Blankaart (Leipzig, 1690). Wellcome L0006636.

system of beliefs about pollution, sin, and death. Indeed, as the perusal of the uncanny illustration from *De humani corporis fabrica libri septem* (Figure 2.2) reveals, the medical profession itself in the Renaissance was becoming intimately associated with sin in the form of criminal bodies: the burgeoning field of human dissection relied on a ready supply of corpses from the gallows.

The viewing of Vesalian anatomies, like the act of "seeing disease" is, as Sander Gilman's important work has argued, "socially coded in many complicated ways."[8] In this chapter, I intend to explore and amplify ways of "seeing" the diseased body in the Renaissance through examining two very different but common afflictions—bubonic plague and kidney stones. As will become clear, in this period moral politics and socio-religious issues were frequently central components of "dis-ease."

BUBONIC PLAGUE

It hath a Preheminence above all others: And none being able to match it, for Violence, Strength, Incertainty, Suttlety, Catching, Universality, and Desolation, it is called *the Sicknesse*.[9]

The playwright Thomas Dekker describes at length the "blew wounds," bodies like "speckled marble," ulcerous running sores in groins and armpits, and the "carbuncles" and "tokens" that gave this particular disease "Preheminence" in the horror stakes. The title-page of one of his many pamphlets, *A Rod for Run-awayes* (Figure 2.3), graphically captures the cultural imagery of early modern plague, which was frequently personified as Death laying siege to the city while anyone with sufficient means simply "ran away"—fled in terror. As his account reveals, the signs and symptoms of "the Sicknesse" were shrouded in mystery and evoked immense fear.[10] Supernatural explanations were bound to be rife in these circumstances, and even the most pragmatic accounts dwelt on its mysterious aspects. Thomas Paynell asked, for example, "Why that some do die and peryshe of the foresayde sycknesse, and some not: and beynge in the sayde same citie or house, why one dothe dye, and another dyeth not?"[11]

Because of the suffering it caused, the fear it inspired, and its mysterious qualities, plague was a particularly powerful polemical vehicle in premodern Europe. Even the painful, stinking, and unsightly purple "buboes" or "tokens" that appeared on the bodies of victims (Figure 2.4 depicts St. Roch symbolically displaying his bubo), which all plague writers describe in painstaking detail, seemed to confirm that it was a blow or stroke meted out somehow by an angry deity.[12]

As medical anthropologists have demonstrated, the human response to devastating diseases is inevitably to ask "Why me?" "Why now?" Further, "Am I myself to blame?" or "Am I the victim of an attack from outside?"[13] In the

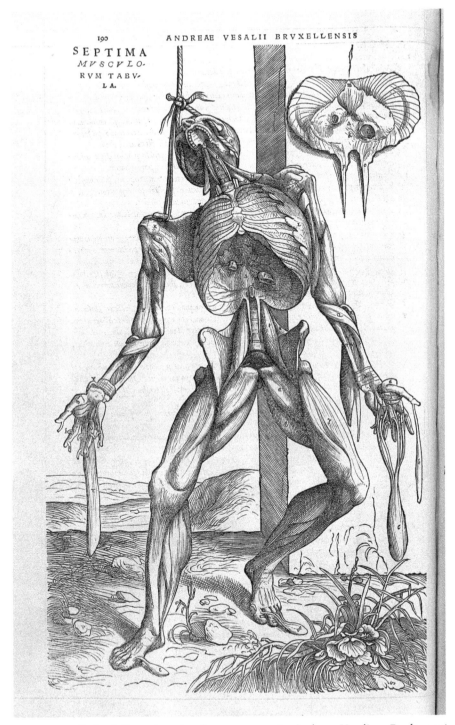

FIGURE 2.2: Skeletal figure hanging from a noose, in Andreas Vesalius, *De humani corporis fabrica libri septem* (Basileae, 1543). Library ref. no. EPB 6560. Rare books. Wellcome L0031739.

A Rod for Run-awayes.

Gods Tokens,

Of his fearefull Iudgements, sundry wayes pronounced vpon this City, and on feuerall perfons, both flying from it, and staying in it.

Exprejfed in many dreadfull Examples of fudden Death, falne vpon both young and old, within this City, and the Suburbes, in the Fields, and open Streets, to the terrour of all thofe who liue, and to the warning of thofe who are to dye, to be ready when God Almighty fhall bee pleafed to call them.

By T н o.D.

Printed at London for *Iobn Trundle*, and are to be fold at his Shop in Smithfield. 1625.

"A ROD FOR RUN-AWAYES"

The title page of one of Thomas Dekker's plague pamphlets, 1625. The plague was almost continually present in London until late in the seventeenth century, but in some years, the so-called plague years, the disease broke out in a violent epidemic; 1625 was one of these plague years. In his pamphlet, "A Rod for Run-awayes," Dekker describes the conditions in London during the epidemic. The illustration on the title page shows the wrath of God descending as lightning from the clouds, and in the center death stands represented as a skeleton. On the left are men and women dead in the fields and over them is the inscription, "Wee dye"; on the right is a group of people fleeing from the plague and in response to their words, "Wee fly," death answers with, "I follow." The people of the suburban districts realized the truth of death's "I follow" and attempted to prevent the infected Londoners from contaminating their towns, as is shown by the armed men marked with the inscription, "Keep out."

FIGURE 2.3: Title page, Thomas Dekker, *A Rod for Run-awayes* (London, 1625). Wellcome M0001927.

FIGURE 2.4: St. Roch represented showing a bubonic sore under the guise of a pilgrim. A citizen of Montpellier who devoted his life to serving those stricken with plague. Museum no. R 4756/ 1937. Wellcome M0014127.

absence of scientific knowledge, plausible stories fill the gap, imposing meaning and textual order on an intractable medical problem. Interpretative frameworks of bodily misfortune may offer supernatural or naturalistic explanations
of disease or a combination of both. In the case of bubonic plague the primary
answer to the "Why me, why us?" "Why now?" questions was always God's
wrath. With remarkable consistency, when plague appears in early literature
it is closely associated with divine disapproval of greed for riches—the sin of
avarice—and frequently with consequent exploitation of the poor and social
discord.[14] Two prime examples of this are *Dives and Pauper* and a popular poem
that emerged in 1382, called simply "A Warning to be ware." The latter poem
explicitly connects the Peasant's Revolt of 1381 with fearful pestilence and the
earthquake of 1382, which it attributes to "gret vengaunce and wrake. . . . for
synnes sake."[15] The poem proceeds to elaborate on the prime sin, which is greed
for money: for money the verse declares, many would betray their own father
and mother. The poem might have functioned as a warning, or even as a threat,
alerting the wealthy lords to the social repercussions that would ensue if they
did not take the demands of the commoners seriously. Moralistic plague writings attributing blame inevitably have a tendency to inscribe social tensions.

If the primary cause of plague epidemics for all writers in this period is divine
rage, the secondary causes—the agencies or instruments of his displeasure—are
interestingly various: supernatural, natural, or a mixture of both. In the case
of this mysterious disease, which, since the late nineteenth century, has been
identified as a bacterial infection spread by the fleas of the black rat, there was
no one agreed upon conceptual framework for making sense of it. The medical
tracts from the Renaissance reveal that all the clues about how plague was actually spread were circulating in print but were simply baffling. Thus in 1603, the
physician Thomas Lodge was thrown by the capacity of fur collars to infect and
remarked on the flea bites on plague victims, while several plague pamphlets,
including his, recorded how rats left their holes during an epidemic.[16] Many
writers lingered perplexedly on the infectious nature of wool and cotton fabrics
and warned against sharing beds and clothes: presumably fleas were partial to
these. Rather ironically Thomas Nashe chastised those of his fellow Londoners
in 1592–1593 who attempted to play down the horror of the epidemic by jest-
ing about it as a "flea-byting affair," one that would cease with "the season of
the yeere."[17]

It was generally agreed through experience that filthy, stinking, and over-
crowded environments were particularly attractive to the infection and that
plague was more prevalent among the dirty poor. These observations were undoubtedly accurate, for as we now know there are two other forms of plague
besides the bubonic—pneumonic and septicemic—both of which are encouraged by close human contact and filth. The former is spread by droplet infection and the latter by body lice and human fleas; all three forms owe their

genesis to the *Yersinia pestis* bacterium.[18] Indeed, it is ironic with the benefit of hindsight that Daniel Defoe's protagonist, H.F., in *A Journal of a Plague Year*, expresses deep contempt for the "ignorance" and "enthusiasm" of those who maintained that the infection was carried by "Insects, and invisible Creatures," which emit "acute Poisons, or poisonous Ovae, or Eggs, which mingle themselves with the Blood." He also questions the truth of a report that breathing on a piece of glass and then looking at it through a microscope will reveal the strange forms of "living Creatures": it would take many more years before the theories of these "enthusiasts" (in fact, the early exponents of germ theory) were verified.[19]

In the early modern period, three prominent medical theories jostled, often side-by-side, for precedence. Corrupt, poisoned air ("miasma"), stars, and contagion were all cited as "natural instruments" promulgating and spreading God's plague. Thus, William Bullein's *Governement of Healthe* quoted Galen on "distempered air" and described "certain stars called infortunates . . . whose influence bringeth corruption . . . and pestilence."[20] Notably, however, Thomas Dekker rejected both theories and another prevalent one favored by the authorities in 1603—that of infection passing from body to body by close contact or touch—favoring instead a biblical explanation: God's smiting angel. This is the type of "delusion" or "Error of the times" that Defoe attacks a century later, through his relation in *A Journal of the Plague Year* of H.F's embarrassing encounter with a group of poor people busy discerning "An Angel cloth'd in white" with his "fiery" sword in the clouds.[21] Dekker, meanwhile, reserves his scoffing indignation for what was to emerge as Defoe's favored theory, contagion:

> Can we believe that one mans breath
> Infected, and being blowne from him,
> His poyson should to others swim.
> For then who breath'd upon the first?[22]

Indeed, he rejects all the contemporary medical theories and treatments in favor of religious explanations and cures: in his view, prayers, repentance, and charitable behavior toward the poor are the only remedies for plague. Dekker's solely supernatural interpretative model may, however, have been partly strategic. He was vehemently opposed to the policy of shutting up plague victims and their families in their houses, maintaining that it was a cruel practice only effectively enforced against the poor. If plague was understood as noncontagious, such prejudicial segregation policies were rendered pointless.

But other kinds of supernatural theories that would function powerfully to undermine the notion of contagion and the government's shutting-up policy were gaining ground in the early seventeenth century. Some disciples of the physician Paracelsus, and subsequently J. B. Van Helmont, encouraged the notion that fear and the imagination could attract plague infection into a victim.[23] For Van Helmont the key to disease lay in the "Archeus," the vital spirits of man, and the Archeus could receive the image of plague, thus producing infection.[24] The spirits had to be "kept up" for the plague to be kept out, and "shutting up," which was closely associated with fear, was thus highly detrimental to preservation. Indeed, the central controversy in the 1665 plague was between chemical physicians who stressed such novel mechanisms and Galenists who favored the ancient model of "natural instruments" combined with imbalance of the humors. Interestingly, Defoe's *Journal* ignores this 1665 debate, focusing instead on refuting the age-old theory of "miasma" (this "whimsy" that it was "all in the Air") in favor of contagion—the tension between these two constituted the fulcrum of the medical, political, and popular controversy in 1722—the period in which he was writing.[25]

While sin, poverty and filth, miasma, smiting angels, and contagion frequently rubbed shoulders in an attempt to make sense of and control bubonic plague, the social face of the chronic disease of renal stones was, as we shall now see, intriguingly different.

THE STONE

> For several years now I had been a prey to the stone, a most troublesome and dangerous ailment.[26]

With characteristic humor, Erasmus of Rotterdam announced in his Adage "Festina lente" (1508) that it was during his stay in Venice (that notorious cradle of Renaissance luxury) that "I had . . . an encounter with a trouble I had not met before, the stone" (*vesical calculi*)—"a most troublesome and dangerous ailment."[27] This was to prove an ominous introduction: Erasmus's "trouble" continued to torment him for the rest of his life. Strangely, however, such unsolicited and dangerous encounters (they could be fatal) were by no means rare in early modern Europe. In fact, Erasmus was in excellent company: among his fellow sufferers were a cast of early modern intellectual and political luminaries, including Michel de Montaigne, Thomas Linacre, Samuel Pepys, John Dryden, Oliver Cromwell, Isaac Newton, and Robert Walpole.[28] As Montaigne wryly boasted in his essay "Of Experience," "I have fallen into the commonest ailment of men of my time of life. On all sides I see them afflicted with the same type of disease, and their society is honorable for me, since it preferably attacks the great; it is essentially noble and dignified."[29] There was certainly a shared perception that "the stone," together with its frequent companion, gout (podagra), favored the affluent and well-to-do. Like gout, the stone was considered a fashionable illness (a "patrician malady") and this was undoubtedly why these two diseases were so frequently written about.[30] The stone regularly haunted the pages of early modern letters and even inspired a short musical composition: Marin Marais's fascinating and haunting, "Le Tableau de l'Operation de la Taille," a "piece de caractere" that articulated the terrifying, life-threatening, and excruciating experience of its surgical removal.[31] Indeed, it was an affliction that, for a variety of reasons that will be explored over the following pages, gave birth to particularly graphic and illuminating representations and to novel medical hypotheses.

Claiming two literary giants, Erasmus and Montaigne, and one loquacious diarist, Pepys, among its victims, the stone's tormenting proclivities were destined for notoriety. Renal stones cause ureteric colic—acute pain radiating from the lower back to the groin to the genitals (often accompanied by sweating, nausea, and vomiting)—described in a twenty-first century medical textbook as one of the most severe pains known to humans.[32] As Elaine Scarry's seminal study, *The Body in Pain*, has foregrounded, intense pain makes the subject acutely aware of his or her body: the soma is no longer a background medium of foregrounded action, and the individual can become immersed in a world of pain whose parameters consist of a highly restricted body image "shot through with stabbing, piercing, searing agony."[33] Erasmus's numerous letters to friends and doctors detailing his sufferings convey a certain relish on the part of this exceptionally gifted penman for the considerable literary challenge of translating such bodily agony and discomfort, as well as intense fear, into words. His vivid evocation of a biting "dragon" occupying his "poor little body," in a letter written to Melanchthon shortly after his move to Freiburg in July 1530, is a notable example:

> This is now the fourth month of my illness. First I suffered from colic, then from vomiting . . . This poor little body of mine does not get on well with doctors. All the medicines they gave me did me harm. The colic was followed by an ulcer, or more accurately, by a hard swelling which first extended all along the lower right groin. Then it centred on the pit of my stomach, almost like a dragon with its teeth biting my navel while the rest of its body was writhing and its tail stretching towards my loins; when its head was fastened tight it coiled around to the left side of my navel, with its tail almost encircling it. It caused constant, sometimes unbearable pain. I could not eat or sleep or write or read.[34]

In a letter to another friend, Joost Van Gaveren, the stone itself is personified as a tyrant inflicting gruesome tortures repeatedly; a fate that is worse, Erasmus declares, than the singular capital punishment of being dismembered while still alive: "What is this [torture by stone] but tasting of death again and again?"[35] He sought God's mercy through prayer, pleading poignantly: "deliver me, if possible, from this evil which is in me."[36]

By contrast with Erasmus's densely metaphoric and moralistic descriptions, Michel de Montaigne's accounts of his affliction tend to lack rhetorical embellishment and are characteristically stoical and clinically sparse and direct. Indeed, during his travels to Italy between 1580 and 1581 (via Switzerland, Germany, and Austria), he often remained on horseback, seeking to encourage the stone's descent through rough, jolting movement as long as he could grip the reins of his mount:

> In this condition I left Siena; and the colic seized me again, and lasted three or four hours. At the end of this time I clearly perceived, from an extreme pain in the groin, the prick, and the ass, that the stone had descended.[37]

But pain is a culturally framed experience, affecting individuals in diverse ways, and Samuel Pepys's diary entries find him far from stoical, periodically crying, roaring, trembling, farting, clutching his "cods" [scrotum], and performing all kinds of bodily contortions during his acute attacks of the stone:

> 11 October 1661: All day in bed with a cataplasme [? plaster] to my Codd.[38]
>
> 2 August 1662: I was afeard to ride, because of my paine in my cods.[39]
>
> 14 May 1664: After dinner, my pain increasing, I was forced to go to bed; and by and by my pain ris to be as great for an hour or two as ever I remember it was in any fit of the stone, both in the lower part of my belly and in my back also . . . At last, after two hours lying thus in most extraordinary anguish, crying and roaring, I know not whether it was my great sweating that [made] me do it, but by getting by chance among my other tumblings, upon my knees, my pain began to grow less and so continued less and less, till in an hour after I was in very little pain, but could break no wind nor make any water.[40]

He did, however, endure lesser sufferings with considerable fortitude: his diary bears witness to his continuing to engage in an impressive range of work and pleasure activities, while experiencing extended periods of discomfort and weakness in his lower back. On the third and fourth of June 1664, for example, suffering "constant akeing" in his back, which he had experienced for six days, he criss-crossed London traveling from his home to his office, to the Exchange, to Whitehall, to St. James, engaging in a cramped schedule of business meetings and social gatherings.

Although these afflicted men experienced and responded to renal colic in diverse ways, they would likely have shared an understanding of what caused it: a surprising amount was known and written about the conception and birth of renal stones, and several physicians published treatises of helpful knowledge and advice. The Dutch doctor, Johan Van Beverwijck, was one of these. Drawing on the writings of Greek, Roman, and contemporary physicians, as well on his extensive personal experiences both as a practitioner and as a sufferer from *vesical calculi*, Beverwijck's *Treatise on the Stone* is comprehensive and fascinating. In common with most early modern authorities on the subject, Beverwijck agrees that the "immediate cause" of the formation of gravel and stones in the kidneys is an "earthy substance" mixed with salt and water, which, remaining there—or in the bladder—too long in a "torpid state," forms the "kernel" or nucleus of larger stones.[41] This earthy substance is in all sorts of soil and is ingested, together with salts, from vegetables, fish, and meat. Sometimes it is present in food in greater abundance and this factor, or "the weakness of the digestive, separating or expulsive power," leads to its being driven into the blood to the kidneys where it forms sand. In healthy people "the gravely substance is evacuated with the motion"—only "less healthy" persons "grow sand." Furthermore, "flabby and torpid kidneys" fail to expel the sand efficiently. People of a "stony" disposition tend to pass milky, plastery urine, gravel, and sometimes, more ominously, blood. He also notes that the shape and size of the urinary organs determine the ease with which gravel and stones can be passed: the "narrowness of the lower part of the kidneys near the ureters" is particularly significant. The shorter, wider female urethra enables women to expel stones more easily.[42]

Beverwijck recognized a strong hereditary element at work in this disease:

People who have this disposition from their parents, are gravely by nature and have inherited this evil together with the good. For fathers and grandfathers, as GALEN rightly says, procreate weak bodies if their seed is bad, whereby children easily get the defects of their ancestors. If somebody has inherited such tendency in his kidneys, he will hardly escape the gravel.[43]

Beverwijck's mother, grandfather, and brother all suffered from the stone. Both Pepys and Montaigne had relatives who grew stones, and the essayist was clearly ruminating on this when he wrote "Of the Resemblance of Children to Fathers." His father had not developed the disease until he was 67; until then he had been remarkably healthy, yet the inherited stony propensity eventually caused his excruciatingly painful death. As Beverwijck's treatise reveals, hereditary factors, diet, poor digestion, and diseased kidneys were all implicated in the process of stone formation. Strikingly, the same culprits are lined up for inspection in twenty-first century medical textbooks.[44]

For obvious reasons, early modern chemical practitioners were particularly interested in this ailment, and the noted medical innovator, Johannes Baptista Van Helmont, appropriated the stone to underpin and illustrate his novel theory of disease:

> For in the Stone, a Disease, it is most material and manifest; but the stone is not the Disease, but the primary Lithiasis or Stony affect, and the true Disease is the Idea itself, radically implanted in the powers of the Archeus of the kidneys or bladder.[45]

The Archeus was the chemical and spiritual governor of the body in Helmont's bodily schema. His English disciple George Thomson expanded on his master's theory in vivid and emotive prose (in Helmontian medicine, like its Paracelsian forerunner, seeds of disease were the disastrous consequence of the fall):

> The Petrifying Imaginary Seed, closely seated in the Archeus, is that which first laid the Foundation of the Stone . . . As long as there remains the stonifying seed, or invisible Beginning, the Person before rid of this hard Concretion, may ere long (if the Idea, the Principle of the Congealing be not absolutely brought to nought, or blotted out) be vexed, tortured or crucified with the like deformed Matter again . . . That the Archeus should be put into such a disturbance or Passion through any disorders in Diet . . . as to frame within its own Bowels such a dreadful, unhandsome Substance, is to be lamented.[46]

Here the victim is "tortured or crucified"—martyred, Christ-like—by the "hard Concretion" born of his activated "stonifying seed." This theory had the virtue of seeming adequately to explain the pronounced hereditary nature of the disease.

As Thomson's account reveals, in chemical medicine, as in Galenic medicine, diet and digestion were construed as of seminal importance to the growth of stones. Treatises on the illness thus contain long lists of treacherous foods. Milk, especially of the "thick and coarse" type was notorious, as was cheese, particularly "old cheese."[47] Also implicated were undercooked bread, various unripe fruits, including medlars, pears, quinces, and sweet chestnuts. Old or smoked pork and beef, smoked or salt fish, waterbirds, wild swans, geese, ducks, and hard-boiled eggs were all to be avoided. Beverwijck concluded, however, that the "stone chiefly grows from wine, which is unripe, sour, brown, thick or too sweet, and most of all from must or wine cultivated where the soil is stony."[48] "Young and turbid beer" and any food cooked in "turbid, muddy or snow-water" were hazardous.[49] Other negative influences were constipation, unhappiness, and infrequent micturition.[50] Rich people were construed as being more prone to stones because they ate white bread ("common wheaten bread" was preferable)[51] and slept on feather beds, which heated the kidneys.[52] Idleness was also implicated, although, rather paradoxically, other "further and external causes" conducive to stone formation included shocks, jumping, long riding, and all sorts of strong movements of a jolting character. Immoderate sexual intercourse causing "inflammation of the loins" was also a noted factor.[53] A "stony propensity" was thus loosely linked to the sins of luxury and undoubtedly this was one reason why Erasmus reassured the physician John Francis (1526) that he had not got "gonorrhoea," and that he had adopted "a moderate way of living."[54] In several letters he stressed that it was overwork, especially standing at his desk, writing straight after dinner, which caused his ailment:

> I surmise that I know the cause of this illness. For more than twenty years already I am in the habit of standing when I do my writing and hardly ever sit down, except when I take my midday-meal or dinner.[55]

In a letter to the famous chemical physician, Theophrastus Paracelsus (Basel, March 1527), he again stressed his industry and commitment to work:

> I have no time for the next few days to be doctored, or to be ill, or to die, so overwhelmed am I with scholarly work. But if there is anything which can alleviate the trouble without weakening the body, I beg you to inform me.[56]

Physicians tended to prescribe mild purgatives to keep the bowels loose and a diet low in earthy substances with the addition of foods that might stimulate the kidneys to expel gravel: as Beverwijck stresses, "the enemy must be checked before he captures."[57] Wine or beer "should always be clear and thin," as should water used in cooking food. Moderate exercise is advisable, "the mind should be made cheerful," and mild purgatives such as manna, cassia and diacatholicon should be taken regularly.[58] Water should be passed frequently: "nothing is so bad as retaining the water a long time."[59] Thin Rhine wines and the mineral water of Spa also "drive out gravel,"[60] but there is nothing like Venetian turpentine (a "nut" a day with sugar) for cleansing the kidneys.[61]

The writings of our three literary men reveal much about their favored regimens for countering their "stony propensity." Notably, all were avid consumers of wine. Erasmus only drank beer in the absence of wine and seems to have regarded good mature Burgundy taken in moderation as a medicine, especially when "diluted with a decoction of liquorice" or other "water of sweet–root" and sugar.[62] According to the noted humanist, Scaliger, Erasmus had been given to drunkenness in his Venetian days when he first encountered problems with stone.[63] One of his colloquies, "A Fish Diet," laments the serious illnesses caused by eating fish—"fever, headache, vomiting, the stone."[64] Forced by his host to eat fish, the interlocutor Eros heads for home clutching a bottle of almond milk and raisins (treatment for stones) yet, in spite of this rapid preventive action, "the stone made itself evident and he was in bed a whole month."[65] Bad, young wine and fish were definitely not on the menu for Erasmus, then. Rejecting his doctor's suggestion of warm baths at Baden, and dismissing his physicians altogether and putting his trust in God, he adopted a regimen of general temperance, of avoiding misty weather, which made his symptoms worse, of "passing water often for fear it will petrify in the bladder,"[66] and, of course, of old Burgundy wine—in moderation (he assures his readers).

Montaigne favored mature wine, too, and, judging from his frequent complaints of headaches in the *Journal*, he might have drunk rather a lot of it.[67] But he was not averse to taking mineral water either, and during his travels he visited numerous spas consuming vast quantities ("each morning at seven he drank nine glasses of water"),[68] plunging his recalcitrant body into hot mineral baths, and undergoing strenuous horseback rides in an apparent effort to free up its rickety, obstructed conduits ("to drive out the matter, dissipate and scatter it").[69] To some extent this worked; he passed numerous stones and much gravel in these months.

Pepys's approach to his ailment was far more haphazard. He was always anxious about catching cold, or overheating, and actively avoided horseback when suffering his "old pains." Like Montaigne he kept his bowels loose with preparations like cassia, and he regularly took turpentine, but he swore by something else, too—a "Hares-foot" he wore against his body to ward off wind and colic.[70] He regularly consulted physicians but, as on October 23, 1663, when he was prescribed a "Glister," "hony," and "Lyncett oyle" by Mr. Hollyard, he would rush hastily on to more pleasant, distracting venues, like that of "Mr. Rawlinsons" to inspect his newly crested bottles "filled with wine, about five or six dozen."[71] Like Erasmus and Montaigne, Pepys's favorite medicine was, undoubtedly, good mature wine.

As we learn from the accounts of all these writers, preventive regimens frequently failed and then drastic surgery to remove the offending stone(s) might be contemplated. John Evelyn's diary provides us with a fascinating insight into what this horrific operation entailed. On May 3, 1650, his curiosity led him to the "Hospital of the Charitie" in Paris where he witnessed five lithotomies. This is how one man, aged 40, was trussed up like a chicken in a lithotomy position and then had a stone "bigger than a turkeys Egg" removed from his bladder:

> The sick creature was strip'd to his shirt, & bound armes & thighs to an
> high Chaire, two men holding his shoulders fast down: then the chirur-

gion with a crooked Instrument prob'd til he hit on the stone, then without stirring the probe which had a small chanell in it, for the edge of the Lancet to run in, without wounding any other part, he made Incision thro the Scrotum [probably, rather, the perineum] about an Inch in length, then he put in his forefingers, to get the stone as neere the orifice of the wound as he could, then with another Instrument like a Crane's neck he pull'd it out with incredible torture to the Patient, especially at his after raking so unmercifully up & down with a third instrument, to find any other stones that may possibly be left behind: the effusion of blood is greate.[72]

Following surgery, a silver pipe was inserted into his penis for the passage of urine, and the wound was sewn up. Evelyn tells us that if the patient made it thus far the subsequent dangers from fever and gangrene were great.

As a young man, in 1658, Pepys chose to undergo the operation, carefully selecting a skillful surgeon, Thomas Hollier, from St. Thomas's and Bart's hospital. He was one of the fortunate survivors of this life-threatening procedure, and after 35 days on bedrest he pronounced himself fully recovered, holding "stone feasts" yearly thereafter to celebrate his wonderful good fortune.[73] Some years later, Evelyn records his encounter with Mr. Pepys's impressive stone:

I went that evening to *Lond*, to carry Mr. Pepys to my Bro: (now exceedingly afflicted with the Stone in the bladder), who himselfe had been successfully cut; & carried the stone (which was as big as a tennis-ball) to show him, and encourage his resolution to go thro' the operation.[74]

Many victims, understandably, chose not to "be cut" and these, like Dr. John Wilkins, famous member of the Royal Society, were often prescribed unpalatable potions (lithontriptics) and poultices with alleged stone-breaking propensities: favorite ingredients included dried shark's brain, crushed beetles, scorpions, ashes of hare, eyes of lobster, and oyster shells.[75] Unable to pass urine, Wilkins subsequently died and, after his postmortem failed to reveal obstructing stones, his learned fellows concluded that he had probably been poisoned by his physicians' "remedies."[76] A few years later the scientist Robert Hooke carried out experiments on his own gravelly urine under a primitive microscope. Finding that the crystals he discovered there could be dissolved with, among other chemicals, "Oyle of Vitriol" (concentrated sulphuric acid), amazingly, he contemplated this as a potential remedy for stone.[77]

Far more palatable medicines were, however, just over the horizon. In the late seventeenth century a promising new therapy emerged—the imbibing of tea, coffee, and beer between meals. In the words of the famous draper Antoni Van Leeuwenhoek:

I think, that to drink tea or coffee promises much . . . , because therby much liquid passes to the kidneys; and, further, that we should drink plentifully of beer, between dinner and supper time, and not follow the example of those who boast that they drink very little in the day, and none at all at supper.[78]

Stressing the need to "drink in sufficient quantity," Leeuwenhoek lamented the ridiculous "manner of living" of some of his gentlemen acquaintances and patients who took pride in strict bodily control to the unhealthy extreme of denying the call of thirst.[79] A master surgeon from the same period commented, significantly, on how "calculous people usually are thirsty."[80] It is interesting that the first line of treatment for "stony" people today is to increase their intake of (nonalcoholic) fluids to a minimum of three liters a day.[81] Indeed, it is quite probable that, as Leeuwenhoek's letter implied, dehydration was a major contributing factor to the high level of stone disease in early modern Europe. This might explain, too, why the incidence remains problematically high in parts of the developing world today where clean water is not freely available (in northern European countries this disease declined from the eighteenth century; only 1–2 percent of people now develop stones).[82] Sadly, Erasmus's, Montaigne's, and Pepys's shared affection for the old wine "remedy" would only have enhanced any bodily tendency toward dehydration, and they just missed the new European fashion for taking tea and coffee liberally between meals. Although the latter would not have cured their disease, the regular higher fluid intake it ensured might have reduced their stony propensity and alleviated some of their distressing symptoms.[83] Each eventually died of a range of debilitating ailments, including chronic damage and infection of the renal tracts through recurrent painful encounters with a, by now, familiar enemy—the tyrannical stone.[84]

As will be clear from these vivid and sometimes disturbing descriptions of plague and renal stones in the early modern period, the understanding of the malfunctioning body in pain was constituted through a web of associations and myths—"dis-ease" and "helth" cannot be extricated from the labyrinth of high-affect discourses in which then, as now, issues of health and illness tend to be enmeshed.

CHAPTER THREE

Sexuality

Of Man, Woman, and Beastly Business

KATHERINE CRAWFORD

In a popular manual about pregnancy and medical issues related to pregnancy, Jacques Guillemeau noted, "Some women when they be with child hate the companie of their husbands: which quality is said also to be in brute beasts when they be great with yong. . . . And surely there be certaine times and seasons of the yeare proper for brute beasts to couple, but man (as Pliny saith) hath neither time nor season limited him, neither day nor howre appointed him, that so he might have his desire at all times: which hath been thus ordained by nature, as being more fit, and necessary for man to multiplie in his kind he being (the lively image of God, and made to behold his glory) then for brute beasts, which were created onely for the use of man."[1] Guillemeau's struggle to distinguish man from beast in matters of sex reflected a number of impulses that marked Renaissance thinking about embodied sexuality. The religious logic that had long dominated Western thinking is remaindered in his second parenthetical comment, but the earlier invocation of Pliny indicated the developing conceptualization of sexuality that accompanied the Renaissance. Understandings of sexuality in the Renaissance changed with the influx, assimilation, and adaptation of information about antiquity. The Christian model of sexual sin did not disappear; it did lose some of its power over notions of sexual behavior.

At the same time, the need to distinguish humans from animals took on new urgency. This chapter, then, uses the recurrent theme of the need to distinguish humans from beasts in sexual terms as a prism through which to examine how understandings of sexuality changed in the Renaissance. What people did sexually did not change all that much—there are only so many possibilities, and in the absence of proof about what individuals did, I assume that many of them engaged in some of those possibilities at some point—but how sex was understood changed significantly. The study of antiquity revealed that neither civilization nor culture nor the status of humans could be explained by the advent of Christianity. Antiquity undermined ideological certainty in one way, but it also offered a new epistemology of sexuality in which human rationality and understanding of the past created the cultural context for sex. This chapter considers salient examples of how the Renaissance culture of sexuality addressed the human role as rational beings in defining the parameters of sex. Through discussions of aspects of textual criticism, developments in medical knowledge and its diffusion, and a central philosophical movement, this chapter will show how we learned that we needed to understand humans as rational sexual actors (and thus distinct from beasts).

This both is and is not how the story usually goes. Recent discussion of the history of sexuality has been dominated by Michel Foucault's account of sex as discourse, in which the Renaissance does not really matter in and of itself.[2] The story Foucault told challenged a number of truisms. For Foucault, frankness about sex in the early modern period was replaced by sexual prudery such that Victorians talked about sex all the time under the guise of condemning it. Seventeenth-century Europeans were playful and shameless; bourgeois Victorians made all sorts of rules about sex. Catholic confessional practices started to take hold in the eighteenth century, the state policed sexuality more effectively, and Enlightenment concerns about birth rates, death rates, and manpower capacity made interest in sexual practices widespread. Confession was especially important in that Europeans were exhorted to think critically about their sexual practices, to articulate them, and then to repress those deemed unproductive or "bad." Confession required that every desire, every action be transformed into language so that it could be mastered by the self. Policing of sexuality turned into mechanisms that defined the self in sexual terms. These habits led to the idea that it was necessary to expel unacceptable forms of sexuality, often defined as unproductive in economic terms. Healthy, affluent married couples who produced children were economically beneficial; those who were healthy and affluent and did not reproduce or who were unhealthy (physically or mentally) and/or poor were increasingly regarded as not merely unproductive but also as detrimental to civilized society. Sexual irregularity in these terms marked a person's being. Because sex was constructed in language and seen as integral to the self, it became a matter of identity. A "homosexual" as a person defined by sexual practices was a modern category to Foucault, as was a "heterosexual."

That is the broad outline of Foucault's narrative. This chapter takes up his focus on discourse and language but eschews his modernist account of identity and presents aspects of the ways the Renaissance reshaped understandings of sexuality Foucault omitted.[3] The Renaissance as a rediscovery of antiquity is significant in this argument: Renaissance Europeans worked out new approaches to the past and toward sexuality through their understanding of ancient texts and the cultures that produced them. Consider for instance Ovid's *Metamorphoses*. From late antiquity through the Middle Ages, Ovid's stories were interpreted as Christian allegories. Christian belief that sex was a distraction from God; that marriage was to control lust, but it was never as good as virginity or celibacy; and that confession of sexual sin was central to spiritual health determined how Ovid was read. This might seem odd to a modern reader who has encountered just plain Ovid. The *Metamorphoses* features a wide range of eroticism, some of which is cast in a negative light (sisters who desire their brothers, gods who rape or trick women), but Ovid also celebrates sexual desire (Ganymede is raised to the heavens, as is Callisto, in her way). Ovid's eroticism was precisely what brought him the unfavorable attention of early Christians like Clement of Alexandria, who denounced Ovid as immoral. Nonetheless, Clement maintained that Ovid's allegorized tales represented hidden truths that had to wait until Christianity came along to be fully revealed.[4] This idea encouraged reading Ovid for the coded ways his tales might reflect divine wisdom. Eusebius, Boethius, Lactantius, and Fulgentius were among the more prominent early interpreters in this mode, and their insistence on allegory remained central to reading Ovid through the Middle Ages.[5] Over time, the *Metamorphoses* was subjected to extensive Christianized reading, and the eroticism of the stories became central to the moral readings to which Ovid's tales were subjected.

The interpretation of Ovid took some time to develop. Medieval exegetes gradually developed the practice of reading texts on four levels or "senses." These were considered different categories of truth, and although the terminology varied, the basic divisions had distinct valences. The natural (also called physical or literal) described properties or phenomena. The historical sense allowed reading in terms of past events. The tropological sense was the moral interpretation, in which, for instance, characters stood for vices or virtues. The allegorical sense was Christian interpretation, and commentators sometimes called this the "spiritual" reading.[6] Allegorical readings of Ovid interpreted eroticism in conformity with Christian sexual ethics. The *Allegoriae super Ovidii Metamorphosin* (ca. 1175) by Arnulf d'Orléans reads the three apples Venus gives to Hippomenes to divert Atalanta as wisdom, beauty, and nobility. Arnulf describes these as Hippomenes's inner virtues, and as such, they are acceptable reasons for her to fall in love with him. More often, female characters are symbols of lascivious desire of various kinds. Some, like Myrrha, the incestuously inclined daughter of Cinyras, are obviously sexually depraved. Others, like Orpheus's beloved wife Eurydice, are more ambiguous. Identified as "good judgment" initially, Eurydice dies from the snake bite because she has immersed herself in pleasure. In contrast, several male characters (Acteon, Orpheus, Hercules, and Aeneas, among others) are described as prefigurations of Christ.[7] Even more elaborate is the entirely allegorized reading of the *Ovide moralisé*, a French poetic rendering of the *Metamorphoses* with commentary pointing out the proper Christian readings. For Iphis and Ianthe, the general lesson is a denunciation of desire as, "Against right and against nature," ("Contre droit et contre nature") if it is "carnal."[8] The allegorical interpretation describes Iphis's mother, Telethusa, as the Church, and Iphis as the sinful female soul regenerated by baptism. Through the efforts of the Church, Iphis/the soul is nourished, and God takes pity on the poor soul, raising it up to heaven. God's grace and glory are revealed in the salvation of the sinner, as signified by the miraculous transformation of Iphis's body from female to male.[9] In medieval versions, the overriding interest was in rendering Ovid useful for Christian consumption. Allegorists, unconcerned about anachronism or historical context, turned a story about the complications of transgender sexual identity into a Christian salvation lesson.[10]

Although moral readings of Ovid continued during the Renaissance, efforts focused on locating Ovid's meaning in his historical context, allowing the sexual content to emerge. Humanists like Raphael Regius, whose analyses of the *Metamorphoses* influenced generations of critics, situated Ovid's sexual and erotic content in terms of social values of ancient Rome. For Regius, Ovid's moral lessons might still apply to Christian society, but historical information and humanist erudition located the larger moral purpose in time and place. The duty of the commentator was to bring out the relevance of the text through rhetorical, grammatical, historical, artistic, and philosophical analysis. The most significant and obvious shift in Renaissance usage of Ovid was the privileging of historical interpretation at the expense of allegory. The impetus for this shift came from humanist interest in the circumstances of antiquity. Although it took several generations, Renaissance humanists developed the ability to analyze historical change in language based on knowledge of material conditions, events, and the chronology of the past. Tropological interpretation then gave humanists a broader purpose, often structured around a text's pedagogical uses, such as training students in (pagan) poetics. Conversely, humanists scoffed at the propensity toward anachronism that marked allegorical reading. Their attacks on allegorists found an ally in the Catholic Church at the Council of Trent, which condemned allegorical Christianization of pagan texts in 1559.[11]

Renaissance priorities gave Ovid's sexual stories a decidedly different shape. For Regius, the birth of the giants was the result of demons copulating with women, a basically natural explanation with euhemerist overtones.[12] When he comments on the story of Pygmalion, Regius draws on his own Italian gender values in his explication: "*Nam virgines bene educate prae pudore in virorum conspectu vix audent se movere*" [For well-educated girls hardly dare to move themselves, because of modesty in the sight of men]. The explication of the tale of Salmacis and Hermaphroditus includes material from Cicero's *De officiis*, Juvenal, Pliny, Horace, and Vitruvius. Regius explains words like "remolliat" (renders soft and effeminate) and "eheruet" (takes away strength and vigor) as part of his analysis of Hermaphroditus's effeminacy.[13] Orpheus' visit to Hades to retrieve Eurydice is accompanied by information about ancient beliefs, including the location of the door to Hell. Regius later added information about Hecate, Persephone, the meanings of serpents that appear in ancient fables, accounts of Latona, Jupiter, Cerberus, and the geography of Thrace. In good humanist fashion (if a bit breezily), Regius invoked the authority of Diodorus, Homer, and Horace on various aspects of Orpheus' significance in ancient religion and music.[14] Allegory gave way to history and to providing factual, rhetorical, and literary context in order to understand Ovid more on his own terms than in service of Christian mores.

Despite his insistence that meaning was made in language, Foucault all but ignored the development of sexual discourse in the Renaissance. And yet, meaning was made in language in new ways. The humanist emphasis on contextual explication and the rejection of ahistorical allegory rendered sex in increasingly rational terms. This move was not just in commentaries on Ovid. Renaissance humanists used their textual practices not just to recover but also to recuperate sexually problematic ancient authors. Generations of humanists, including Pierio Valeriano, Marc-Antoine de Muret, Achilles Statius, and Joseph Scaliger, emended, commented on, and glossed Catullus's poems, often in the face of ecclesiastical disapproval. Lost for the most part for a thousand years, the project of reconstituting Catullus's corpus and context resulted in, as Julia Haig Gaisser observes, a Renaissance Catullus who was in part a sensualist devoted to sexual pleasure.[15] Plato's trajectory was rather different. Never entirely forgotten, Plato fell into abeyance until knowledge of Greek returned in force to the West. At that point, the desire to deploy Plato as a civic humanist in service of secular political projects ran into concerns about his moral content. As James Hankins points out, translators and commentators had to either omit or explain away Plato's ideas about male–male love and marital communism.[16] Through the knowledge of antiquity, language and thought about sexuality expanded in volume and complexity. Cultural encounters with the poetics and philosophy of ancient Greece and Rome offered glimpses of sophisticated societies that treated sexuality differently than did Christianity.

Renaissance humanists accordingly reconsidered Christian imperatives in light of exposure to antiquity. Several pondered the idea that marital sex might be a good thing. They argued that marriage, and the sexual union at the heart of it, produced intimacy, children, and even love. Ambrogio Traversari celebrated chastity by translating St. John Chrysostom's *On Virginity* and dedicating it to Pope Eugene IV in 1434. Venetian humanist Ermolao Barbaro (1453–1493) wrote that celibacy enabled the practice of philosophy. Sex and family were distracting. Others were less certain. Some hazarded that marriage was a morally superior condition to celibacy and virginity as sexualities. For Alexander Niccholes, virginity is better, "But the greatest authority we have in praise of Marriage, is the union of Christ with his Church compared unto it."[17] Bernardino of Siena (1380–1444) denounced bachelorhood as miserable, insisting that single men would resort to sodomy and ought to be barred from office until married. Marital sex, in Bernardino's view, should displace the development of unnatural sexual habits. Bernardino's arguments were the result of demographic disasters caused by plague and war. More abstractly, Bernardino raised the question of where marriage fit relative to celibacy, as did Poggio Bracciolini's *On Whether an Old Man should Marry* (1437), which similarly advocated marriage because celibacy often failed. His endorsement of marriage was grudging:

> In what way can one who abstains from marriage avoid being an adulterer or a fornicator or [avoid] becoming more attached to another more detestable vice? Nor do you show me a life of continence, and there are indeed very few people who embrace such virtue. Therefore, one should embrace the married state, since it is a more virtuous life.[18]

Others, such as Coluccio Salutati, denounced celibacy in more emphatic terms: "Nature, or rather God, the author of nature, put genitalia in the human body so that humans might procreate and preserve the human race . . . Since Nature has bound all men to procreate, we call the sterile 'defective' and 'half-men' as they lack what is natural."[19] In his *Encomium matrimonii* (1518), Erasmus argued against clerical celibacy, noting that the Church allowed concubinage by fining clerics in irregular situations. He also invoked biblical injunctions to procreate. The mix of pragmatism and moral language indicates that Christian ideas continued to shape understandings of sexuality, but humanists worked toward a logic that made sense of sex in rational terms.

This was not a simple matter. Deciding what was acceptable sexually was not always entirely clear. For some, sex outside of marriage was by its nature bestial. Thomas Cogan opined that God ordained "mankind should not couple together without difference, after the manner of bruite beasts, but beeing joined in lawfull marriage, which estate was established between man and woman." Cogan went on to condemn fornication, adultery, "buggery," and incest as damaging to marriage and its primary purpose of providing, "mutuall society, helpe and comfort."[20] Philip Stubbs regarded "mutual coition betwixt man and woman" outside of marriage as a "horrible vice of whoredom."[21] For Stubbs, and indeed for most early modern communities, human social order depended on sex contained and organized by marriage. What separated human beings from animals was social organization, which was defined in part by marriage and sex kept safely within it.

The exposure to antiquity, however, deepened concerns that sexuality was at its core malleable, and thus, unpredictable. Central to sexual instability was the conception of the human body inherited from Greek and Roman physicians. Historian Thomas Laqueur argues that a one-sex model of human sexuality based on the Greek physician Galen (ca. 129–210) dominated European thought until the late eighteenth century. The male body was the standard, Laqueur maintains, with the female body imagined as an inferior or inverted version of the male, until the cultural need to imagine men and women as distinct and incommensurate sexes took hold in the Enlightenment.[22] While Laqueur has been challenged on several fronts,[23] the view of human anatomy he described was both compelling and unnerving in the early modern context. Particularly useful in terms of the anxiety about corporeal sexual instability was the idea that the vital fluids—blood, sperm, milk, and fat—were all thought to be the same basic material. Chyle, the underlying form of all the four, would change its properties depending on degrees of hot or cold, dry or wet of the body. Medical practitioners considered menstrual blood to be excess, and they thought women who did not excrete it monthly were at risk for diseases caused by retention. Pregnant women did not need to excrete blood because they converted the excess into milk. The other key sexual excretion was ejaculation, and experts described sperm as blood refined into foamy form. Science and medicine defined the sexed body itself by its emissions, seeing distinctions between male and female emissions as matters of degree, rather than absolute differences.

Given the malleability of human physiology, science and medicine did not provide stability about human sexuality, prompting renewed attention to drawing lines between man and beast at law. In Buckinghamshire in 1519, Richard Mayne's attraction to a mare got him accused of carnal copulation.[24] The Parlement of Bordeaux condemned Anthoine de Mars to be burned to death for copulating with "a beast." The magistrates ordered the animal to be strangled, a more merciful death than the burning Anthoine was to suffer, but death was required because the animal had gone "against human nature" in accepting Anthoine's advances. Further cases, in 1528 (no animal specified but the violation of human nature was again invoked) and 1601 (an "abomination" involving a woman and a dog), turned on maintaining the species barrier.[25] Jurist Laurens Bouchel called bestiality the worst crime, and explained that the animal had to be killed because it was the instrument whereby the person violated the divisions of nature. Bouchel and other legal commentators framed their reasoning by reference to ancient jurisprudence.[26]

In part, Renaissance observers feared bestiality because they feared it would result in monstrous offspring. With the breakdown of Catholic religious hegemony, the belief that "wonders" or "marvels" were signs from God made the appearance of monstrous births deeply unsettling. As historians Katharine Park and Lorraine Daston argue, after 1500, the association of monstrosity with the wrath of God gave way to explanations in terms of nature and natural variation.[27] But the change in perspective was gradual and congruent with the shift toward imagining sexuality in rational terms.

Stories abounded and circulated widely about humanlike creatures with animal attributes: in 1540, the monster of Cracow, which had multiple dog heads; in 1512, the monster of Ravenna, which had a horn on its head, wings instead of arms, and a single birdlike leg; lion bodies with human heads; humans with lesser animal characteristics (various hirsute creatures, men with hooves instead of hands and feet or webbed fingers and toes); and human bodies topped with various animal heads (ducks, pigs, or the 1561 baby with cow ears).[28]

Some were difficult to categorize, such as the monster of Freyberg, born in 1522. Called the monk-calf of Saxony, Protestants defined the creature as a monstrous critique of the Catholic Church.[29] Not only could sexual reproduction go terribly awry, but even more, sexual deformity figured in a number of cases of the blurry line between man and beast. Sexual monstrosity encouraged various modes of stratification. Pamphleteers and commentators depicted the monster of Ravenna as a hermaphrodite. French interpretation insisted that the genitals signified sodomy, which French commentators maintained was an Italian vice.[30] Other texts dwelt at length on various conjoined twins as markers of failed, flawed, or portentious generation.[31] In the Renaissance, fascination and repulsion, pleasure and horror mingled in the sexual products that were monsters, moving only slowly to their place in the development of modern natural history.[32]

The gradual rendering of monstrosities into a scheme of natural categories helped highlight categories of unnatural sex. In his narrative of New World exploration, Francisco Lopez de Gomara (1511–ca. 1566) recounts the sodomitical tendencies of the natives and warns:

> The female Indians in general are not as small as our women, but they seem to be, because they do not wear high slippers, much less shoes or heels. There are many of the men who wear tight and short camisoles, having very small sleeves . . . They are great sodomites, and if they are crammed with this vice, it is because of the necklaces they wear, where we wear chains, there they represent the God Priapus humping, and two men one on the other . . . The sodomites wear their hair like a crown, and for

en
od
:e,
pe
ix,
u:
)re
m.
ati
:fu
&
im
)o:
ra
tet
;o:

l res terrenas refpectum:fexum

FIGURE 3.1: The Monster of Ravenna from *De conceptu et generatione hominis, et iis quae circa haec potissimum consyderantur libri sex* (Tiguri: Christophorus Froschverus 1554, fol. 51. Wellcome L0031585.

this reason, one calls them crowned. The girls who protect their virginity often go to war with bows and arrows; they give chase, and can, without any fear of penalty, kill those who claim their honor. They take the children of their enemies, because they are the most tender to eat.[33]

FIGURE 3.2: Human monstrous bodies from F. Licetus, *De monstrorum caussus, natura, et differentiis libri duo*, 2nd ed. (Padus: P. Frambottus, 1634). Wellcome L0007744.

For Gomara, natives in the New World fail to maintain sexual order. Each group does it differently and wrong, and the results are, to Gomara's mind, horrifying. The slippage from effeminacy to transvestism to sodomy for men and from virginity to masculinity to cannibalism for women describes trajectories of unnatural sex for Gomara. Henri Estienne recounted several examples of sodomy and then elided it with bestiality: "Because following this, it is certain that Sodomy ought to be included in this category, and without other dispute, brute beasts render us convinced" [Car suyvant cela, il est certain que la Sodomie doit estre comprise sous ce titre: et sans autrement en disputer, les bestes brutes nous en rendent convaincus].[34] Estienne used sodomy polemically to attack Catholics during the French Wars of Religion. Indeed, the frequent association of sexual deviance with heresy—both being largely in the eye of the beholder—became proverbial in the Renaissance. In other words, another layer of disorder associated sodomy with religious strife.

As the responses to sodomy indicate, many associated unregulated sex with social chaos. Increasingly, interventions worked to imagine sex in regulated and productive ways. This was especially true in the broadening of medical discourse about sexuality. The proliferation of manuals offering sexual advice, often cast as medicinal or scientific information, reflects the move toward a rational view of sex.

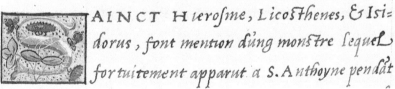

AINCT Hierosme, Licosthenes, & Isi=
dorus, font mention d'ung monstre lequel
fortuitement apparut a S. Anthoyne pendât

FIGURE 3.3: Hirsute man appears to St. Anthony, "Monstre lequel foretuitement apparut a S. Anthoyne pendant qu'il faisot sa penitence au desert," in Pierre Boaistuau,
Histoires prodigieuses (1560). Wellcome L0025537.

Giovanni Marinello spoke explicitly about the penis (including shape, size,
and possible defects) and testicles (focusing on defects). He expostulated about
female anatomy, extensively describing female reproductive ailments centered
on the uterus. Among the most common systemic problems were unbalanced
humors.[35] Going back to the ancient Hippocratic corpus and Galen, physicians
described the human body as a collection of fluids, or humors. The four basic

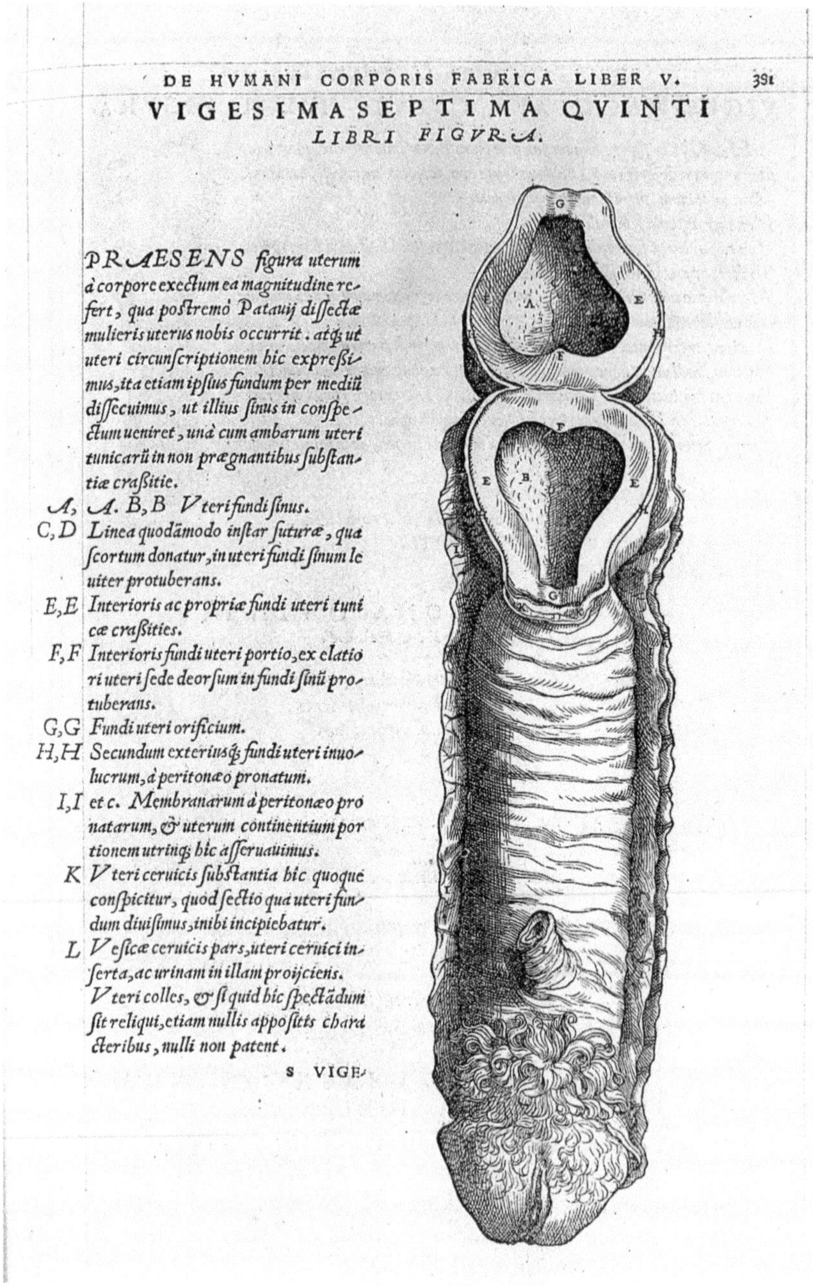
DE HVMANI CORPORIS FABRICA LIBER V. 391

VIGESIMASEPTIMA QVINTI LIBRI FIGVRA.

PRAESENS figura uterum à corpore exectum ea magnitudine refert, qua postremò Patauij dissectæ mulieris uterus nobis occurrit. atq; ut uteri circunscriptionem hic expressimus, ita etiam ipsius fundum per mediū dissecuimus, ut illius sinus in conspectum ueniret, unà cum ambarum uteri tunicarū in non prægnantibus substantiæ craßitie.

A, A. B, B Vteri fundi sinus.

C, D Linea quodāmodo instar suturæ, qua scortum donatur, in uteri fundi sinum leuiter protuberans.

E, E Interioris ac propriæ fundi uteri tunicæ craßities.

F, F Interioris fundi uteri portio, ex elatiori uteri sede deorsum in fundi sinū protuberans.

G, G Fundi uteri orificium.

H, H Secundum exteriusq; fundi uteri inuolucrum, à peritonæo pronatum.

I, I etc. Membranarum à peritonæo pronatarum, & uterum continentium portionem utrinq; hic asseruauimus.

K Vteri ceruicis substantia hic quoque conspicitur, quod sectio qua uteri fundum diuisimus, inibi incipiebatur.

L Vesicæ ceruicis pars, uteri ceruici inserta, ac urinam in illam proijciens.

Vteri colles, & si quid hic spectādum sit reliqui, etiam nullis appositis characteribus, nulli non patent.

S VIGE-

FIGURE 3.4: Uterus resembling a penis. Andreas Vesalius, *De humani corporis fabrica* (Basel: Oporinus, 1543), liber v, fig. 27. Wellcome L0015865.

humors—blood, phlegm, choler, and melancholy—were variable in heat and moisture. They had to be kept in the right proportions or, medical wisdom maintained, the body became ill or susceptible to outside forces. Environment and diet affected the humors, each of which had particular characteristics (e.g., blood was hot and wet; choler was hot and dry). Gender determined the humoral environment. Experts considered men dryer and hotter, and women colder and moister. Humoral difference was thought to explain physiology. Manuals insisted that women were smaller, had long hair, and had internal sex organs because they were colder. Although the female body was necessary for gestation, commentators opined that it was incapable of contributing more than mere matter during pregnancy. By the same logic, men were larger, often bald (their hair "burned" off because of their greater constitutional heat), and their genitals were exterior to the body. Humors also determined generation and pregnancy. Men produced seed, and going back at least to Aristotle, only male generative material contributed to the soul, which experts agreed was the most important part of a human being. Despite what seems now to be obvious gender bias in the conceptualization of the humoral body, its explanatory power was such that it influenced ideas about sexual physiology until well after the Renaissance.

What made the Renaissance treatment of the humoral body new was the attempt not just to present sexual physiology in rational terms but also to do so to a wider reading public. Writing in the vernacular, Michele Savonarola emphasized the rational aspect of sex. The act itself is, "very useful." Sex was part of the excretion process, and it should be moderate in quality and quantity: "Venus, then, or the venereal act, or concubinage, or conjunction, when it is between the masculine and the feminine with the seminal member and the natural member should be moderate . . . [this allows] useful excretion to be of great advantage." [Venere adunque, ò l'atto venereo, ò il coito, ò il concubito, ò il congiungimento, che si fa tra il maschio, & la femina col membro seminale, & col membro naturale, se sarà moderato . . . lo escremento utile, sarà di gran giovamento.] Moderation ensures the health of the sexual organs. Some sex, experts insisted was a good thing because it allowed for the expulsion of excess humors and restored proper balance. Moderation in sex resulted in healthier offspring.[36] Guillemeau warned that too much intercourse could introduce excessive moisture and cause an unborn child to come loose from the uterus.[37] Eucharius Roesslin described the categories of infertility in women: "In women three or foure generall causes by the which the conception may bee impedite an let: overmuch calidity or hear of the matrix, overmuch coldnesse, overmuch humidity, or moystnesse, and overmuch drinesse."[38] He devoted much of the fourth part of his treatise to identifying which condition prevailed, along with remedies and recipes to correct it. The sexed body in these texts was not determined primarily in terms of its relation to sin but, rather, as an object that could be in or out of balance for more or less mechanical reasons.

Cases of hermaphroditism reflected assumptions about corporeal sexual instability and cultural discomfort with the malleability of sex. Put simply, hermaphrodites brought the line between man and woman into question. In 1587,

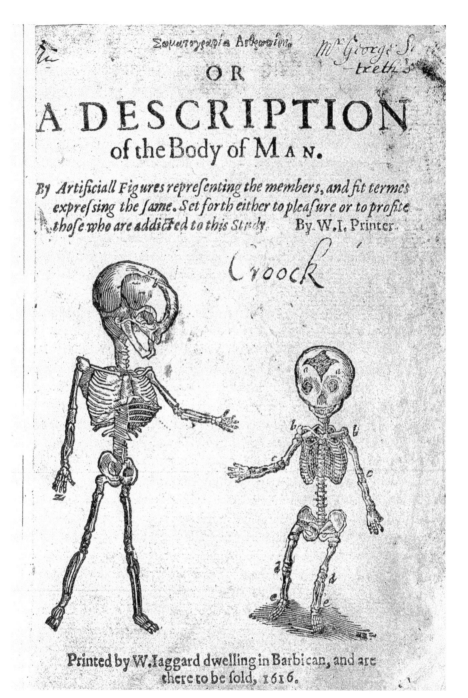

FIGURE 3.5: Title page of Helkiah Crooke, *Somatographia anthropine. Or a description of the body of man* (London: W. Jaggard, 1616). Wellcome L0001604.

Ortego Velázquez denounced Eleno de Céspedes to the Toledo Inquisition for having two sexes. Eleno, at that point married and a successful surgeon, had been born a female slave. As Elena, she had married and had a child, but Eleno claimed to the court that the stress of childbirth had resulted in the emergence of his male genitals. Adopting the language of ancient medical authorities who maintained that sex might change in the direction of greater perfection (that is, from female to male), Eleno contended he was a man. The court disagreed. The court convicted Eleno of sodomy and of using artificial means to deceive his wife. The conviction came down despite Eleno having received permission to marry and having passed muster with a royal physician.[39] In 1612, surgeon Jacques Duval published his account of Marie/Marin le Marcis, another person born female, who later took up sexual relations with a woman and claimed to be a man. The Parlement of Rouen initially convicted Marie/Marin. The court was swayed by the testimony of Jean Riolan, a physician who discounted the possibility of either a sex change or hermaphroditism. He considered Marie to be a sexual deviant, a lusty woman with unnatural appetites. Riolan never examined Marie/Marin, but Duval did. With the condemned set to be executed, Duval's physical examination found that Marin had a small penis. The judges reconsidered, and permitted Marin to live as a man. Duval's account includes a brief glimpse of Marin ten years later as a bearded tailor, capable of procreation with his wife.[40] In both cases, the hermaphroditic body caused unease both because of its changing gender and sex and its combination of the two.

All of which is to say that humanity was neither clearly different from animals nor particularly stable. What set man apart (and it was usually man and not woman) was his ability to display rational thought. Unfortunately, sexual desire often undermined rationality. Cultural remains of the anxiety about sex as a rational practice abound, beginning with articulations that desire easily and often misleads. The author of the *Batchelars Banquet* warned against falling too hard, too fast: "The next humor that is incident to a woman, is when the husband having entred very young into Lovs pound, and there fettered himself by his too much folly, for a vaine hope of ticklish delights which lasted but for a yeeare or two, hath matched himselfe with a very froward and perverse woman (of which sort there are too many) whose whole desire is to be mistresse and to weare the breeches, or at least to beare as great a sway as himselfe."[41] The husband, overburdened because his desire has led to children he must now support, becomes unpleasant or mean to his wife. She loses interest in him and is vulnerable to the seductive wiles of other men. Niccholes excoriated those who allow lust to determine choice: "Some undergoe this curse in steed of blessing, merely for lust chusing their wives most unfitly, as Adulteresses, and such are said to marry by the eye, looking no further then a carnall beauty is distinguished." In his advice on how to pick a wife, he admits that beauty facilitates attraction, but he is wary about getting carried away.

A wife ought to be appraised with the same objective eye that sizes up a horse before buying it.[42] Like many, Niccholes stops short when it comes to advice on how to prevent lust from being mistaken for love.

The Renaissance solution of sorts (which was, as we shall see, imperfect and incomplete) was to draw on the tradition going back to antiquity that deemed man a rational animal. Through reason, man might assert his special, unbestial qualities and use sex for positive good. Sexuality, in this scheme, lost some of its associations with disorder and became a way to achieve higher spiritual aims. One version of imagining sex in rational terms came through the infusion of Florentine Neoplatonism into European thought. Central to Plato's thought was the concept that true love proceeded by stages from love of particular objects to comprehension of the idea of love.[43] Especially in his *Commentary on the Symposium*, or *De amore* (1469), Marsilio Ficino (1433–1499) elaborated on the relationship among beauty, desire, love, and sex. Ficino defined love as the soul's desire for union, or rather reunion, with the divine. Souls desire to return to their creator, God, which they do through the contemplation of beauty.[44] Ficino describes this as noncorporeal and nonsexual at times,[45] but he also repeatedly conflates sexual attraction and transcendent desire in his corpus.[46] Ficino posits that the contemplation of beauty can achieve rational and transcendent noncorporeal love: "No corporeal nature enflames love. Only true beauty retains it. That is, it can have nothing of the corporeal."[47] Agathon's speech from the *Symposium* as analyzed in the *De amore* presumes that external beauty reflects inner, spiritual beauty, and so being attracted to it is a rational act. Real beauty is attracted to the "Angelic Mind," one of Ficino's five ontological hypostases (the others are One, Soul, Quality, and Body).[48] Attraction to the Angelic Mind enables the ascent of the soul to God.

The way this works is that the lover can learn to comprehend, in hierarchical order, love for souls, laws and customs, all branches of knowledge, the science of beauty, and finally, absolute beauty itself.[49] Ascent is marked by degrees of agency. The body cannot move itself, but it can rise to the soul. The soul can move itself but has an intelligence inferior to that of the Angelic Mind. The Angelic Mind has an appetite to return to God, and "The first conversion toward God is the birth of Love."[50] The Angelic Mind tries to reach God through beauty, which is determined by desire: "When we say love, understand this to mean the desire for beauty."[51] The highest good—God—has the potency of cognition and is whole rather than composite.[52] Beauty is infused by God in all beings, and the quest for beauty through sympathetic attraction is the mechanism for salvation of the soul.

The practical problems in this model were extensive. The senses can be taken in by physical beauty, causing love to be directed toward an inappropriate object. For Ficino desire is always for more perfect beings. This means Ficino imagines transcendent love in homosexual and homosocial terms. But later Neoplatonism saw love in almost exclusively heterosexual terms. Whether external beauty reflects internal beauty Ficino left opaque. Ficino concedes that desire for embodied beauty can overwhelm the senses and the intellect, rendering the human will unreliable.[53] The pleasures of taste and touch can be so violent that they dislodge the intellect from its proper state. Ficino claims that this is corporeal love, distinct from spiritual love, which he insists hates and shuns intemperance as contrary to beauty. So whither desire after all?

Lest this seem unduly esoteric, Neoplatonism circulated not only in philosophical texts but also in various forms at noble courts, in poetry, and on stage. Dialogues like Pietro Bembo's *Gli Asolani*, Baldesar Castiglione's *The Book of the Courtier*, and Leone Ebreo's *Dialoghi d'amore* reworked Ficino's ideas for princely audiences.[54] Giovanni Pico della Mirandola developed an alternate version of the ladder of love,[55] while Symphorien Champier and Mario Equicola in different ways struggled to translate Ficino's tenets into practical advice on love.[56] Louys Le Roy presented his reading of Plato's *Symposium*, which is heavily indebted to Ficino, to Francis II and Mary Queen of Scots as a wedding present in 1558. Marguerite de Navarre incorporated Neoplatonic love theory into the commentaries of *The Heptameron*, and her protégés, Antoine Héroët and Jean de la Haye wrote Neoplatonic poems and translated Ficino, respectively.[57] French poets, including Guillaume des Periers, Gilles Corrozet, Guy Le Fèvre de la Boderie, Pontus de Tyard, Maurice Scève, Pernette du Guillet, Louise Labé, Joachim du Bellay, Charles de Sainte-Marthe, Philippe Desportes, and Pierre de Ronsard, invoked Neoplatonic love in their poetry. John Ford's Giovanni in *'Tis Pity She's A Whore* used Neoplatonic logic to justify his love for his sister, Annabella.[58] This very partial list gives a sense of the widespread appeal of Ficino's vision of love as rational contemplation of divine beauty. Moreover, Neoplatonism contributed to the revaluing of sex in Renaissance thought. Beauty and desire were philosophical and spiritual positives in Ficino's scheme. Christian rejection of corporeal desire was tempered by the idea that desire was intrinsic to achieving salvific love of God.

As this brief account of aspects of Neoplatonism indicates, Renaissance sexuality still took seriously questions about sex in relation to salvation. Renaissance humanism did not replace the structure of sexual sin that dominated through the Middle Ages. Gradually, new ways of imagining sexuality and new questions about it came to the fore in large part because of attempts to engage with antiquity on its own terms. The infusion of knowledge about and interest in the ancient past raised questions about the relationship of sexuality to nature. The recurrent anxiety about the animality of sex was a manifestation of this concern. The ability to understand human actions in rational terms accordingly shaped articulations of sexuality. God, Christian revelation, and the ecclesiastical structure of sexual discipline shared space with Renaissance sexuality as a human matter. While the language of nature appeared regularly,

Renaissance references to it with respect to sexuality were of a different quality than later Enlightenment understandings. Not yet the competitive mix of reason and sentiment that marked Enlightenment sexuality, Renaissance versions represented attempts to imagine a place for distinctly human sexuality. It turns out that such imaginings could be substantially more coercive than Foucault's roseate view of premodern sexuality. The science of sex pronounced rational accounts of sex and dysfunction that defined so-called normal practices. The categorization of monsters, whether as objects of revulsion or curiosity, stratified sexual mishaps. The philosphical move toward rationalizing desire threatened failure on every rung of the ladder of love. In a paradox typical of the period, Renaissance sexuality both trusted in human rationality and used it in punitively normatizing ways.

CHAPTER FOUR

The Body in/as Text

Medical Knowledge and Technologies
in the Renaissance

SUSAN BROOMHALL

This chapter focuses on the ways in which the human body was constructed by discourses of medical knowledge and contemporary technologies in Renaissance Europe. Transformations in the understanding of the body and the development of new tools with which it could be studied were to have a profound impact not only upon health and healing but also upon ideas concerning faith and rationality, literature and the arts, and cultural relations in this era. At the same time the changing culture of European society during the Renaissance held important consequences for perceptions of the body and how it could be understood through medicine. The Renaissance was a time of immense challenge to medical ideas about the human body, as practitioners debated publicly the authority of the body in fashioning medical knowledge. How diverse communities developed answers to that question in relation to rival groups, and the impact of these transformations, is the focus of this chapter. It explores the insights borne of the interaction between contemporary culture and medical knowledge and technologies for understanding the human body, and how the culturally coded body itself was critical in the development of medical knowledge and technologies.

In the past twenty years, scholars have contributed much to the expansion of our vision of Renaissance medical knowledge. Such a term might once have been applied strictly to the universities and the textual transmission of their theories and practice. Certainly, the medicine of the faculties had a significant impact on how the human body could be treated and conceptualized, particularly within its elite, literate readership. Its knowledge has also remained in sources most apparent to modern scholars, such as manuscripts, printed texts and illustrations, account books, correspondence, and journals of practice. Studying such materials, scholars have recognized the rich diversity of viewpoints expressed within the university tradition, reminding us that individual imagination or experience was not limited by a shared training. The views, practices, and conceptual hierarchy underpinning the knowledge of other practitioners have been less easily recovered. Those who increasingly accepted the medical hierarchy or supervision established by university-trained physicians, such as apothecaries, surgeons, and midwives, also moved to record their debates through the written word during this period. The voices of these generally guild-organized practitioners have demonstrated their disagreements about corporeal knowledge with physicians and unauthorized practitioners, challenges to other medical guilds, and occasionally divisions within their own "fraternities."

Scholars have also looked elsewhere to uncover the ideas and practices of health care practitioners guided by faith-based systems. Some, such as hospital nursing religious orders or exorcists, are recorded as authorized carers for particular states of the human body within the Catholic Church. Other ideas about healing the body remain in the archives of witchcraft trials and investigations for beatification as victims and promoters sought to encourage acceptance of their understanding of the body's susceptibility to supernatural intervention. Legal records have revealed the therapeutic theories and practices of practitioners deemed illegal by authorizing communities in disputes over local licensing and trading infractions. By reading against the grain of often unlikely sources, scholars have been able to widen our knowledge of Renaissance corporeal ideas, meanings of health and disease, and how bodily states could be managed. The innovative search for evidence and its revelations enable identification of commonalities and distinctions between sets of medical knowledge and the technologies that enhanced them, as this chapter will explore.

THE UNSTABLE CORPUS: MEDICINE AS A FIELD OF KNOWLEDGE

Knowledge about the body in sickness and in health was always a contested and unbounded entity, rather than a discrete set of ideas developed and transmitted through the academies and their texts.[1] Unlike the other faculties of law, theology, and the arts, university medicine in the Renaissance was not an abstract field of study, constructed solely out of textual knowledge and rational analysis and therefore limited to those with access to these intellectual supports. Medical knowledge at all levels was also grounded in another source: the body. This fact made medicine more than a set of abstract theories; it gave it a concrete materiality through which new ideas could be identified or verified. It also opened up the possibility for people everywhere to hold an opinion about how the body functioned. Universal access to the body offered everyone an opportunity to observe and to develop theories about the behavior of their bodies and those they saw around them. The lived experience of the body in health and illness was, for many, a valid criterion for establishing bodily knowledge.[2] Thus, while all forms of medical knowledge shared at least one common text, that text, the body, led to a plurality of voices about health and illness, rather than the creation of a single knowledge community.

The body was an ambiguous producer of meaning for medicine. Individual interpretation of bodies was guided by the ideas of surrounding knowledge communities and by direct and personal experience of one's own body in sickness and health. It followed that the Renaissance medical imagination, particularly beyond the universities in which a recognized corpus of textual evidence had become a powerful tool of validation and coherence, would be fractured by the multiplicity of possible personal experiences and explanations from them. All of these direct readings of the body absorbed the discourses that define corporeal meaning and understanding. Yet university medicine also participated in the development of the regimes of power that determined these discourses by validating and adapting them.[3] As both the source for medical understanding and the site upon which medical knowledge was rendered valid, the body as a readable object, and as the performed embodiment of lived experience, was thus unstable, uncertain, and subject to continual negotiations of meaning between knowledge communities.[4]

Bodies as a source of knowledge about sickness and health were also deeply problematic because their codification and interpretation were not stable between different eras or indeed between distinct groups at the Renaissance. Bodies were read through the lens of historically specific gendered and racial assumptions as well as ideas of social status that were connected to understanding of bodily functions, capacity, and susceptibility to illness. This could render it difficult to confirm observations from other eras, locations, or groups. Shifting ideas about appropriate roles for men and women of different social status and ethnic groups would limit the construction of a coherent or universal set of observations and theories about their meanings.

Furthermore, medical logic was always subject to the unpredictability of distinct bodies, individual bodies that behaved differently. Treatments were thus uncertain, a fact that could lead a practitioner into ruin or prosecution. The contrast between patient assumptions about healers and healing, and the results obtained, led many practitioners before courts and tribunals, which were frequently composed of local physicians.[5] The body as an instrument for testing medical knowledge was unreliable, a factor that led to complex theoretical explanatory models (whether divine, diabolical, or abstract scientific traditions).

The interpretations drawn from bodily experience by different knowledge communities were thus unsurprisingly diverse. Two practitioners attending the same patient could form very different diagnoses and therapeutic regimens based on their intellectual frameworks. What was definitive information about the maintenance or restoration of health to a vast sector of society might be the physician's popular belief: false, misguided, or potentially dangerous. Similarly, medical logic derived from an inaccessible Latin tome might be perceived as irrelevant, and even harmful when translated into concoctions of exotic ingredients, in the eyes of a rural herbalist treating her patient with tried-and-tested simples. In other respects, however, sets of knowledge about the body did demonstrate some commonalities, born of shared foundational belief structures (for Christians at least) or the pervasive conceptual influence of humoral theory. Sometimes, actual medical practices were similar between rival practitioners, although the notions that underpinned them might not be recognized by the others' respective communities. Medical knowledge was not a contained set of ideas but a flexible and shifting corpus operated by different groups with varied degrees of authority in particular situations and for specific contexts.

In addition, what constituted health was, then as much as now, particular to class, gender, as well as ethnic and individual determination, so that even the focus of a corpus of knowledge intended to heal the body might contain surprising diversity. The restoration of health sought by an exorcist was quite distinct to that which a physician might be aiming to achieve. The aims of medical knowledge could be particular to discrete knowledge communities. For many practitioners, medicine enabled the articulation of a vision of the human body, particularly one constructed in relation to God. A wide variety of practitioners argued that understanding the intricacies of the human body was a way of revealing the mastery of God's work. Practical or clinical anatomists often presented their scholarly conclusions of the body's interior in this fashion.[6] In their knowledge of healing, practitioners also argued that they enabled demonstration of the power and mercy of God. The famous claim of the military then royal surgeon Ambroise Paré, "I treated it, God cured it" would be replicated as a defense by other therapists subject to legal investigation across the country.[7]

However, physicians also claimed a second theoretical purpose to their medical knowledge. While many might claim to know the human body, its ailments, and treatments, physicians were most vocal in differentiating the purpose of their knowledge to understand the human body and how and why it functioned as it did. Analyzing the causes and curing the symptoms of illness were argued as critical justifications for the superiority of their medical knowledge. Medical knowledge formed within the university knowledge framework, with its reliance on the learned classical tradition and a strong theoretical focus on cause, therefore served another purpose, which was to bring it closer in line with the abstract forms of knowledge created in other faculties, such as the arts, theology, and law, and to distinguish itself from the manual kinesthetic traditions of other medical work, such as surgery or midwifery.[8] In reality, a number of Renaissance physicians derived a significant proportion of their patronage by producing astrological charts for wealthy clients. Heinrich Cornelius Agrippa's refusal to acquiesce to his patron's request for horoscope prognostications (which he claimed would degrade the purpose of his medical training) spelled the end of his service for Louise de Savoie, mother of Francis I.[9] As the example of Agrippa indicates, the aims of medical knowledge and the uses to which it might be put was itself a matter of debate between different providers as well as between therapists and their clients.

THE LEAKY CORPUS: REGULATING THE TRANSMISSION OF MEDICAL KNOWLEDGE

The Renaissance was a time when practitioners were increasingly vocal about the diversity of their knowledge of the body, while other groups were simultaneously seeking means to control the extent of medical knowledge on display. Therapists voiced concerns about the susceptibility of un(in)formed minds to dangerous knowledge of their bodies, such as the English physician John Cotta in his *Short Discoverie of the Unobserved Dangers of severall sorts of ignorant and unconsiderate practisers of physicke in England* (1612). Physicians claimed, with the legitimation of royal, church, and civic authorities, that their control of medical theory and practice was critical to protect the public from false or unsafe practitioners.[10] Therapists called to the universities and courts who were unable to cite canonical wisdom of the body were condemned. Yet the opportunities for an ever broader sector of society to access medical knowledge were increasing through the period of the Renaissance with the increasing impact of communication networks, transport, and print technologies.

Bodies themselves were crucial to this phenomenon as one of the most important sites through which medical knowledge could be displayed and transmitted. One way in which knowledge about the body in sickness and in health could be gained was by observing practitioners at work. Indeed many guild and nonauthorized medical providers trained primarily or wholly by this means.[11] In the same way patients could observe treatments provided to them or those around them, asking, or surmising their own ideas for the rationale behind these therapies. Testimonials of traveling salespeople in the marketplace to the power of a particular cure might have added another dimension to general understandings about the body, and must have also responded to perceptions of bodily capacity for healing. The physical appearance of bodies also carried indications of disease impact and survival rates—the evidence of smallpox or measles scars might bear witness to local disease burdens but also to the low mortality rate of those illnesses.[12] Beauty itself was one indicator of bodily health, as guides to whiter teeth, sweet breath, shining hair, and other beauty regimens found in vernacular printed manuals, often written by physicians, attested.[13] Rhinoplasty and prosthetic limbs attested physically to ideas about the normative body and advances in surgical techniques, published by surgeons Gaspare Tagliocozzi and Paré among others.[14]

Transmission of bodily knowledge was also regulated by the Church in addition to, as well as in conjunction with, medical faculties. Before the events of the Reformation, most communities possessed relics, spiritual locations, and catalogs of saints to which one, with the encouragement of the Church, might turn for assistance with particular medical conditions. The intervention of St. Roch, for example, could assist plague victims as the representation in Figure 4.1, informed by contemporary university medicine, signaled to viewers.

Christians could enhance their knowledge of the human body and its susceptibility to supernatural forces by reading or hearing of miraculous healing events that occurred in relation to relics, sites of worship, or new forms of prayer. Some circulated in popular pamphlets and prints. Ideas about the body also developed through the practices of the Catholic Church. Through the collection of miracles judged fit to be recorded for beatification trials, both local communities and the Church negotiated ideas about what was possible for the human body, what was curable through nature, or what was only possible through divine or diabolical influence.[15] Such knowledge of the body evolved in dynamic relation to the Church and people's interpretation of broader medical information, which spanned the range from Latin university texts in the hands of Church officials to the views of local wise women who spoke of their observations of people's responses of similar illnesses, injuries, or events.

The mass organization of bodies in medical contexts also provided a means by which medical theory and practice might be conveyed to the lay community. Public health measures such as quarantine in times of plague and for lepers and syphilitics, as well as the more mundane requirements for street cleaning or disposal of human waste, spread notions broadly among urban populations about the human body's susceptibility to disease by contact and inhalation. Expectations about good health might be fueled by public health interventions only for particular afflictions (plague but not intestinal worms) and that promoted a vision of acceptable levels of disease impact.[16] Mortality rates for

FIGURE 4.1: St. Roch with his dog, indicating a plague bubo on his groin. Line engraving by P. L. after Bartolomeo Passarotti, ca. 1580. Wellcome L0022461.

children remained alarmingly high by modern standards, caused by social practices such as wet nursing that worked against transmission of maternal immunity, and concerns reflected in a nascent literature of infant and child health by physicians Gabriel Miron, Simon de Vallambert, and midwife Louise Bourgeois.[17]

Organization of patients within Renaissance hospitals also articulated, however secondary to cure of the soul, a medical knowledge that housed bodies upon distinctions of gender as much as illness, and reflected the authorized hierarchy of the medical world in the labor of its personnel from nursing sisters and charitable women to barbers, surgeons, apothecaries, and an occasional visit from a physician.[18] Hospitals' increasingly visible role in urban culture, as locations for civic charitable donations and art patronage, also spoke to contemporaries about the importance for care of bodies of the poor and the rich. Their architectural design, artwork, and other decoration signaled to a broad public the shifting religious and civic perceptions of charitable medicine and documented transformations in Renaissance cure of the body. The frescoed walls of the Ospedale di Santo Spirito in Rome, as shown in Figure 4.3, documented the significance of a broadening pharmacopoeia (here cinchona, important in the treatment of malaria) through contact with the Americas. The providentialism of some Protestant physicians merged morality, medicine, and bodily affliction.[19] The impact of such doctrines may be seen in the selective support of poor relief provided by the diaconate of the Dutch Reformed Church and of the municipalities in the Dutch Republic, signaling the importance of confessional identities in the assistance of bodily needs.[20]

FIGURE 4.2: A hospital ward in the Hotel Dieu de Paris, facsimile after a sixteenth-century original. Woodcut, from *Les edifices hospitaliers* by C. Tollet. 2nd ed. (Paris, 1892). General Collections. Wellcome M0004486.

Other changes occurring in medical knowledge could be charted in textual sources of scholarly manuscripts and printed texts. Within the Renaissance universities, the intellectual impact of humanist concerns for philology and a return to original sources was leaving a palpable impression on medical thought. The critical texts of the university syllabus were being cast in a different light by the new information provided by Greek texts, not least the value of Galen himself. The Hellenization of sixteenth-century university medical education signaled a paradigmatic shift from the Arabic tradition of medieval medical theory.[21] While the academy began to question from which texts medical knowledge should be established, the bodies of medical authors became important in the judgment about the knowledge they conveyed. As the Ottomans pressed on the edge of Christian Europe, Arabic sources, even those by Christian authors, were losing sway in the university community.[22] A focus on the ethnicity and geographic origins of authors shifted textual knowledge toward the classical world, and medical sources from Ancient Greece and Rome were reinserted in the so-called new medicine of the academy.

Debates between the advocates of the old and the new medicine raged in letters and published texts, even within their illustrations, where the weight of

FIGURE 4.3: Copy of fresco in the Ospedale di Santo Spirito in Rome, *The Count of Chinchon Receives Cinchona*. Oil. Twentieth century. Wellcome L0016709.

canonical wisdom could be demonstrated to the reader, as in Figure 4.4. Those
in favor of the new humanist medicine declared themselves by their participation in translating the revived authors into Latin or vernaculars or writing
commentaries about the value of these sources. In Paris, the medical humanist
Johann Guinther von Andernach, a staunch supporter and member of the university's traditional Galenist fraternity, published a Latin translation of Galen's
anatomical work as *De Anatomicis Administrationibus* (1531). In Lyon in the
1520s and 1530s, the practicing humanist physician Symphorien Champier
was an ardent and vocal advocate for the new syllabus, expressed through
his works in Latin and even French, such as the *Myrouel des Appoticaires*
(1533).[23] The act of translation not only signified the alignment of its authors
in the internal debates of the faculties, but it also introduced the university's
medical knowledge to a wider literate readership.

Although some within the academy debated which texts should form
the foundations of the medical syllabus and its theories and practices, other
graduates observed that various other therapists might have practical lessons
from their observations and experiences that could complement or enhance
the canonical texts. Phillippus Aureolus Theophrastus Bombastus von Hohenheim (Paracelsus), a contentious student of the medical faculties, even
advocated that physicians consult with villagers, empirics, and wise women
who might have bodily knowledge to benefit the academy. Paracelsus's ideas
were widely condemned within the universities during his lifetime, but some
aspects of his natural philosophy (as Paracelsianism) would reignite widespread university debate after his death.[24]

The technologies of print had a profound impact on knowledge transfer,
on the rate of transmission, and in the visualization of knowledge about the
body. In the first instance, it was generally the knowledge community of the
universities who used print to circulate their thoughts across national boundaries, beyond faculty politics, and for posterity. The concurrent development of
technologies to support sophisticated visual representations of the body, from
woodcuts to detailed metal engravings, enhanced the transmission of precise
medical information from anatomical images to ligature techniques. The rapid
exchange of ideas between physicians made possible through print, and accessible to others able to purchase the texts, far exceeded the slow circulation
and compilation of bodily knowledge in manuscript form. Moreover, academic
theories of anatomy and disease were widely transmitted within a Latinate
community of scholars through the exchange of letters, manuscripts, and, later,
printed texts. Latin enabled ideas to jump local linguistic distinctions and to
define a shared university schema for the production, verification, and content
of Renaissance medicine across Europe.[25]

The fault lines of bodily knowledge to be revealed to the wider public in
the vernacular, or obscured by Latin, was clearly gendered. Parisian physician

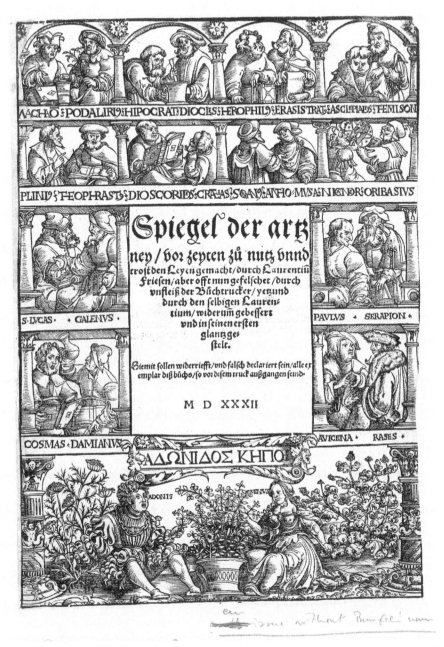

FIGURE 4.4: Detail of a woodcut depicting ancient herbalists and scholars of medicinal lore. "Herophilus and Erasistratus." The whole woodcut (Galen, Pliny, Hippocrates etc.); and Venus and Adonis in the gardens of Adonis by Lorenz Fries (Balthazar Beek-Strasburg, 1532). Wellcome L0040792.

André Le Fournier's *La decoration dhumaine nature & aornement des dames*
(1533), a manual on beauty and good health, shifted to Latin when he referred
to gynecological matters. Later in the century, Laurent Joubert's detractors argued that young girls could be corrupted by knowledge of their own reproductive functions through his 1578 vernacular text, *Les Erreurs populaires*.[26] Few
women were of course schooled in Latin. In such texts, Latin thus became a
gendered bar to bodily discovery and a signifier of a kind of learned masculine
incorruptibility.

In rural areas, there were few physicians to consult, a fact admitted even
by physicians themselves. Some published first-aid texts to assist countrymen
and women in the case of minor injuries or manageable conditions. A proliferation of market-driven self-help manuals, from Charles Estienne and Jean
Liébault's household medicine *La Maison rustique* (1564) to Michel de Notredame's practical cures for the plague in *Excellent et moult utile opuscule à
tous* (1556), also offered university theories for mass consumption. The quality
of the medical information that circulated in print was widely contested, especially by physicians, as a broader number of therapists proposed public communication of their techniques through the printed word and image. Use of
such disparate information was determined by readers, a fact that horrified
some practitioners. Laurent Joubert, chancellor of the Montpellier Medical
University, proposed to help lay readers make sense of the mass of information
by explaining the "errors of the people" while validating other popular views
of the body where they matched his own. Such texts may have sought to control the practices and ideas of the populace but they also provided access to the
tools for others to contest medical ideas into the future.[27]

The broadening intellectual framework of humanism sparked a number of
well-known physicians to participate in broader cultural domains, especially
literature.[28] In doing so, they transmitted information from their training to a
wider readership, sometimes in surprisingly influential ways. Girolamo Fracastoro's 1530 poetic analysis of a seemingly new sexually transmitted disease would be long remembered after the name of the central protagonist, the
shepherd Syphilis. Rabelais' 1546 *Tiers Livre*, the third installment in one of
the most widely reprinted fictional series in sixteenth-century France, included
extensive comic discussion of contemporary university debate that transmitted
a complex smattering of medical ideas to the reading public. Rabelais' depiction of a lively, excessive, unstable, and grotesque body contrasted with the
controlled classical body represented in the academy.[29]

Beyond regulating the tendency of their own fraternity to disseminate their
knowledge in print, the vernacular, and varied popular genres, physicians were
also concerned by the extent of "other" medical voices in print. Physicians'
involvement in the training and licensing of surgeons, apothecaries, and midwives asserted their own superior bodily knowledge, but it came at a price.

As guild-trained practitioners became increasingly integrated into the university training system, they mimicked physicians' textual transmission of their knowledge and highlighted their affiliation with the faculty as a form of accreditation of their knowledge. Increasingly, however, surgeons and midwives found in the medium of print a means to situate their distinct learning by text and experience alongside the theories of the physicians in the public eye. Other forms of medical knowledge in print were created anonymously and collated by printers seeking a regular income in the popular genres of the shepherd's almanac and the revelation of university or mystical formulae such as the texts of Pseudo-Albertus Magnus. Printers and publishers spurred by novelty and marketability, and reader interest in health, encouraged a plurality of medical opinion, forcing physicians to renegotiate their claims to superiority in a textual environment.

Pamphlet literature also proposed teratological analyses of bodies, fueling a fascination for explanations of deformed fetuses and infants or "monstrous births," as shown in Figure 4.5.[30] These frequently provided woodcuts offered visual and alphabetic access to contemplation of the miraculous possibilities of the human body. Many such texts were anonymous, and many developed their analyses within the context of Catholic and Protestant apocalyptic literature, but occasionally the voice of a medical provider could be heard in these discussions. The Zurich physician and playwright Jakob Rueff engaged with text and image in inventive ways, producing an illustrated book for a broad German-speaking readership, the 1554 *Ein schön lustig Trostbüchle von den Empfengknussen und Geburten der Menschen*, on the generation and birth of children. The work offered a unique analysis of monstrous births that was not located within a cosmological framework typical of contemporary teratology but rather grounded in the natural world.[31]

The increased transportation networks, volume of trade across Europe, and increased literacy during the Renaissance also increased communication technologies beyond print. Exchange of medical information was developed between different knowledge communities by a range of interactions. Elite patients who sought medical consultation by correspondence accessed a variety of scholarly opinions on their illnesses, introducing them to the ideas and internal controversies of the faculties. The advent of the scientific community of letters may have been encouraged by print, which created the impetus to develop a closed mechanism for information exchange. Laypeople who could read and write exchanged ideas about their own cures and favored practitioners in letters, spreading knowledge about different practices far and wide across the breadth of Europe and beyond.[32] By the end of the period, many people were conscious of diverse communities of bodily knowledge and had been exposed to a variety of conceptual frameworks for developing their own logic with which to make sense of the experience of their bodies in sickness and health.

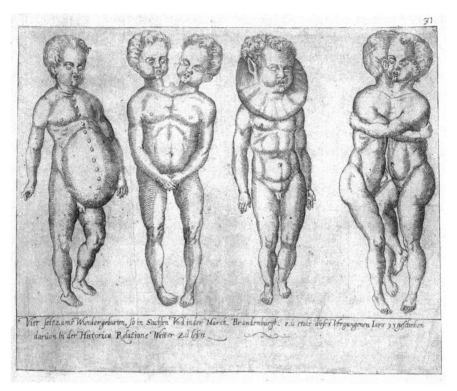

FIGURE 4.5: Four grotesquely deformed children. Line engraving, 21 × 26.7 cm. Iconographic Collections. Library reference no. ICV No7702. Wellcome V0007482.

AUTHORITATIVE BODIES: VALIDATING
MEDICAL EPISTEMOLOGIES

The matter of who could create medical knowledge (and who accepted or
regulated it) in the Renaissance was not straightforward. The evidence increasingly points toward plurality both in terms of the makers of corporeal ideas
and of those who chose to accredit its authority by their patronage. The authority of medical knowledge constantly needed to be renegotiated with other
practitioners and even within their groups, but most particularly with patients.
Primarily, it was patients and their families, neighbors, and community, who
determined from whom it was acceptable to ask for assistance, although these
choices were also limited by the availability of both providers and texts.

Fundamental to the determination of authority about medical epistemolo-
gies were mechanisms to assess its validity. Patients might consider a range
of factors to judge the quality of corporeal knowledge, including its cost,
previous experiences, the experiences of family and friends, their faith, and
whether the language and appearance of the practitioner concurred with expectations developed in discussions and from the observation and illustra-

FIGURE 4.6: Quack displaying his wares to a crowd. In the Rijksmuseum, Amsterdam. Pencil drawing, seventeenth century, by Willem Buytenwegh. Wellcome L0009785.

tion of such people in texts. Thus, the bodies of medical providers themselves became critical signifiers of the validity of their knowledge. The performance of corporeal knowledge provided by the charlatan in the marketplace fused authority with entertainment through a powerful combination of language, action, and appearance (see Figure 4.6).[33]

The physical manifestation and demeanor of varied medical practitioners were distinctive. A physician dressed in the long robe of the faculty was widely recognized, and his language and therapeutic logic might reflect the Latin and Greek textual authorities upon which his knowledge was based. His remedies based on analysis of the underlying cause of an illness would set him apart from those providers who proposed a therapy derived from long experience alone. Rhetoric and eloquence were emphasized within the academy as an important part of (the patient's confidence in) the physician's presentation of his medical theory.[34] Presentation of medical authority through the physical body could also be transferred into print, where author portraits reassured readers by their clothing and regard of their right to respect.[35]

By contrast, for others it was the experience of their own bodies that commanded esteem as a practitioner. Birth order, style of delivery, birth on a particularly auspicious day, or divine intervention provided mystical corporeal authority that was vital to the validity of medical therapists who used astrological and supernatural insights to form their diagnosis.[36] Personal, direct observation of techniques, as well as experience of their usage, were the hallmarks of validity for many practitioners. At the vanguard of advances in surgery, as well as their dissemination, was Paré, the French royal surgeon whose claims to truth in his illustrated French-language *Anatomie universelle du corps humain* (1561) stemmed from his experience as a surgeon to the French army. Following in Paré's footsteps was Louise Bourgeois, the first midwife to outline her obstetrical techniques in print, some of which consciously contradicted the standard university authority, Galen. Bourgeois could know what she did of the human body, specifically the female and infant body, she argued explicitly, because of her long experience with women of all classes and deliveries from the poorer mothers at the Hotel-Dieu in Paris to the society elite.[37] Significantly, she, like Paré, was also empowered to speak publicly of her bodily knowledge because of her proximity to the royal body. Bourgeois first published while she was the attendant midwife to Marie de Médici, Henri IV's consort.

Gendered bodies carried different kinds of medical authority. Women struggled to gain recognition from elite male communities for their experience or theories outside of contexts to which they were biologically connected. Although women were likely to have been responsible for most primary medical care and assistance in the home, this did not translate into authority to speak in printed forums (where physicians produced volumes to help women) or to recognition of their complementary labor by local faculties.[38] On the other hand, women, even those who were not trained or active providers, were assumed to have innate nursing skills.[39] Midwives were often accredited in part because they were mothers and because of their practical experience. Female nursing and midwifery activities are both displayed in Figure 4.7. Accordingly, male accoucheurs fought to assert themselves as valid alternative practitioners of obstetrical care, sometimes insisting upon feminine qualities to assert their legitimacy.[40]

Non-Christian bodies were also subject to scrutiny about the kinds of medical knowledge they could offer to Renaissance Europe. Jewish practitioners, male and female, had long been banned by medical and civic authorities from medicinal authority over Christian bodies. Yet Jewish and converso healers

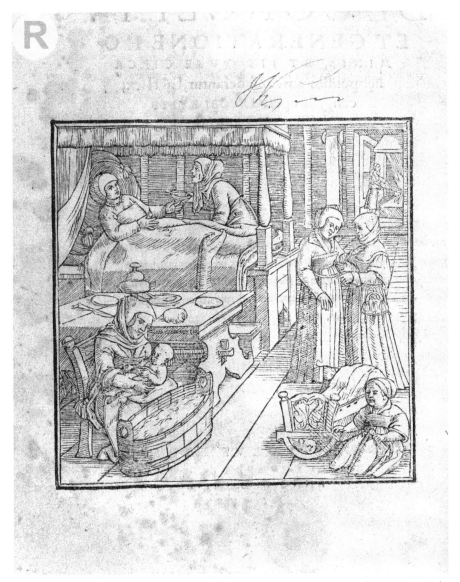

FIGURE 4.7: Rueff, Jacobus (1500–1558). *De conceptu et generatione hominis* (Zurich: C. Froschover, 1554). Frontispiece illustration showing birth scene. Collection: Rare Books. Library reference no. EPB5612. Wellcome L0032614.

fared far better within the medical establishment of Spain than did Morisco practitioners.[41] Through travel and trade, Christians were increasingly exposed to alternative medical theories and practices. Initial analysis of medicinal plants and their properties in the Americas was often conducted by those who saw products to sell: tobacco and chocolate among them.[42] Merchant traders such as Richard Hakluyt and John Frampton encouraged the consumption of new goods in Europe by publishing and translating information on

American medicinal plants, such as Hakluyt's *Discourse of Western Planting*, 1584. Observations of merchants, missionaries, travelers, and physicians about non-European bodily practices (from use of drugs to inoculation) meant that such ideas could return to Europe outside of academic contexts, and from European communities who perhaps held only tangential claims to authority about bodily knowledge.

THE BODY AS SOURCE: CREATING NEW KNOWLEDGE

Bodies were critical to the validation of medical authority, but their place in creating new knowledge was changing during the Renaissance. Distinctions in the practical and theoretical rationale underpinning medical knowledge between groups led naturally to divergence in the kinds of things known, knowable, and worthy of being known by medicine. A surgeon, midwife, oculist, lithotomist, or herbal practitioner might claim that successful treatment stemmed from careful and experienced observation of the human body, with perhaps some access to textual supports.[43] Within university medicine, the transition to permit contemporary personal observation of a theorized body to contribute to medical knowledge was a slow but profound development. As observation and experience of bodies became valid sources of new knowledge for the academy, practitioners explored new fields of analysis.

The developments of clinical anatomy at this period, merging theory, direct observation, and the manual techniques of surgery, would have deep and resounding impact within and beyond the learned academies. University dissection had occurred from the fourteenth century, typically as a ritual confirming the rational analysis of the practitioner. Now clinical anatomists, Andreas Vesalius most vocally, demanded new status for their investigations, arguing its power to create new knowledge about the body. In 1543, Vesalius who had studied at Louvain and at the Medical Faculty in Paris, published his extravagant, lavishly illustrated, and incendiary text, *De corporis humanii fabrica* (see Figure 4.8).

Vesalius's anatomical study offered no less than to denounce the false theories of Galen, privileging (culturally conditioned) direct and personal observation of the body's structure over textual tradition. Yet, informed by humanism, the work of some anatomists could be seen as an attempt to replicate the anatomical programs of the ancients.[44] Even within the images created by clinical anatomy, the call to antiquity was ever present. Perhaps because of its very innovation, anatomical illustration insisted on a classical heritage in the features and dress of its bodies on display and in the physical context of classical ruins in which they were frequently located. The anatomists' desire to value personal and direct observation of the human body could be reconciled to the antiquarian desire to see the sites of the past and the humanist desire to read the original source.[45] The negotiation of the place of the body and ancient commentaries within aca-

FIGURE 4.8: Andreas Vesalius by Jan Stephan van Calcar. From *Andreae Vesalii Suorum de humani corporis fabrica librorum epitome* (Ex officina J. Oporini, Basileae, 1543). Size: atlas folio. Collection: Rare Books. Library reference no. EPB6565/F. Wellcome L0046304.

demic corporeal knowledge also paralleled contemporary confessional debates concerning the role of scripture and biblical commentaries.

In another way, the challenges to religious beliefs presented by reformists profoundly restructured knowledge about the body. In areas that accepted

Protestantism in its varied forms, the body quickly became far less susceptible to divine supernatural influences, such as intervention of saints, although
female bodies still remained potentially open to diabolical forces. The intel-
ligibility of the Protestant body in religious terms decreased, as it became
comprehensible in "scientific" terms. Such changes to religious doctrine affected ritual healing beliefs and practices even in communities that remained
Catholic. Here doctrinal clarifications of the Catholic Reformation led to
new understandings of the body in illness, particularly illnesses that could
be identified as divine or diabolic. Authorized practitioners, such as exorcists, would be required to treat such diseases.[46] Yet some physicians, such
as Andrés de Laguna and Johann Weyer, would publicly reject supernatural
foundations to witchcraft and argue that the beliefs resided only in the minds
of witches and their victims. Scientific experimental method, developing the
transformations of the age, combined observation as a tool for theoretical
insight and as an instrument of practical experiment, and it would seemingly
have invalidated other "untestable" forms of bodily knowledge that could not
be subject to a system that privileged sight over other senses.[47] On the surface,
the experimental method may have looked similar to the practices of other
empirical practitioners, but distinctions remained in the scientists' theoretical
apparatus of the hypothesis and final thesis, and in their inability to detect the
same in the intellectual systems of other practitioners deemed inferior.

Clinical anatomy represented an attempt to control the unstable body by
conquering its landscape, as if by naming it, it could be known and controlled. The
analysis of clinical anatomy may have been restricted generally to a Latin readership, but its visual presentation could be interpreted by the paying clientele of the
anatomy theater and illustrations seen by a broader public. The legible body was
now something more than simply external manifestations and symptoms. Indeed,
although the anatomical illustration stripped away skin in dramatic écorché figures, as though discarding the last barrier to the hidden corpus of knowledge of
the body's internal structure, it was also transforming the understanding of skin
as a barrier to the university medical knowledge.[48] Diagnostic techniques using
the external products of the body—its sweat, urine, blood—were progressively
ridiculed in art and literature. Moreover, through anatomy, the body's concrete
parts were increasingly understood to have discrete, if coordinated and complementary, systems. The new focus on them would fracture the coherence of humoral theory and move away from its holistic approach to therapy.

Bodies could simultaneously provide new medical knowledge and confirm
cultural assumptions by their responses to new diseases introduced into Renaissance Europe. Their impact would expose the body to knowledge analysis
by the varied groups who witnessed the devastation they wrought. Syphilis
would be understood by many as a gendered and socially segregated disease.[49]
A surgeon such as William Clowes could describe the elite as unfortunate but

the poor as deserving victims of their fate.[50] Explanations for the impact of measles in the Americas would, like similar explanations for syphilis in Europe, be layered with racial and religious analyses, as much as revelations about the body itself. In apocalyptic strains, both diseases bore witness to the degeneracy of non-Christian peoples. The susceptibility of fragile indigenous American bodies to commonplace European diseases, and thus to diabolical control, exhorted European Christians to endeavor in their missions.

Knowledge about the capacity of bodies to respond in illness was also undergoing transformation. As a result of the new materia medica, tobacco, quinine, and sarsaparilla all entered into the European pharmacopoeia, providing new kinds of treatment and ideas about somatic possibilities, as shown in Figure 4.9. Diseased bodies en masse, together with print, also provided new data to be analyzed with new methodologies. The Bills of Mortality published for London were interpreted by John Graunt in *Natural and Political Observations Made upon the Bills of Mortality* (1662) in ways that would open up new understanding of bodily responses to disease (see Figure 4.10).

Yet other parts of the body did not appear readily accessible to developing medical science. Clinical anatomists read female genitalia through the lens of the normative male body, confirming text-based conclusions from antiquity that women were an inferior, imperfect version of men.[51] Yet female genitalia were still problematic; a confusing, complicated, and imperfect parallel for equivalent male reproductive organs. Early modern gynecological knowledge presented difficulties for physicians because it could not be personally experienced or understood. Its seeming instability and impenetrability to male analysis was reflected in an author such as Rabelais, whose comic grotesque style returns time and again to the female body to examine, exaggerate, and ultimately to reject it in favor of the more clearly comprehensible bodies of his male protagonists.[52]

NORMATIVE BODIES: COMPARATIVE BIOLOGIES

Ultimately, for all the knowledge it was capable of creating and reproducing, the human body was only knowable in relation to other bodies. Learned medical texts provide a particularly compelling set of assumptions about the hierarchy of bodies. The human body was understood by scholarly commentators in relation to two entities: God and animals. As God's most perfect creation, many viewed the human body as a unique achievement to be understood in isolation from other living creatures. As anatomists such as André du Laurens would argue, clinical anatomy provided a window into the majesty of divine wisdom.[53] Yet however much the material body might be known, the will of God could not, rendering unknown how the body might respond in the particular circumstances of a specific individual.

FIGURE 4.9: Indigenous people gathering cinnamon bark. From *Les oeuvres*, Ambroise Paré G. Buon, Paris, 1579. Collection: Rare Books. Wellcome L0006010.

If anatomy presented a view of divine majesty, it was because the human body was his most perfect creation. Du Laurens devoted a whole chapter of his 1600 anatomical history to the topic of why human bodies were superior to those of other animals.[54] Comparative anatomy advanced by reference to both Aristotle and Galen in the works of pioneers Volcher Coiter and Girolamo Fabrizi d'Acquapendente among others. Popular reproductive knowledge may well have come from observation of animal behavior and husbandry, comparisons which may not have been helpful for conception and contraceptive uses.

The normative human body of the faculties at least was male, youthful, European, and in good health. Monstrous bodies were perceived across both

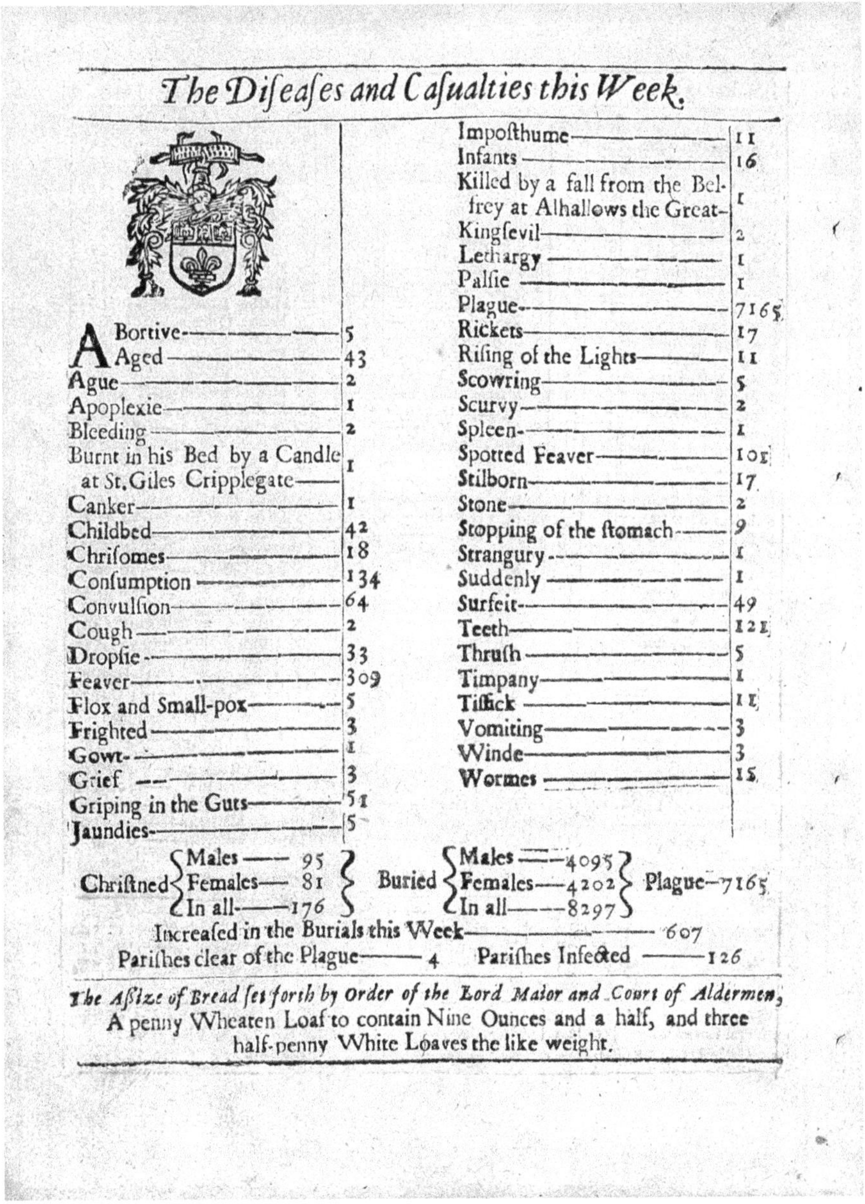

The Diſeaſes and Caſualties this Week.

Abortive	5	Impoſthume	11
Aged	43	Infants	16
Ague	2	Killed by a fall from the Bel-frey at Alhallows the Great	1
Apoplexie	1	Kingſevil	2
Bleeding	2	Lethargy	1
Burnt in his Bed by a Candle at St. Giles Cripplegate	1	Palſie	1
Canker	1	Plague	7165
Childbed	42	Rickets	17
Chriſomes	18	Riſing of the Lights	11
Conſumption	134	Scowring	5
Convulſion	64	Scurvy	2
Cough	2	Spleen	1
Dropſie	33	Spotted Feaver	101
Feaver	309	Stilborn	17
Flox and Small-pox	5	Stone	2
Frighted	3	Stopping of the ſtomach	9
Gowt	1	Strangury	1
Grief	3	Suddenly	1
Griping in the Guts	51	Surfeit	49
Jaundies	5	Teeth	121
		Thruſh	5
		Timpany	1
		Tiſſick	11
		Vomiting	3
		Winde	3
		Wormes	15

Chriſtned { Males 95, Females 81, In all 176 } Buried { Males 4095, Females 4202, In all 8297 } Plague 7165

Increaſed in the Burials this Week 607

Pariſhes clear of the Plague 4 Pariſhes Infected 126

The Aſſize of Bread ſet forth by Order of the Lord Maior and Court of Aldermen,
A penny Wheaten Loaf to contain Nine Ounces and a half, and three half-penny White Loaves the like weight.

FIGURE 4.10: *The Diseases and Casualties this Week*, London 39, from 12th to 19th September, 1665—recto. From *London's dreadful visitation: or, a collection of all the Bills of Mortality for this present year* . . . (Printed and are to be sold by E. Cotes. London 1665). Wellcome L0030701.

learned and popular traditions as prodigious (as in Figure 4.11).[55] The body of the hermaphrodite could be both creative and indicative of notions of sex and gender, as well as religious and political ideologies.[56] The female body was never the source for knowledge about the human body within the academic

discourse of medicine, but only for her own different functions, primarily reproductive.[57] Sexual function and reproductivity were equally critical to rendering the normative male body. The masculine entity against which "other" was
defined was proudly described and depicted in the prime of his reproductive

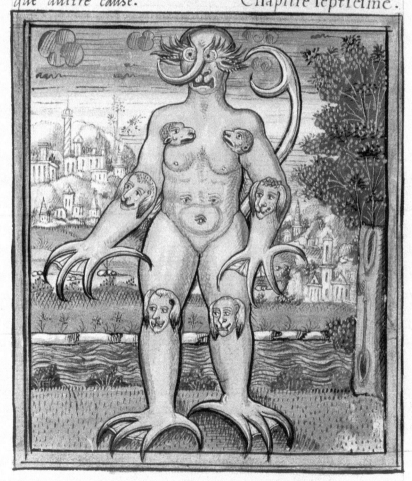

estonnement son enfantement dépuis à rendu tesmoignage).
Voy semblablement Cardan en ses liures de la Subtilité, ou il
confesse que l'estonnement a peu aider a lier ces deux enfans
ensemble, mais il dît qu'il fault qu'il y ait eu encore quel=
que aultre cause.   Chapitre septiesme.

FIGURE 4.11: *Creature monstrueuse engendree de parens honnorables.* Monstrous creature born to honorable parents, 1560. From *Histoires prodigieuses* by Pierre Boaistuau.
Wellcome L0025532.

functions, youthful, vigorous, and sexually mature. The physical body's decrepitude was rarely illustrated in university medical texts but fear of an aging human form found voice and vision in the age-related gendered and sexualized discourse of contemporary witchcraft.

Non-European bodies attested to the strength and power of the normative European body. Explorers reported back to their patrons that the bodies of American peoples were physically weaker than those of Europeans because the former were unable to withstand diseases that had long been childhood illnesses in Europe. The relatively hairless bodies of the Americas gendered them feminine, contrasting the masculine vigor of Renaissance Europe and may even have influenced Renaissance European popularity of the beard as elite men sought ways to demonstrate their bodily superiority.[58]

The identity of the normative male body was, moreover, obedient. Bodies were frequently shown participating in their own dissection, by lifting their skin to reveal inner secrets.[59] In such illustrations, it was "the body" that came to life, eschewing the specific character of the individual who owned it. The victim of execution was rendered compliant, contributing to the advancement of anatomical knowledge and reintegrated into the respectable community.[60] Within clinical anatomy, individuality was, it seems, progressively effaced from the physical body.

CONCLUSION

Knowledge about the human body in sickness and health is inherently unstable. Bodies as a source of knowledge would always be problematic to control but could also provide useful diversity and innovation for medical knowledge. At the Renaissance a series of events, technologies, and cultural transformations complicated the dynamics of knowledge production about the body. What medical knowledge was—and who could make, voice, disseminate, practice, and embody it—underwent profound transformations during this period. The authority of the body as a legible text was a critical debate in the university sphere, and its ramifications would have implications for healers in many domains beyond. Events of the period and its developing technologies participated in this discussion as they continually confirmed the unknowability of bodies, as well as offering new and challenging ways to interpret them. Such instability in discourses about the human body and the cultural disruptions they caused may also have led to reinforcement of other sets of cultural understandings, such as gendered hierarchies, which, despite all the new possibilities of evidence, discovery, tools, and ways of thinking, appeared to transition effortlessly to new forms of knowledge and their ontological systems, relocating themselves almost unchanged within new structures of thought, in equally powerful and influential ways.

CHAPTER FIVE

The Common Body

Renaissance Popular Beliefs

KAREN RABER

In the presence of its murderer, a body's wounds begin to bleed anew. Children's malleable bodies must be carefully fashioned by binding in swaddling cloths, lest their unreliable skin fail to maintain an appropriate shape and they become malformed. It is possible to contract the pox (syphilis) from dirty sheets. Plague is a visitation from God for one's sins. Lice are produced when perspiration is obstructed. The king can cure scrofula by his touch. Witches can cause disease with spells. The stars can generate or foretell plagues. Women who are lustful or commit sinful acts bear monstrous children.

These statements represent some popular ideas about the early modern body and its behaviors. The period 1500–1700 marks a tidal shift in the ways that the body was probed, examined, charted, and diagnosed by science and medicine—yet the beliefs listed here remain fairly constant among the broader population of the same period. Despite the professionalization of medicine and the advent of new theories about wellness and disease, despite the advances of the new science in discovering the internal workings of many bodily systems, despite changes in the social and economic situation of Renaissance bodies, popular lore persisted, proving resilient in the face of change. This chapter investigates the nature of the popular lore about bodies and posits some possible reasons for its staying power.

At one time it may have been tempting to see the kinds of beliefs represented in the examples cited above as superstitions embraced by the illiterate masses, battled over and gradually replaced with more rational, scientific views belonging to the educated elite, which ultimately trickled down to become the broad standard for all. Such a view dichotomizing high and low culture and their respective belief systems, and structured around a teleological narrative of progress, has been successfully deconstructed by scholars in history, anthropology, literature, and the history of medicine and science. Rather than a binary model that assumes a top-down imposition of reason on unreason, current thought on early modern body lore argues for a dialectic in which the exchange of beliefs and ideas is complex and multidirectional. Where the body and its illnesses are concerned, recent criticism in the history of medicine and science adopts the view that "the boundaries between learned and popular medicine were porous and blurred."[1] For Doreen Evenden Nagy, "the assumption that there existed [in the Renaissance] a body of superior, scientific medical knowledge accessible only to educated, trained professionals is untenable, as is a conception of popular medicine as being in essence illiterate folk remedies."[2] Margaret Healy warns that systems for explaining the early modern body were diverse, competitive, and often conflated; in addition, she notes that critics' post-Cartesian "desire to separate knowledge into distinct disciplinary categories" obscures the reality of how the Renaissance treated bodies: the "majority of interpreters" of the body and its ailments, for example, may not have been "learned physicians" but rather artists, writers, clerics, or even ordinary people.[3] In a related vein, Mary Lindeman suggests the limits of an iatrocentric history of the body, written by professionals and so disposed to treat popular beliefs as foolish superstitions or antiquarian curiosities.[4] And although it may be fair to describe movement from the imagination of the body as an open system to a new view, or desire, that fashions it as a closed, impermeable, and discrete entity, this movement does not necessarily constitute or confirm a whiggish or teleological narrative about medicine or the history of the body.

I would add that even the tendency to view history of medicine, science, and the body through a pre-Cartesian versus post-Cartesian binary creates a rupture where there may be none—Descartes' methods and conclusions build on aspects of change already present in European thought, though their influence is neither immediate nor complete for centuries afterward. Nevertheless, changes are at work in the Renaissance that, even if they do not conform to a neat narrative of a sudden and complete paradigm shift, do influence popular constructions of the body. How to balance a (potentially) triumphal narrative of historical rupture against evidence of continuity and non-linear historical change is especially important when dealing with accounts of the body that attend to oral, folk, and other traditions that percolate inconsistently throughout the cultural record we inherit.

The term "popular" that defines the purview of this chapter introduces a category that is not entirely stable or simple to define. It belongs to the "I know it when I see it" system of classification, and is the product of retrospective his-

FIGURE 5.1: *The Village Doctress Distilling Eyewater*, Thomas Rowlandson, ca. 1800. Although from a later period than the Renaissance, this plate indicates the contempt in which wise women or medical amateurs were held: the plate depicts an elderly woman urinating in a funnel and a younger woman and a cat all doing the same; the urine is intended for treating the eye. Ironically, urine's sterilizing properties may have made it a less harmful treatment for many conditions than the remedies prescribed by physicians. Wellcome Collection L0040282.

torical mapping. Few people in the Renaissance thought of themselves either as being part of, or diverging from "popular" ideas about the body. We could indeed borrow the high/low dichotomizing that marks popular as non-elite, or the lay/professional binary that has popular align with ignorance of the new science, but as I've indicated, part of the purpose of this chapter is to suggest that what is popular is precisely what is not reducible to such binaries—that popular

FIGURE 5.2: *Charlatans at Work*, from William Clowes, *A Briefe and Necessary Treatise touching the disease called Morbus Gallicus*, 1585. Wellcome Collection M0010402.

is common, in the sense of commonly available and commonly espoused. Most early moderns believed some or all of the statements with which we began, whether or not they were educated, regardless of class status, and sometimes in spite of all apparent scientific evidence to the contrary. Popular beliefs about the body thus transcended social and other forms of division, a fact that is central to my argument. Popular beliefs, in fact, structured the possible routes for the inroads of empirical science and medical authority when and where they did encroach. John Henry has argued that most people outside the medical profession in the Renaissance had extensive knowledge of the body and its functioning and a good grasp of medical theories,[5] which led to real resistance to the professionalization of medicine. If so, the Renaissance may have experienced the last historical gasp of such broadly available participation in medical expertise, one of the consequences of which was the tendency for professional physicians and scientists to defer to popular attitudes and ideas.

Much excellent recent work has been accomplished on the relationship of the body to science, culture, literature, and politics.[6] However, with a few notable exceptions, most of these analyses tend to privilege either subjectivity per se or the connection between the subject and a political order.[7] In this discussion of *popular* body lore, which I understand to imply a different orientation that privileges connection between groups and institutions rather than conflict, I will instead suggest that the Renaissance is a moment when the body is being redefined in ways that will eventually render it anti-communal, individualized, the material for use in a labor and consumer economy—being quite literally denatured and deracinated, cut off from an organic relationship with the environment, alienated from com-

munitarian, holistic models of its interaction with nature and its peers. The tenac-
ity of popular beliefs is a kind of register of (often inchoate) cultural awareness
that there are costs, individual and communal, involved in such a redefintion.

Medieval popular lore constructed a body whose most significant division in
the cosmos was from God; its most significant social division had to do with the
relative holiness of certain individuals.[8] The medieval body was continuous with
and a microcosm of God's creation. When Piero Camporesi notes in more than
one place the tendency to think of the body as "organic" in the sense of analogically related to natural things like plants and trees, he is registering the last expression of an essentially medieval perspective.[9] In a sense, the environment and the
community of fellow humans are versions of the same: the pre-Renaissance popular body was defined by its organic connection, or communion, with both, because both were its natural contexts. Even the negatives of early body lore tended
to reinforce the sense of community: all people suffered the same vulnerability to
the assaults of the world on a fragile, often treacherous physical existence.

The Renaissance, with its investment in the new science, what we now
identify as Cartesianism, and the rationalization of the body via professional
discourses of medicine, slowly, fitfully, but inexorably divides the body from
the natural world and from the body's human peers, until the body can finally
be divided from the self, imagined as a separate abstraction unlocatable in the
flesh. Norbert Elias notes that the Renaissance body was becoming "severed
from all other people and things 'outside' by the 'wall of the body,'" a process
registered in the development of manners.[10] The mind–body abyss that Cartesianism supposedly introduces erases what Michael Schoenfeldt calls the "seductive coherence" and "experiential suppleness" of the Galenic system, which
dominated popular understanding of the body from classical times even into
the eighteenth century.[11] In distinction to an Enlightenment sense of the self as
divided from the body, Schoenfeldt returns to Galenism to remind us that early
modern interiority was embodied, profiting from regulation and self-policing.[12]
But ultimately, Schoenfeldt's work suggests, Renaissance discourses about the
body will help subject it to an Enlightenment biopolitics that serves a modern,
rational, capitalist cosmos in which it is an insignificant cog, a machine that in
turn serves the machine of governance and commerce, alienated from its soul
and the souls of others. Renaissance popular beliefs, as supple and difficult
to dislodge as the Galenic system that underwrites many of them, remained a
drag on the process, and a reminder that it had a potentially high cost.

Discussing popular beliefs could logically take us into virtually all the areas
covered by other chapters in this volume—health and disease, cultural constructions of the body, gendered and racially or socially marked bodies, birth
and death, sexual behavior, medicine and its interventions, and so on. In the
interests of brevity, and to elicit a specific set of ideas about the body, I limit
what follows to several topics: the sensory body, the alimentary body, popular

remedies, and the dying or deceased body. These topics are also meant to offer a set of popular ideas that overlap with the material offered in other chapters, and to balance that with popular beliefs in domains not necessarily central to other parts of this volume. For the most part, although not exclusively, I discuss body lore in Renaissance England, rather than all of Europe. And although this chapter does not specifically analyze the medical profession, it necessarily touches on the role of medicine, because, as I have suggested, professional and popular overlapped. In addition, it is often only through the written texts left by the literate elites, including physicians, that we see the variety of popular lore reflected.

THE SENSORY BODY

What was it like to inhabit the early modern body? Historical documents and literature suggest that the body's experience of its own existence in the period involved a constant struggle to find an equilibrium within the body itself, and between the body and the world outside; the body suffered a bewildering array of discomforts and illnesses and submitted to practices that rendered everyday life a form of near-constant torture for some. At birth the body began to die, to become materially and spiritually corrupt, and to disintegrate (literally, to fall to pieces in some cases where disease was involved); the whole of life was a war against the inevitable. Ulinka Rublack points out that the early modern body was "situated in the continually-changing context of a relationship to the world whose precise effect was never stable or predictable, so that one simply had to submit to it.[13] Popular beliefs about the body both contributed to, and attempted to palliate, this experience.

The foundation of most thought about the body in the period, regardless of any other factors that might influence such ideas, was the Galenic humoral system that attributed the body's behavior to four essential elements: black bile, yellow bile, blood, and phlegm.[14] Dating from classical times, humoral theory involved a complex set of interactions between the balance of each of these bodily humors, with diet, activity, and even climate.

The basic types involved a larger proportion of one element, so that an excess of blood led to a sanguine temperament, black bile a melancholic one, yellow bile a choleric one, and phlegm a phlegmatic one; however, various degrees of heat and cold also had to be factored into the equation to determine exact proportions (women were, for example, colder of nature than men, but men could become colder on balance due to certain activities), as did other influences like moisture versus dryness, time of day, season, country versus city life, climate of place of birth or current dwelling, and so on. Because most disease was assumed to be the consequence of imbalances in the humors, constant monitoring of this very complicated set of conditions was necessary to guaran-

COMPLESSIONI
CO LERICO PER IL FVOCO.

VN giouane magro di color gialliccio, & con fguardo fiero, che effen-
do quafi nudo tenghi con la deftra mano vna fpada nuda, ftando
con prontezza di voler combattere.

Da

FIGURE 5.3: Figure representing one of the four temperaments or humors ("Complessioni"), in this case the choleric, associated with fire. From Cesare Ripa, *Iconologia* (ca.
1610). Wellcome Collection L0003189.

tee good health (and perhaps provided a more understandable way to explain
the much greater prevalence of bad health among the mass of people).

Further, for early moderns, even those with a strong sense of the mechanical
or physical origins of illness, the status of the body correlated with the state
of the soul. Early modern beliefs about physical illness, for instance, equated
physical symptoms with mental or moral states of being: pain in the body might
be a reflection of psychic distress or moral disorder within the individual; physical expressions of disease could register with the suffering individuals or their
caregivers as the gradual revelation of consequences from past behaviors or
events, while distortions in physical topography (ranging from birth defects to
moles or warts or simple bumps and curves) were assumed to reflect a perverse

FLEMMATICO PER L'ACQVA.

FIGURE 5.4: Figure representing one of the four temperaments or humors ("Complessioni"), a seated man with tortoise, representing the phlegmatic humor. From Cesare Ripa, *Iconologia* (ca. 1610). Wellcome Collection L0003191.

interior landscape. At the same time, moral conditions within the community could spur disease—hence, a common understanding of visitations of plague saw outbreaks as punishment for communal sin or social degeneration. Such a belief system had implications for the relationship between the individual and the group. Early modern body lore reflects a dominant concept of the body as a social instrument and a consequent profound anxiety over the body's ability to disrupt the social and spiritual health of the community.

The body of early modern popular imagination was open and vulnerable in part because it was experienced as transected by the senses. The very organs that helped it negotiate the world gave opportunities to invasion, corruption, or assault: the senses control the interaction with the two important aspects of the world that I am suggesting were experiencing a seismic shift in cultural construction relative to the status of the individual: the environment and other people. Medieval ideas about the senses involved a "much more open process, not just a form of transmission of information about objects, but one which enabled tangible qualities and, indeed, spiritual or intangible qualities to be passed from one party or object to another."[15] The Renaissance continued to experience the senses as a point of exchange with the environment and with other humans in it, although this experience was under revision. With the Enlightenment, sensory interaction with the world would begin to be reduced to a one-way process through which input is received only.

Eyes, most famously, were the potential target of terrible wounds. Vision was believed to happen by the active projection of rays from the eyes that entered into the receiving organs. Helkiah Crooke summarizes the thoughts of the classical philosophers on the subject of optics, some of whom assert that "certaine beames do issue from the Eyes and reach unto that which is to be seene," while others imagine that the eye sees by reception of some thing, likely an element of spirit that enters and imprints itself on the brain.[16] John Donne registers the physicality involved in seeing via projected rays in "The Extasie": "Our eye-beames twisted and did thred / Our eyes, upon one double string."[17] Light was the medium for vision, and light had moral connotations (light being associated with God, darkness with evil, and twilight with confusion). Eyes were also apertures that connected directly with the soul, and were at the same time windows to it, so that close scrutiny of them could deliver up information about their owner. Crooke agrees with this popular idea in *Microcosmographia*:

> the eyes are the mirror or Looking-glasse of the Soule. The Eyes wonder at a thing, they love it, they desire it, they are the bewrayers of love, anger, rage, mercie, revenge: in a worde, the eyes are fitted and composed to all the affections of the minde, expressing the very Image thereof in such a manner, and they may seeme to be even another Soule, & therefore when we kisse the eyes, we seem to reach and dive even to the verie soule.[18]

Vision could equally be turned inward, associating vision with insight into one's own moral state. Famously, Milton borrowed the classical idea of antiquity that blindness, rather than a sign of God's punishment for sin, was a means to better insight: although the "Holy Light" of heaven "revisit'st not these eyes" in *Paradise Lost*, Milton nonetheless hopes he is as a result "equall'd" with "blind Thamyris and Blind Maeonides, / And Tiresias and Phineas Prophets old."[19] The aggressive transmissive powers of the eye are evidenced in the folk belief in the "evil eye," or the idea that ill will could be conveyed by staring too long at someone or some thing; Alan Dundes speculates that this function of the eye is always connected to effects of drying, dessication, and withering.[20] Francis Bacon notes that envy is also produced by action of the eye: "We see likewise, the Scripture calleth envy an evil eye; and the astrologers, call the evil influences of the stars, evil aspects; so that still there seemeth to be acknowledged, in the act of envy, an ejaculation or irradiation of the eye."[21] Even in the happier event of falling in love, the eye was an aggressor that attacked and sucked the life spirit from unwary objects: "Love's arrows" were "not . . . a mere metaphor" but were "equipped with invisible pneumatic tips able to inflict severe damage on the person shot," while a lover ensnared by the vision of a beautiful woman could leak "so much spirit mixed with blood that his pneumatic organism is weakened and his blood thickens."[22] The very disconnect between the immateriality of sight and its demonstrable capacity to make physical effects begged for material explanations or made its actions appear a powerful mystery, potentially sorcerous in nature.

Active, aggressive sight had several implications for how people thought of themselves in the world and in their relationships to others. Although the examples cited earlier seem exclusively to convey anxiety about the vulnerability of the body, they must also be recognized to indicate multiple sites of agency. Donne's lovers are rendered coequal in passivity, as their eyes conduct a dance of connection; love's "arrows" darted from the eyes of the beloved to capture the soul of the gazer are a warning of the age-old suspicion of women's capacity to seduce and ensnare witless suitors, yes, but they are also the instrument of a common pain, however misogynistically expressed, that unites the male experiences of courtship. The evil eye tradition reminds watchers that their own aggressive gaze can be met and defeated by sorcerous intrusion, while the idea of the eye as "window to the soul" that "bewrays" the emotions within promises to allow a certain access to the internal thoughts of others. Needless to say, as such beliefs are replaced by scientific models of vision, the potential to change the world by looking on it diminishes. The idea that the eye can shape the world around it is instead made metaphorical. After this, any claim to transform the world through individual acts of vision will come to seem dangerously solipsistic.

The ears also opened the body, and the soul it housed, to external assault and influence. Sound reached the ear through waves of air hitting the ear: from classical times to the Renaissance, the belief was that inside the ear was a pocket of internal air that transmitted sound to the spirits within. As Bruce R. Smith points out, when Renaissance anatomists discovered the series of small bones that transmit sound waves, they were "clever enough . . . to reconcile these discoveries with the received idea of an aerated fluid as the main conductor of sensation.[23] Thus, Crooke, after noting the ancients' explanation of hearing produced in the "infancie" of anatomy, writes, "Indeede I esteeme this Ayre to be very necessary unto Hearing . . . but I can never perswade my selfe that it is the principall organ of Hearing."[24] The idea that sound is produced by a "blow" upon the air that reaches the ear led to the representation of sounds ranging from speech to city noise as the cause of great discomfort, even pain. Ben Jonson's character Morose in *Epicoene, or the Silent Woman* (1609) cannot tolerate even the slightest sound, and removes himself entirely from the social world of the city that produces intrusive sounds. His nephew, Ned Clerimont, describes his "disease" with the world:

> But now, by reason of the sickness, the perpetuity of ringing [bells] has made him devise a room, with double walls, and treble ceilings; the windows close shut and caulk'd: and there he lives by candlelight. He turn'd away a man, last week, for having a pair of new shoes that creak'd. And this fellow waits on him now in tennis-court socks, or slippers soled with wool: and they talk each to other in a trunk [sound-dampening tube].[25]

If we take the report of a foreign visitor to England at face value, then Morose's sensitivity might not seem so strange. Paul Hentzra reported in 1598 that the English were "vastly fond of great noises that fill the ear, such as the firing of cannon and the ringing of bells."[26] Morose, however, is clearly a cautionary figure whose extreme withdrawal from the community is condemned and, indeed, punished in the play through his mistaken marriage to a "silent woman" (who turns out to be neither silent, nor a woman). As hearing moved from its function in the Middle Ages as a vehicle for transcendent sounds that were reminders of heavenly realms, to its modern role as the means for torture through admission of the cacophony of daily life, what is lost is the recognition that admitting sound is part of the social life—and health—of the human community, precisely Jonson's point in skewering Morose.

Smell rendered the body a porous, osmotic entity, and perhaps surprisingly so did touch. A belief in the potency of "airs" to determine health and illness led to a literature about where to build a home, what kind of air to seek in general, and how to refer to healthy air to heal oneself. City air was more noxious than country air, but popular ideas also dictated that one's native air was more important to health than any gross division of the world into good and bad air; thus, if a person came from a certain part of the country, the air of that place was most conducive to his or her health, while being displaced into foreign air could cause disease. Specific diseases like the plague were believed to have bad air as one possible cause, sometimes influenced originally by the planets. "Putridity engendered putridity," a basic idea that was assimilable to the new empirical science.[27] "The air of a place," writes Alain Corbin,

> was a frightening mixture of the smoke, sulfurs, and aqueous, volatile, oily and saline vapors that the earth gave off . . . the stinking exhalations that emerged from the swamps, minute insects and their eggs, spermatic animalcules and, far worse, the contagious miasmas that rose from decomposing bodies.[28]

In such conditions, it is not surprising that smell had a physical, material dimension for early moderns. Odors are impossible to keep out of the body, and thus enhance the fear and anxiety over the body's openness to infection and corruption; in this fashion, however, odor can also stand for the condition of the human soul, impossible to make impervious to sin. If the stench given off by plague victims' bodies reinforced a sense of direct connection between disease and odor, that smell in turn signified the "reek of sin" made physical.[29] A distinction was made between "honest" bad odors, like the regular foul odors of animal or even human excrement, and "evil" bad odors that emanated from or acted upon human bodies and souls. Advice from one author about diet notes that abstinence results in "freedom from dishonest Belchings," presumably those full of bad smell.[30] Farts too figured in the popular imagination as signs of internal sin or ill will, attested to by the role of the fart in two of Chaucer's *Canterbury Tales*: in the Miller's Tale, the fart is a "blow" at a jealous husband from an adulterous lover, while in the Summoner's Tale it is a sign of clerical corruption.[31] Significantly, the medieval tradition of sanctified humans giving off heavenly scents did not persist into the early modern period in popular lore with the same regularity as did the tradition of evil smell. We might interpret this as the success of a version of the body that requires its fellow humans to wall themselves off from the stink of others, a body that is attempting to justify the need to bar itself from communion-via-scent with its peers. Apparently no one wanted any longer to open the nostrils and breathe in the odor of another on the chance that doing so would be a pleasurable and salvatory experience.

Constance Classen categorizes touch as the sense most associated with women, "imagined to be essentially feminine in nature: nurturing, seductive, dissolute in its merging of self and other" in opposition to the "dominating, rational" and so masculine faculty of sight.[32] Touch involves an organ that escapes localization—the skin, which is everywhere on the body: Crooke writes that "al other senses are retrained within some small Organ about the brayne, but the Touching is diffused through the whole body."[33] Crooke celebrates the sense of touch as preeminent—it is, he says, the sense most responsible for our conception, we have it even in the womb when other senses lie dormant, and it saves us often from mishap. However, Carla Mazzio argues that the sense of touch is subversive of rational systems of categorization to "disable ways of calculating knowledge" because it bypasses the more "rational" senses.[34] In its best role, the skin helped balance the body's internal fluids through sweating, and traditional medicine used touching and stroking as part of the healing art. The king's "laying on of hands" to cure scrofula participated in this positive dynamic of touch.[35]

William Clowes cautioned that the cure of scrofulous tumors should not be undertaken by "chirurgeons," lest their attentions drive the tumor to become

FIGURE 5.5: King Charles II touching for the King's Evil. From R. White, *Adeno-choiradelogia* (1684). Wellcome Collection M0010336.

cancerous. Instead, "Divine and holy curation" is required, and provided best by the king's touch.[36]

Despite this, however, and Crooke's efforts to rehabilitate the skin's reputation, it remained a source of apprehension. Georges Vigarello calls it "the wide open skin" that did not adequately hold the inside of the body together—hence the fear that children would not grow properly if not swaddled or if given too many baths.[37] The pores of the skin came in for suspicion too, because lice and fleas could be imagined growing out of them, "the result of uncontrolled exhalations": "They came out from the skin like certain maggots."[38] The sensory body of popular belief is a body in transition from its fully porous medieval incarnation, to its closed and walled-off condition in modernity. Even the fears and anxieties produced by popular views on the body's sensory connections to the natural environment and others affirmed the body's potential unity with creation—few images could inspire confidence in that belief as that of the skin itself generating insect life.

BELLY LORE

Early modern Europe experienced what is now often referred to as "the Little Ice Age," nearly four centuries following the Medieval Warm Period when temperatures made survival difficult for most of Europe's populations. England did notably better than other European countries, largely because of its relatively early embrace of farming practices and crops that responded to the new climate; however, most people had an insufficient diet for most of their lives, and occasional outright famine reinforced a sense of the relationship between humans and their environment as a war waged over the body's needs.[39] The agricultural effects of the Little Ice Age, compounded by the extreme complexity of humoral medical theories and patterns of social distinction, may also explain some paradoxical popular attitudes toward the consumption of food. The "famous Italian" whose discourse on diet was included in *The Temperate Man* notes the decline in abundance from the classical era, "that nowadays the store of victuals is so much abated and the price inhaunced" so that he concludes "undoubtedly that scarcity . . . can proceed from nothing but our excessive Gluttony."[40] On the one hand, food is associated with the celebration of the community. In Bahktin's terms, "man's encounter with the world in the act of eating is joyful, triumphant; he triumphs over the world, devours it, without being devoured himself."[41] This carnivalesque form of consumption by the "grotesque" or popular body is counterbalanced, however, by acts of extreme control asserted over individual acts of eating, which elevated abstinence and regimen to supreme values. Such a paradox is often understood in terms of a history that opposes the communal and collective world of the masses either to the social elites, possessors of a classical and well-policed body, or to the social norms, resisted through cultural expressions of the grotesque.[42] But if such a paradox is read, not against literary evidence or ideas about social class, but against the climate record, both approaches make some sense as cooperative expressions: abundance and the ability to consume in unfettered ways was indeed something to be celebrated fulsomely, not least because such abundance was nearly always temporary and perishable; but in a world that experienced dearth as a normal state, abstinence and self-regulation served the community as much as they seemed to serve the individual—avoiding gluttony was, as the "famous Italian" hints, a necessary path to restoring access to food for all.[43] Indeed, during one period of bad harvest, the Jacobean authorities in England issued edicts to curb eating in order to prevent further food shortages,[44] an act of policing satirized by Thomas Middleton in *A Chaste Maid in Cheapside* (1613): government representatives "prow[l] the streets and confiscate[e] foodstuffs," "arrest the dead corpse[s] of poor calves and sheep" (2.2.63).[45] However foolish such government action might have seemed at the time, it attests to the degree of social impact caused by individual choices about diet.

As a rule, popular ideas about eating among early moderns themselves came from the Galenic system and the general idea that a balance or equilibrium within the body needed to be maintained through constant vigilance. Dietary advice in the Renaissance is so ubiquitous and so voluminous that it would be impossible to excerpt any proportion of it here. The goal of all the advice, however, was the same: to adjust diet to a degree of compatibility with the basic humoral makeup of the body in order to produce an ideal mean, a stasis point of temperance, when all factors are precisely balanced for optimum health. Further, as Schoenfeldt observes, "Galenic physiology, Classical ethics, and Protestant theology" cooperated to represent "all acts of ingestion and excretion as very literal acts of self-fashioning."[46] Thus, the importance of what was consumed and how it was consumed was of constant concern on moral, physical, spiritual and social grounds. Inattention to diet, or indulgence of any sort, could lead to illness and premature death. "Not only is every act of eating [in the Renaissance] a performance of our fragile dependence upon the recalcitrant grace of a predatory world; the processes of corporeal assimilation demand the continual intervention of divine grace to accomplish the miracle of digestion."[47]

"Eating was considered a shameful act" both because it signified the triumph of the corrupt flesh via the appetites, and because it required processes that resembled the lowly domestic chores of cooking.[48] Popular and medical ideas about digestion borrowed from the visual, commonly experienced chemistry of the kitchen, so that one's intestines were believed to boil, bubble, and concoct raw materials into bodily fluids. If it was preferable in the Catholic world to assume a rigid posture of bodily denial that attempted to transform the body into a "sealed jar," especially through extreme fasting, Protestant England translated such a desire into a more rational medicalized aspiration to keep the body strictly at a point of equilibrium through moderation or intervention via weights, measures, analogical theories about types of food, and vomits spurred by purgatives.[49] Instead of a sealed jar, the body was a labyrinth, an alembic, requiring evacuation to match its consumption. Viscount St. Albans, quoted in a preface to *The Temperate Man*, notes that "a slender Diet . . . answerable to the severest Rules of Monastical life" prolongs life, but if one can't achieve such a thing, keeping food "equal and after one constant proportion" will help. The author of the 1676 treatise *Rules for Health* goes so far as to recommend a "weighing chair," that will show by its descent that the diner seated in it has eaten the precise amount required for his health, only a tad more extreme than Leonardo Lessius's advice to weigh food before eating.[50] Lessius argues that the "catarrhs, coughs, headaches, pains of the stomach, fevers" other men ascribe to various causes are in fact all due to poor diet.[51] Ludovico Cornaro's *Treatise of Temperance and Sobriety* claims that it was his "infirmities" that first persuaded its author to "leave Intemperance, to which I was much addicted."[52] Indeed, the call to moderation in diet is so powerful that even fat and gluttonous Ben Jonson, imagining a feast of expansive proportions for thirty-five lines of poetry, feels compelled to observe in the thirty-sixth line: "Of this we will sup free, but moderately."[53] Some applied themselves too much to the antidotes for food—purges and vomits, which were intended to evacuate offending excesses or get rid of excess moisture or other elements to return the body to temporary stasis. William Vaughan calls a "vomit" the "wholesomest kind of physick" because that "which a purgation leaveth behind it, a vomite doth root out."[54]

Individual dietary behaviors connected to the fate of the community at large. As Cornaro argues, diet is a linchpin of order: he blames family arguments and the deep melancholies felt as a result by other family members on lack of a temperate diet.[55] Schoenfelt describes the attention paid to diet in spiritual autobiographies: "Tempering what enters the body, and sweating out one's inevitable excess, become essential components of the spiritual life of the subject which was in turn linked to the spiritual health of all mankind."[56] Spenser's treatment of the virtue of Temperance in his *Faerie Queene* ends at the Castle of Alma, where the castle of the body includes a "bustling" gut that regulates and evacuates as if it were a good Protestant imbued with a solid work ethic.[57] Cornaro also claims that "Order makes arts easie, and armies victorious, and retains and confirms kingdoms, cities, and families in peace."[58] The spiritual health of individuals, the spiritual and social health of the community, are all bound up in a dynamic that requires knowledge, vigilance, and active remediation of the belly's condition.

Suspicion of food came in part from the view that food was already in a process of decay. Cheese, for instance, "increases the existing putrefaction in the dark meanders of the intestines and the recesses of the human bowels"; as a result, it was the suspected cause even of contagious diseases like plague, and on the basis of analogical theories, seemed suitable for use as a poultice or medicine.[59] The appearance of worms in food that had begun to rot seemed to link it to the fate of human flesh; the significance of worms was that much more powerful given that they cohabited within the body of humans and they waited to eat the body after death. Hamlet's reminder that the body is "food for worms" applied not only at or after death, or in sermons, but in everyday treatments for parasites, those creatures that eat instead of being eaten by human beings. Most Renaissance treatises on physick include remedies for worms, usually emetics or purgatives of some harshness. John Tanner includes the judgment that worms are associated with gluttony, because worms are "bred of such Nourishment as easily putrifieth" and so tend to show up most in "Children and such as are Gluttonous."[60] The potential foulness of the body's interior registered in descriptions of the effects of worms in bad breath, swollen stomach, foul wind, and so on—all reminders not only that worms were eating the body from within or stealing its nourishment, but also that they were themselves voiding into the gut, excrement within excrement. Part of a circular material process of eating, living, and dying, the body was a burdensome reminder that nature would not be denied her meat.

The alimentary body would eventually be cut off from its Galenic past. By the late eighteenth and early nineteenth centuries, fatness and thinness would no longer be perceived as the effects of imbalances, nor would individual gluttony be interpreted as social subversion. Whether or not this change constitutes progress, however, requires one to consider the effects of the new dietary regimens on individual self-conception and account for the disciplinary division of diet from other forms of medical interest in the body. The pathologizing of fatness and thinness in later periods suggests that change was change, not progress.

HEALING BY NATURE

One of the most common areas to find folklore and beliefs well-represented is also a field that has garnered relatively little attention from theorists or social historians in the same way that, for example, anatomies or dietary regimens have, although it is an important focus of some medical and scientific literature: the creation of remedies—potions, pills, possets, and other treatments—is a crucial part of body lore. Early modern illnesses can be very roughly divided into three categories: those that are generated by imbalances in the humors (thus, internal), those that are normal infections commonly experienced by many, and plagues or epidemics that are seen as unusual and distinct from other infections. Each of these categories of illness involved vastly different cultural perceptions of the body and its cultural place: brilliant work has been done to distinguish how epidemic disease affected the way early modern society characterized national identity, for instance, or how plague challenged traditional interpretations of local or specific sin and its role in bringing down divine retribution on a few unfortunates.[61] Equally, the critical literature on the medical profession has accounted for economic influences on where, how, and by whom differing levels and types of illness were treated. But one constant in all these accounts seems to require further examination: regardless of what new forces influenced the cultural understanding of any particular type of illness, a persistent, basic reliance on remedies specifically from herbal, domestic, sources remains the same.

There was ample reason for the lore of herbalists and folk practitioners to fall out of favor in the period, given the rise of a professionalized medical community and the association of herbal lore with country ignorance, women, domestic knowledge, and witchcraft so it is all the more surprising that herbal and folk treatments remained so popular for so long. Learned medicine and herbal lore were divided by many things: learned medical knowledge claimed the classical texts and traditions as its own, while herbal lore grew out of oral, often local or familial transmission and custom; learned and herbal lore relied on different notions of what constituted the holistic treatment of the individual. Perhaps the two most important sources of division, however, have to do with the basis of herbal remedies in the domestic household and the natural environment, and the important role of women in creating and dispensing them. Thomas Lodge recommends a variety of potions and pills as restoratives to prevent plague, mainly composed of rosewater, rhubarb, or other natural and common ingredients, but cautions that they should be under the "direction . . . [from] a learned and diligent Physition and not according to the fancie of foolish chare women and ignorant practisers."[62] Laurent Joubert's 1578 *Erreurs Populaires* castigates folk cures in a section titled "Those who know a little about medicine are worse for the sick than those who know nothing at all":

> There are people who know nothing at all about medicine nor anything about arguments or reasoning (such are ignorant women). They do not even know how to read or write but imposed on us their observations and their rules. Just because they know how to do a pottage, a sling, restoratives, a barley soup, a proper bed, a patient's hair, and a few simple remedies for the mange, burns, the falling of the uvula, worms, the suffocation of the uterus, etc., they think they know everything and undertake numerous cures, according to their own humour and fantasy . . . It would be much better if people attending the patient know absolutely nothing other than how to follow the physician's orders.[63]

Joubert does not, in fact, deny that these "ignorant women" can cure with their simple remedies, only that they overstep the limits of their knowledge when they interfere with the business of the learned doctor. However, in the teeth of written diatribes against old wives' tales about the efficacy of herbal and other homegrown treatments, and in the face of the growing control over drugs by licensed apothecaries, popular faith in traditional potions took a remarkably long time to wane—indeed, it is possible to argue that it never fully did. Nor did the professional medical community entirely reject herbals: most physicians were experienced in some herbal lore, and they proffered herbal and chemical remedies. Nearly half of William Bullein's *The Government of Health* is a recitation of the properties of herbs, foodstuffs, and flowers for the regular diet and for the treatment of specific conditions; likewise, books of "simples" and herbal preparations, like Nicholas Culpeper's *The English Physician* (1652), William Langham's *The Garden of Health* (1597[1598]), or Robert Turner's *Botonologica* (1664), included extensive lists, uses, and instructions for preserving natural remedies.

Andrew Wear points out, and these works by these purported physicians attest, that there was real reciprocity between the folk/domestic sphere and the professional regarding remedies, so that a doctor's recipes might become part of the domestic property of a housewife, and many country wise women's treatments found their way into official texts.[64] However, professionals (both physicians and apothecaries) often argued that the uneducated lacked the resources (intellectual and material) to preserve or correctly prepare herbal treatments, and there was constant tension between the two camps.

The number of women involved in treating illness and discomfort can only be inferred from the evidence of those represented in history and literature, usually the wealthy elite. Noblewomen not only treated their own households with their homemade preparations but also administered to often far-flung constituencies in their towns, especially if geography or poverty made learned doctors inaccessible. Mary Josselin had skill at the process of distillation, and Lady Grace Mildmay's papers reveal her "medical practice [to have been] extensive, systematic and at the forefront of contemporary medical knowledge."[65] The average housewife of whatever degree was expected to be expert in physick and even surgery; in more literate homes, recipes for remedies might be conveyed in written form, some even published as part of general collections that might include food and medicinal recipes, as Robert Boyle's sister, Lady Ranelagh, did in the later seventeenth century.[66]

Folk and herbal remedies have two characteristics that may have contributed to their effectiveness, as well as to the suspicion in which they were held by subsequent generations of medical science: one plant can serve in a vast array of treatments, and most herbal preparations contain many, sometimes dozens or more, ingredients. Lady Mildmay's medical records include a recipe for

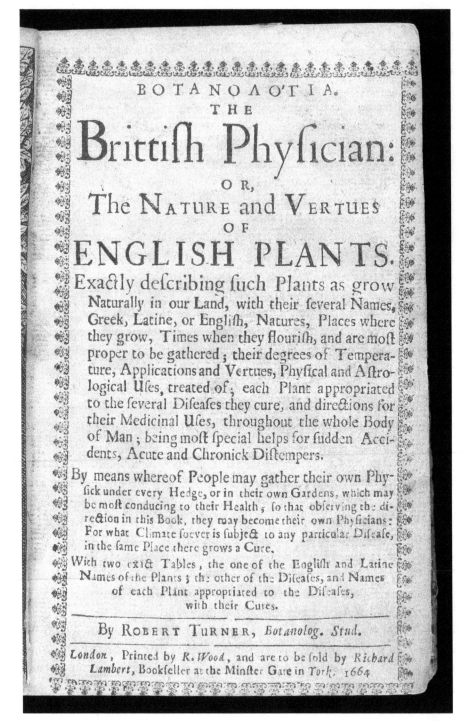

FIGURE 5.6: Title page of Robert Turner's *Botonologica* (1664). Wellcome Collection L00037384.

"A most precious and excellent balm" that includes twenty-four roots, more than twice that many herbs, as well as dozens of seeds, flowers, spices, gums, and other plant materials; and a cordial water for which she gives the recipe also contains dozens of ingredients.[67] Clearly, such a vast pharmacy was unlikely to be affordable to most, nor would the storage of such a variety of ingredients have been easy for anyone but a woman with substantial income; nevertheless, Lady Mildmay's preparatives stand for a notion of natural remedies that speaks to the cornucopia of nature, the plenty to be found in the world that can alleviate human suffering. But this copious pharmacy, and the knowledge that goes into generating herbal remedies from it, is far less well-represented in the official medical literature of the day. Most treatises that include herbal remedies, like Bullein's or Lodge's, offer those that include several ingredients, not dozens or hundreds, indicating perhaps a professional discomfort with variety—the more targeted and limited the remedy, male physicians seem to assume, the more scientific it will appear to their public.

At the same time, herbalists use the same materials for many different complaints, something equally true in oral as in learned tradition. Natural ingredients that crop up frequently for both external and internal application are roses, honey, and rhubarb. Indeed, all three appear to have broad value in the treatment of a spectrum of complaints. Roses have anti-inflammatory and astringent qualities, honey has been found to be a useful antiseptic for wounds and can heal stomach ulcers, and rhubarb was thought "to contribute to the amelioration of an impressive range of complaints," although it was most used for its laxative and gentle purgative effects in the Renaissance.[68] For Gabrielle Hatfield, the idea that a single plant or a single "active principle" in a plant must be responsible for its effectiveness is a modern imposition that undercuts the effectiveness of herbal treatments.[69] Just as plants exist in a complex organic system that requires diversity to succeed, so at the microcosmic level plant remedies may rely on the combination of chemicals within a single plant, or a combination of chemicals in their natural form in several plants, to work on the body correctly. This sense of the body and the plant as complex ecosystems with multiple points of interaction seems to ground Lady Mildmay's practice, as well as that of village folk healers, and even the professional physicians borrowing herbal treatments as their own. Andrew Wear points out that "the linkage . . . between the literate and the oral tradition of herbal remedies gave the former a kind of validation which is different from its association with the world of learning, for it can be seen as having its roots in the countryside, the fount of health and the home of plants." In Wear's reading, the country was "where health resided" (because country people were less often sick) and a reminder of paradise, because it was where God placed healing plants for human use.[70] I would like to put additional pressure on this characterization, and argue that the origins of popular herbal lore were linked with the countryside not merely as a place where people were assumed to be healthier but also with rural places as the locus of a more organic, multivalent conception of the body in the natural environment, a place where humankind's needs for connection with nature and with the benign, healing influences of domestic society were better met. This is not merely a reinstantiation of either our, or the Renaissance's, nostalgia for a paradisal pastoral past. Rather, I would argue that belief in a continuum that links human with human, and plant with plant is the basis of a rudimentary ecological consciousness about the human body that is essential in comprehending, embracing, and believing in herbalism. Human cooperation, plant cooperation, human interaction with nature, plant interaction with the human body—these exchanges appear fundamental to the belief system that accepts herbal lore.

We have, then, two aspects of popular treatments for complaints consistent with what we have seen in popular sensory knowledge and belly lore: attention to the communal nature of the body that is best when cultivated in the context of family and home, and the organic nature of the body that places it in reciprocal exchange with the rest of creation. Here, the domestic world of the household is the communal frame, and plant life the source of profitable interpenetration.

THE PERSISTENT CORPSE

The relationship between the living body and the dead corpse was the source of much popular anxiety as is reflected in folklore and religion, both of which often cooperate to either explain or relieve fears associated with the disposition of the dead. For medieval and early modern men and women, death was a lengthy process that began literally at birth. Walter Cary notes in his 1587 *Briefe Treatise* that "all men are subject unto death and our bodies (as yielding thereunto) always from time to time gather corruption."[71] Just as life is a process of dying more or less slowly, so death is not necessarily concluded at the exact time of expiration: burial rituals included periods of attendance on the corpse, indicate that death was a time of transition, rather than a sudden event. Such periods of attendance might also have been designed to ensure that the deceased truly was dead in a period with limited medical skill at detecting comatose states. The Catholic invention of purgatory was crucial during the Middle Ages in giving the living a sense that their dead were not entirely gone from the world—acts like burning candles, saying masses, and generally praying for the soul of the departed spoke to continuity, rather than rupture. Although the Protestant Reformation challenged these beliefs and attempted to repress the ritualized aspects of the treatment of the dead, success was slow and inconsistent. Shakespeare's Hamlet sees the ghost of his dead father as late as 1601, and the ghost makes clear the possibility that it dwells in purgatory. But even having denounced the concept of purgatory, Protestants still saw ghosts with some regularity: many returned, it was thought, to haunt their murderers, leading to the practice of swearing oaths on a tomb or grave, which presumably guaranteed that anyone violating the oath would be visited by a grisly shade.[72] Phillippe Ariès points out that there was real ambiguity in the moment of death, which raised questions about where life was whether in the material of the body, or in the soul, and if in the body, whether it resided in "the whole body or . . . its elements."[73] Popular belief seized on evidence of the body's persistence, in sounds emanating from graves or tombs, in motions of dead corpses, anything that tended to argue for the role of matter as the seat of life. Thus, popular lore about death and the dead depicted the body's materiality as occasionally and worryingly transcendent.

Early moderns pared down the rituals of burial after the Reformation, limiting most to a simple family ceremony. In the folk beliefs of the period, fears about what, exactly, happened to that buried corpse persisted. Legends about revenants grew out of a fascination with blood, aligned with the idea that the products of the human body could remedy even the most extreme cases of disease. Blood, along with semen, milk, and other bodily fluids, figured in the popular imagination as the sustenance of life, as food: it was, as Camporesi notes, a main sauce in the kitchen.[74] We are, then, back in the realm of belly lore, of the body that emulates a successful kitchen. Blood played an important role in remedies: the "blood of a fresh, delicate man, one well-tempered in his humours" included in a potion was thought to slow the aging process.[75] The use of mummia (the dried by-product of Egyptian or other mummies) in medical preparations was only one more esoteric way in which the dead body could serve the living; children's hair and teeth were also used in treatments, bones could be ground up or their ashes incorporated into cordials—indeed, there was no part of the human body that could not function in therapies, creating a culture of consumption that linked the living with the dead.[76] The perspiration of corpses was thought to be a cure for hemorrhoids and tumors, and "the hand of a cadaver applied to a disease area can heal.[77]

Yet blood was in particular a difficult medium to control—an excess of blood could congeal in the body and create tumors or other disorders. "If obstructions should happen [in any part of the body that naturally purges sweat or urine], all the while filthy masse of noisome humours is thereby kept within the body, and then given violent assault to some of the principall parts," writes William Vaughan.[78] Purging and bloodletting were thus signs of the body's constant need for intervention in its internal balances and were closely tied to such things as strict dietary regimens, again affiliating blood with food.

Thus, it is no surprise to find popular beliefs in dead or corrupted bodies thirsting for the blood of others. Revenants, cannibals, and witches were all seen as craving human blood (witches were supposed to suck the blood of infants, while in turn suckling the devil or its minions with their bodies' milk, thus transferring the important stuff of life from needy souls to evil ones).

FIGURE 5.7: Man wearing a tourniquet letting blood into a bowl. From *Arzneibuch* (ca. 1675). Wellcome Collection L0041074.

Revenants figure in the lore of almost all cultures from antiquity forward; however, in the twelfth century, Walter Map and William of Newburgh both report cases of the dead returned to life and threatening the living, appearing from traces of blood around their mouths to have drunk the blood of the living. In the seventeenth century, Protestant doctrine did not prevent continued stories of revenants, and the tradition ultimately gave rise to the lore of vampires in the nineteenth century. Unlike later vampires, however, Renaissance revenants were usually florid of complexion, healthy, even swollen as if from surfeit of food, fluid, or blood. Common threads that unite revenant lore in England and Europe involve the corpse's need to complete unfinished business with the living, the lack of proper burial or funeral rites, early or unexpected death, or pernicious behavior that continues despite death.[79] In all these cases, the body turns out to be highly susceptible to interference by all sorts of forces that do not permit a quiet or secure rest after death, forces from within— sinfulness, a need for what the living can offer, desire for more time with loved ones, confusion over the event of death, reproach for incomplete burial, and sanctification—and from without, like violence or injustices committed against the deceased. A lingering popular fear that the immaterial soul is somehow ensnared by the material flesh is manifested in these stories of revenants.

Blood also factored into popular practices regarding death and crime. Exposed to the person who committed its murder, a corpse could indicate the crime by bleeding anew from its wounds. In some instances, the practice of using dead bodies to determine guilt or innocence was spontaneous, suggesting that divine providence was at work. In others, the encounter was manufactured as part of the investigation into the death. In 1591, a sensational account of a father's murder of his three children was published in England. Brought into the presence of the unearthed bodies of the three, the bodies begin again to bleed, "which when the Crowner [coroner] sawe, he commanded the partie apprehended to looke upon the children, which hee did and called them by their names." At this point, a miracle occurs, and the children's appearance changes so that they seem alive again, causing the father to confess and repent.[80] The boundary between life and death is a fluid one, and here, as in many other cases of bleeding corpses, the dead may exhibit the characteristics of the living when justice requires it. In this instance as well, the pull of natural and bodily familial ties reasserts itself, so children may be brought to life when they hear their father's voice. Not only does such a story affirm the crime's violation of basic human parental obligations, but it also suggests that such obligations reside in the body and cannot be repressed even in death.

In death the body also reaffirmed a mutual and reciprocal identity with its environment. Food for worms, as we have seen, was a literal truth of the body, and this is the case for burial practices as well, because ordinary burials did not separate the body to any effective degree from the earth in which it was placed—the use of a coffin was new to the period, becoming more common during the seventeenth and early eighteenth centuries, its growing ubiquity corresponding to the increasing importance of ideas about the bounded, separate body.[81] The cemetery of the nineteenth century, with its individual plots and monuments, was largely unknown in the Renaissance. Communal graves for plague victims were but an extreme version of a common fact: bodies were placed one on top of another as space required in parish graveyards throughout England. "A widespread response to a growing shortage of space, especially in the inner city, seems to have been to pack the maximum possible number of corpses into the available ground; various parish registers record the multiple use of graves, with several bodies being placed together."[82] Contra Andrew Marvell's opinion that "the grave's a fine and private place / But none, I think, do there embrace," the Renaissance grave was a highly social place, with worms, other bodies, and the earth comingling comfortably, where the body embraces its own organic origins anew.

CONCLUSIONS

This narrative about the Renaissance body as it was popularly conceived is only one of several that might have been offered here—equally valid, for instance, would be a discussion of how popular beliefs resist the encroachments of science and the alienation of the body from the natural and communal domains of its past through a stubborn attachment to the supernatural, magic, and religion. However, in foregrounding the two areas, I have attempted to read popular beliefs for their organicism, their assumption of an ecology of being.

The fact that the body does not go willingly into its future post-Renaissance incarnation as the dissected, rationalized, examined, disciplined material of an economy of exploitation and economic consumption is demonstrated in the residual traces of these organic versions of the body long after the actual systems and ideas they described disappeared. Their ghosts are reminders of a struggle that is about more than class status, labor, or intellectual hierarchies, a struggle that does necessarily end with the Enlightenment or modernity. Belief in bodies assaulted and probed by aliens, macrobiotic diets, homeopathic remedies, and aromatherapy: these popular beliefs may be the heirs to an older tradition than even their adherents imagine.

CHAPTER SIX

Beauty and Concepts of the Ideal

MARY ROGERS

An interest in praising beauty and striving to achieve it, in both nature and art, has often been thought to differentiate the cultures of Renaissance Italy from those of other centers in contemporary Europe. Whether remarking on the grace of Italy's inhabitants or the quality of its works of art, the admiration for beauty is widespread among those who have left records of their opinions. Many social practices were directed toward nurturing and enhancing such physical beauty, from conception onward. Many artifacts kept in the home in the fifteenth century, especially in bedrooms, incorporated images of handsome young males and females (inside wedding chests, *cassone*) or vigorous male infants (on birth salvers, *deschi da parto*, Figure 6.1), the sight of which were supposed to encourage the procreation of similarly healthy and well-formed offspring.[1] The innocent, golden-haired beauty of young boys would have been admired when portrayed with older family members (as in the portrait by Domenico Ghirlandaio in the Louvre in chapter 10, Figure 10.1) or when participating in civic events. Girls, who were more restricted in their public appearances, would also be scrutinized for their physical and behavioral charms as they reached marriageable age. Once betrothed and wed, their beauty was felt to be appropriately enhanced by fine clothing and jewelry, often carefully recorded by painters.[2] In the sixteenth century, upper-ranking women might receive fulsome admiration for their beauties from their literary hangers-on, as praise for female beauty came to be a major poetic genre. On a less rarefied

FIGURE 6.1: Domenico da Bartolo, *Infants*, Reverse of a *Desco da parto*, 1430s. Venice, Museo del Ca' d'Oro. Photo Credit: Cameraphoto Arte, Venice / Art Resource, ART170885.

level, cleanliness and grace in body and clothing for males and females alike were advocated in the burgeoning literary genre of treatises on behavior, and assisted in real life by a range of material aids, from cosmetics to medicines, hair tweezers, and toothpicks.[3] The advance of mirror technology in the sixteenth century, making possible large, flat, plateglass rather than small, convex examples, must have stimulated additional bodily self-scrutiny.

Only in the later fifteenth or early sixteenth century, however, did many Italian intellectuals move beyond noting and appreciating the presence of beauty in life or art to attempting definitions or extended analyses of the concept of beauty in a general sense, and even more, of the many qualities that might be associated with it, such as grace, vigor, charm, or liveliness. Discussions of the human figure and its beauty, of course, differed according to whether they were dealing with nature or with art, yet were interrelated to the extent that they were informed by similar literary sources that discussed art or aesthetics in general. These came mainly, though not exclusively, from classical antiquity, the Latin rhetorical writings of Cicero and Quintilian being the most important, together with the sections on painting and sculpture in Pliny the Elder's *Natural History* and parts of the architectural treatise of Vitruvius. These had earlier been used in discussions on the merits and achievement of the major Tuscan poets and artists of the fourteenth century.[4] Later in the Renaissance, the influence of Florentine Neoplatonism stimulated lengthy, often rhapsodic, attempts to extol the significance and value of beauty, even in tracts whose tenor is largely Aristotelian. This chapter will outline these strands in Renaissance thinking about beauty, mostly in art theory or criticism, and then suggest how they were developed in a discourse that became richer in its terminology and more widely diffused over the course of the fifteenth and sixteenth centuries. Perfection of form and expressiveness of demeanor are the two qualities most frequently commended in beautiful figures in both art and life, qualities that might be difficult to reconcile. Color and luminosity might also be admired, although there was less of a consensus as to their value, and diverse styles of beauty were always accepted.

Perhaps most fundamental for Italian, though not northern European, aesthetic ideals was the antique-derived belief that beauty was a matter of innate structures or of the arrangement of the whole. Few Italian Renaissance observations about a beautiful figure failed to say it was "well formed" or "justly proportioned," as they were ultimately influenced by general ideals of harmony, balance, and proportion pervading antique writing on literature, art, and the natural world and stemming from Platonic and Aristotelian traditions. Many ancient Greek and Roman texts known in the Renaissance articulated these ideals with varying detail and consistency. Cicero extolled human bodies and the universe where no part is superfluous: the whole form had the perfection not of chance but of a work of art (*De oratore* III, xlv, 179).[5] Aristotle proclaimed in his *Poetics* (which became well-known in translation only in the sixteenth century) that "whatever is beautiful, whether it be a living creature or an object made up of various parts, must necessarily not only have its parts properly ordered, but also be of an appropriate size, for beauty is bound up with size and order."[6] These texts express the antique ideal of *symmetria*, which, rather than its modern sense, meant a proportioned harmony of all the parts, both in themselves and in relation to each other, in a body or structure. For the Aristotelian tradition, which continued to be dominant in the aesthetic and scientific theory of the Renaissance, despite the impact of Platonist thinking, this *symmetria* was a norm immanent in the varying forms of nature and an aesthetic guideline for architects and visual artists.[7]

Most directly influential of all for theorists of art and architecture was the first century A.D. architectural treatise by Vitruvius, whose contents became more carefully studied, and eventually translated, after a less corrupt text was discovered in 1404. In his architectural treatise of ca. 1450, Leon Battista Alberti, the "father of Renaissance art theory," famously defined beauty, after Vitruvius, as "that reasoned harmony of all the parts within a body, so that nothing may be added, taken away or altered, but for the worse" (*De re aedificatoria* VI, 92–93).[8] In his own Book III, Vitruvius, basing himself on Greek sources now largely obscure, had given the proportions of the well-formed human body, which he claimed had been adhered to by the famous painters and sculptors of antiquity. These included the principles that became most firmly established in the Renaissance, that the normative human figure with outstretched arms should fit into both a circle and a square, that the head is one eighth of the whole and the face one tenth, and that the face is divided into three equal parts, the forehead, the nose, and the nostrils to the chin (*De architectura* III, 1).[9] By implication, these proportions should be applicable to any human who is not deformed, but the subsequent book, on temples, moves from a universal standard to describe several different and gendered norms. The architectural orders are related to varying types of the human body, young and old, male and female, each with its own style of beauty differing mainly in column thickness and embellishment, the rugged manly beauty of the Doric order being "naked and unadorned," the feminine beauty of the Corinthian more delicate and ornamented (IV, 1).[10] Appropriate variation in architecture, then, is connected with norms that vary like human beings. This basic principle of diversity was adaptable to the different types of figures found in the painting and sculpture of the Italian Renaissance, as will be seen later in this chapter.

The Vitruvian ideas on proportion might also be combined with another set of ideas from the Aristotelian tradition, those of physiognomy. Though not directly concerned with the beauty of the human form, but rather with the diagnosis of inner psychological characteristics from outward physical form, physiognomy texts, or Renaissance writings influenced by them, frequently connect good proportion with moral worth and deformity with deficiency. The clearest examples in art theory are the treatises on painting by Cennino Cennini (ca. 1437) and on sculpture by Pomponius Gauricus (ca. 1504).[11] Cennino saw women's bodies as entirely lacking in proportion, basing his remarks on ideal proportion, which apparently were influenced by Byzantine and Vitruvian canons, solely on the male.[12] Gauricus, who recommended that artists study physiognomy, outlines women's physical characteristics (smaller head, less muscled legs, softer, more rounded feet and hands) in a way that clearly connects with his categorization of their psychology as timid, indolent, feeble, intemperate, and much else. His preferred male ethnic type is of medium height, with honey-colored skin and moderately wavy hair: midway between the tall, blond northerner, who is considered choleric and stupid, and the smaller, dark, wiry-haired Arab or Sicilian, who is considered dishonest and ill-suited to learning. The best ethnic group is ingenious, thoughtful, modest, and agreeable: flatteringly to himself, this is exemplified by Greeks or Italians.[13] Few writers on aesthetic topics accepted the more elaborate physiognomic systems known in Renaissance Italy, however; although another of its key ideas, the similarity of human, especially male, types to certain animals, both in their physical and their psychological characteristics, seems to have influenced certain artists. Verrocchio, Leonardo, Dürer, and Titian, among others, sometimes intensified the communication of the idealized character of a warrior, ruler, scholar, or artist by giving the countenances of their sitters heroically leonine, or loftily aquiline features.[14] The general notion found in Aristotelian physiognomy theory, and elsewhere in physiological and medical writings, that the outward forms of the body in one way or the other indicated its inward character was all pervasive in the Renaissance, however, and came to be reinforced by the later cult of physical beauty as related to inner grace.

The two rather different passages in Vitruvius (stressing normative proportions, or stressing different styles of beauty) generated various responses. At a simple level, the Vitruvian ratio of one to eight for the face to the whole figure might have encouraged sculptors or painters of earlier fifteenth-century Florence to produce less elongated figures than those in currently fashionable forms of late Gothic. The Byzantine-based, ratio of one to nine, though, was still recommended as a possible alternative throughout the Renaissance by such writers as Cennini, Gauricus, Dolce, and others, and there was general awareness that several canons of proportion had existed in antiquity. Other artist-theorists picked out elements from Vitruvius's norms that accorded with their particular interests. Several architectural treatises included illustrations of the "Vitruvian man" inscribed into a circle and a square, later given classic treatment in the drawing by Leonardo da Vinci (chapter 9, Figure 9.1), which seems to encapsulate the belief, deeply attractive to the Renaissance, that both man and the cosmos were structured according to regular geometry.[15] However, there was also a consciousness of Vitruvius's inadequacies, and the critical and empirical bent of both Leonardo and Albrecht Dürer led them to make their own detailed studies of human proportions. Leonardo stressed the diverse forms beauty could take, based on the variations found in nature: "Beauty of face may be equally fine in different persons, but it is never the same form, and should be made as different as the number of those to whom such beauty belongs" (Codex Urbinas 51v) and repeatedly insisted on the need for a similar variety of bodily types in art.[16] At a time around 1500 when he was enamored with the "secrets of proportion," which he thought generated the greater beauty of the figures in Italian art, Dürer made many drawings of classicizing figures, women and men, with Vitruvian proportions, using compasses and ruler (Figure 6.2).

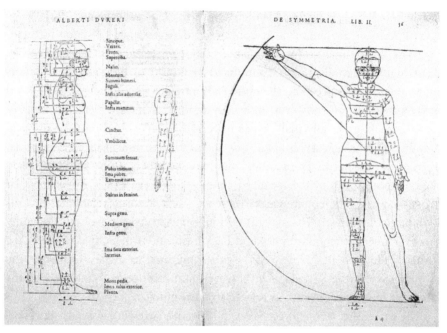

FIGURE 6.2: Proportions of a female figure. From Albrecht Dürer, *De symmetria partium humanorum corporum libri quatuor* (Paris, 1557). Wellcome L0000296.

This type of drawing would later be given artistic form in his famous *Adam and Eve* engraving of 1504.[17] Here the ideal proportions of the first couple are given philosophical weight, as they serve as an indication of the pair's prelapsarian perfection by demonstrating the balance of the four Galenic humors in their bodies. The subtlety of the rendering of bodily features, light effects, and background detail all testify to the need for a fundamental beauty of proportion to be literally "fleshed out" and completed by a richly varied setting in a successful work of art. Over the next decades Dürer made varied explorations, into figures in motion and stasis. Some graphic works suggest his interest in combining Vitruvian proportions with non-classical, non-ideal types (notably the *Great Fortune* engraving of ca. 1502) or non-European ethnic types (the drawing of the West African Catherine of 1521 in the Gallerie degli Uffizi, Florence). These show his evolution of a conception of beauty that, though informed by Vitruvius, went beyond a straightforward differentiation of human types into male and female, old and young and did justice to variety in nature: "there is a great harmony in diversity," he would later write.[18] In his theoretical writings (translated into various languages and widely disseminated throughout the sixteenth century) this resulted in his expansion of existing canons of beauty to thirteen basic types, each beautiful in its own way, and eventually, in his *Vier Bücher von Menschlicher Proportion* of 1519, to his denial that absolute beauty could be understood by humans.[19]

Dürer's labors attracted attention but also a certain amount of snide comment, and most artists in both the fifteenth and sixteenth centuries seem to have rejected as impractical the overly rational, mathematical approaches for achieving beauty. Leonardo's remarks on creating grace or charm in a human figure place greater emphasis on softness, lighting, and the definition of form.[20] No human proportion studies survive by Raphael, perhaps the most influential of sixteenth-century artists. Raphael studied Vitruvius carefully but claimed "Vitruvius helps me, but not much," and made the famous remark that, in the absence of sufficiently beautiful models, he resorted to creating not according to arithmetical guidelines but rather to "a certain idea" in his mind.[21] Only one proportion drawing exists by Michelangelo, who emphasized the artist's "judgement of the eye."[22] This emphasis on the skilled artist's subjective judgement in creating beautiful figures, rather than a following of rules and norms, connects with the widespread acceptance in the sixteenth century that beauty or grace were elusive and mysterious matters.

The belief that beauty was a matter of artistic judgment rather than objective norms could have been connected with the acknowledgment by 1500 that different types of beauty existed in the styles of current artists, each meritorious in their different ways, as in certain late fifteenth-century writings famously analyzed by Baxandall.[23] This was encouraged by the widespread reading of antique texts, especially Pliny the Elder, a crucial source for ideas concerning the visual arts, with its characterizations of the contrasting styles of various antique masters.[24] More fundamentally, the frontal, static nature of rules of figural proportions derived from columns in architecture was limiting for working artists needing to present figures from various angles, moving in action or in emotion, as both Leonardo's writings and some of Dürer's diagrams acknowledged.[25] Furthermore, it was art that was animated and lively that had been commended by antique writers and appealed to the Renaissance as being connected with grace, *grazia*, a word with rich overtones but often concerned with beauty as revealed in action and movement.[26] Nonetheless, an emphasis on the importance of elements that were harmoniously shaped and arranged (often loosely defined) continued in art theory and in writing on beauty in other contexts. Most influentially, Giorgio Vasari, although mainly concerned with the analysis of distinctive artistic oeuvres in his *Lives of the Most Excellent Painters, Sculptors, and Architects* (1550, enlarged 1568), also stressed in the preface to his third part the need for *regola*, *ordine*, *misura* (rule, order, measure—qualities clearly associated with the Vitruvian tradition of ordered proportion) and *disegno*, that crucial concept developed to embrace not only the linear definition of bodily forms, but the planning of the entire composition.[27] His definition of the fifth desirable quality, *maniera*, as a stylish license outside the rules allowed himself considerable freedom of maneuver, however.[28]

The general conviction that beauty was grounded in nature, that *symmetria* was both a principle guiding nature and an aesthetic ideal, had to be reconciled with an awareness, increasingly clear in sixteenth-century aesthetic theory, that not everything in nature was beautiful. Philosophical principles could be invoked to explain why existing natural forms might be unsatisfactory models for the artist. In 1548, Paolo Pino defined beauty as "a measured correspondence of members produced by nature without any hindrance from accidental flaws."[29] Pino, who cited Aristotle elsewhere in this section, is here using the traditional Aristotelian distinction between the form of nature, which is perfect, and the accidents, which are flawed, an idea repeated elsewhere, as in Vincenzio Danti's treatise on proportion of 1547.[30] This could explain why the Vitruvian norms might be generally true, though every human being does not conform to them, so the artist needs to do more than copy live models exactly if the artist wishes to achieve true beauty. Resorting to an analogy found in Cicero and elsewhere, and cited over and over again in the Renaissance, Pino stated that the artist needs to select from the best parts in nature, and then judiciously combine them as Zeuxis had from the women of Croton when creating his image of Helen.[31] Thus he would be, in a Renaissance sense, an imitator rather than a copier, an emulator of nature's underlying processes and norms rather than her superficial manifestations. Such a distinction, expressed in Italian through contrasting the verbs *imitare* and *ritrarre*, became increasingly clear in sixteenth-century art theory.[32] Other, more literary-minded writers might also encourage visual artists to look at the beauty created in the "word-paintings" of leading poets: the obvious example is in the *Dialogo della pittura* of Lodovico Dolce of 1557 where the author counsels painters to follow Ariosto's description in his *Orlando Furioso* of the beautiful Alcina.[33] Still others, notably Vasari and after him academic theorists, would counsel guidance from available works by good artists from the recent past or from classical antiquity to attain a desirable *maniera*.[34]

A harmonious arrangement of the parts entailed an aesthetically satisfactory total composition, not just individually well-formed figures. While art theorists' interest in the overall composition partly goes beyond the scope of this volume, it did raise issues relating to the variation or enhancement of beautiful bodies. The concept of beauty as the proportioned harmony of all the parts was extended to the pictorial composition by Alberti, in his *De pictura/Della pittura* of 1435–1436. As is well known, in the first part of his treatise, dealing with perspective (I, 19–21), Alberti recommended the use of a module relating the height of the figure to the architectural surrounds in order to gain tighter compositional control: something like a pictorial form of the architectural notion of *symmetria*.[35] However, in the subsequent book (II, 40), introducing ideas that would have a longer life in Renaissance art theory, he argued for the desirability of richness of detail (*copia et varietas*) as the source of pleasure in art, attained through variety in the types, costumes, and poses of the figures as well as in color. This is another version of the "beauty completed with ornament" idea more explicitly stated in the *De re aedificatoria* (VI, 2, 93–94), an idea adapted from antique rhetorical theory where convincing overall arguments were said to become palatable by refreshing variations in language or sentence length, by "ornamental" figures of speech, or by appropriate gestures to make up an effective oration. Quintilian, indeed, in a passage strongly influencing Alberti, had employed a pictorial analogy, how painters introduce pleasing variation in the clothing and movements of their characters (Quintilian II, xiii, 9–10).[36] However, he and many other antique writers had considered "ornament" to be a double-edged concept, giving splendor and delight when judiciously used, but bringing the dangers of over-ostentation or confusion if taken to excess. Different contexts demanded different levels of ornament, sometimes resulting in a splendid style, sometimes one like that which Renaissance commentators on painting would commend as *puro senza ornato*.[37] Alberti and other Renaissance thinkers, like the classical writers, considered that ornament and variety must be judiciously adjusted to the main subject matter and setting of the painting, in an adaptation of the antique notion of decorum.[38]

Renaissance writers' preferences as to appropriate ornament or variety tended to be shaped by personal or current tastes: Alberti gave as examples of "ornament" windswept hair or drapery that both suggested graceful movement and the form of the figures it contained.[39] Here he is responding to a standard effect in Roman relief sculpture taken up by contemporary artists such as Donatello and Ghiberti, and then becoming very prominent in the style of later Florentine artists such as Pollaiuolo or Botticelli, especially when treating antique subjects such as the birth of Venus (Figure 6.3).

Variety could mean the avoidance of monotony in figure types. Leonardo warned painters to take care lest their fictive figures resemble themselves and counseled the study of many different human types as observed in daily life.[40] In Dolce's treatise, Pietro Aretino, the author's mouthpiece, criticized Michelangelo for his monotonously muscled male nudes, and Raphael and Titian were praised as greater painters by virtue of the *varietà* of their figure types.[41] Classic examples of such praiseworthy variety, both within the figures in a single painting and in the cycle as a whole, might be found in Raphael's Stanza della Segnatura (Figure 6.4) and d'Eliodoro. Titian's mythological scenes for Philip II of Spain provide a rare example of an artist testifying to his search for variety in the form of contrastingly posed nudes. The painter wrote in a letter on September 10, 1554:

> since in the *Danae*, which I have already sent to Your Majesty, one sees all the parts from the front, I wanted variation in this other *poesia* and to reveal the opposite parts, so that the room becomes . . . more graceful to the

FIGURE 6.3: Sandro Botticelli, *The Birth of Venus*, ca. 1480. Florence, Gallerie degli Uffizi. Wikimedia Commons: http://commons.wikimedia.org/wiki/File:Sandro_Botticelli_046.jpg.

FIGURE 6.4: Raffaello Sanzio (Raphael), *Parnassus,* ca. 1510 Vatican Palace, Stanza della Senatura. Photo Credit: Erich Lessing / Art Resource ART214313.

view. Later I will send the story of Perseus and Andromeda, which will
have a viewpoint different from these, and similarly Medea and Jason.[42]

However much mimetic richness was enjoyed, however, it was balanced by ideals of decorum or of economy of means and clarity of structure.

The admiration for variation in the types and poses of the figures leads
one to the other way in which human figures were thought to be beautiful in
the Renaissance: in their expressiveness, whether performing appropriate actions or displaying appropriate emotions. Pliny's criticism had a part to play
in this, such as where the author commends Parrhasius for adding "vivacity
to the features," or Aristeides for being "the first to paint the soul" and give
"expression to the affections of man," but the quality of apparent life was
admired in numerous antique remarks on sculptures and paintings, as part of
the *mimesis* likewise celebrated in its literary theory.[43] When early humanist
writers, writing in both Greek and Latin, praised beautiful, or perhaps more
accurately, striking effects in the visual arts or in nature, they usually focused
on the human figure. Often modeling their passages on ancient examples of *ekphrasis,* they most often commended vivid and dramatic depictions of human
personages, whether in examples from classical antiquity, such as the Meleager
sarcophagus, with its living and dead figures, described by Alberti in Book II,
37 of *De pictura,* or from art of the present day.[44] Writing around 1450, the
Neapolitan Fazio admired the "faces that live" (*vivos vultus*) in the sculpture
of Donatello (chapter 10, Figure 10.9), painted bystanders that caused amusement in a work of Pisanello, or holy figures that induced reverence in those of
Jan van Eyck.[45] Rogier van der Weyden's nude Adam and Eve, then in Ferrara,
were "of the utmost beauty" (*ad summam pulchritudinem*); equally excellent
was the painter's depiction of the Virgin Mary, tears streaming at the death of
her son, yet retaining her dignity. Though neither work was known to the author, Rogier's *Crucifixion* in the Escorial (Figure 6.5), presents similarly strong
yet restrained grief, and van Eyck's *Enthroned Virgin* in the interior of his
Ghent Altarpiece (Figure 6.6) a beauty (*venustate*) like that Fazio found in an
Annunciate Virgin also by van Eyck.[46]

It should be stressed that these earlier fifteenth-century writers made no distinction between artists featuring as "Renaissance" and therefore progressive
in modern art-historical accounts (Donatello, Ghiberti, and Masaccio), artists
who tend to be classified as "late Gothic" (Gentile da Fabriano and Pisanello),
and the Netherlanders, van Eyck and van der Weyden. Whether or not the bodies
of their figures conformed to Vitruvian norms, all were seen as achieving the
other fundamental goal of art: *mimesis* combined with decorum. They engaged
and moved the spectator by making visible character and emotion in ways that
were varied and appropriate to their themes.

FIGURE 6.5: Rogier van der Weyden, *Calvary/Crucifixion*, ca. 1450–55. Real Monastero de San Lorenzo, Escorial, Spain. Photo Credit: Erich Lessing / Art Resource ART214231.

FIGURE 6.6: Jan van Eyck, *Virgin Enthroned*, from *The Adoration of the Mystic Lamb* (*The Ghent Altarpiece*), Cathedral of St. Bavo, Ghent, Belgium. Photo Credit: Scala / Art Resource, ART65707.

Throughout the fifteenth and sixteenth centuries, many of the most no-table descriptions of works of art, whether Italian or otherwise, or whether in letters or within works with more biographical or theoretical structures, essen-tially follow this tradition of *ekphrasis*. Developing a specialized vocabulary of praise, sixteenth-century Italian art theorists repeatedly commend paint-ers, such as Leonardo, Raphael, Andrea del Sarto, Parmigianino, or Titian, whose figures combine beauty of form or grace in action and movement with expressions of appropriate character and emotion, usually called *aria* or *costume*.[47] The poets and muses in the *Parnassus* in Raphael's Vatican Stanza della Segnatura (Figure 6.4), for example, were according to Vasari given "so much beauty in expression (*aria*) and divinity in their figures that grace and life breathe from them."[48] In a letter of ca. 1554–1555 to Alessandro Contarini analyzing Titian's *Venus and Adonis*, Dolce found beauty not only in the graceful and well-proportioned body of the young male but also in his facial expression (*nell aria del viso . . . certa graziosa bellezza*) and his bold (*gagliardo*) yet easy (*facile*) movement.[49] However, it was not only young and slender figures that could have *bellezza*, *grazia*, or *aria*: Michelangelo's *Moses* (Figure 6.7), according to Vasari, had supreme beauty in its face "which has a certain expression (*aria*) of a truly holy and awe-inspiring (*terribilissimo*) prince," and hair, beard, drapery, body, and pose.[50]

FIGURE 6.7: Michelangelo, *Moses,* from tomb of Julius II, ca. 1515, Rome, S. Pietro in Vincoli. Photo Credit: Mary Rogers.

The quality of liveliness suggested in phrases such as the *vivos vultus* of Fazio becomes expressed in Italian in such words as *vivacità*, *prestezza*, or *prontezza* and their cognates.[51] In a minor but clear volume of 1571, Francesco Bocchi expounded the three paramount qualities found in Donatello's *St. George* and in his other sculptures, which as well as *bellezza* were *costume* (the communication of character and emotion through facial expression) and *vivacità*. The latter is defined as a lively movement in forceful action, showing readiness and swiftness with beauty.[52] Following this general tradition, Karel van Mander, the "Dutch Vasari," in his 1604 *Schilderboek*, praised the expressiveness of van Eyck's faces in the *Ghent Altarpiece*, claiming that the Virgin (Figure 6.6) is speaking the words she reads and that the spectator can distinguish the different musical parts the angels sing from their appearance.[53]

These writers varied in how much attention and value they gave to features like costumes, color, lighting, or the painters' touch, and in whether these constituted important aspects of *varietà* and completeness, as in Pino and Dolce, or secondary, minor elements. They varied, too, in whether they clearly located grace, expressiveness, liveliness, and the rest purely in the works of art, or also in the actual or imaginary figures portrayed (a Muse or a Moses), or in the agile hand and person of the artist. Often they conflate two or more of these, or fluctuate between them, so that both Michelangelo and his figures have *terribilità*, both Raphael's sacred and allegorical figures and his person are endowed with grace.[54] As modern critics have pointed out, grace can be not only a physical quality but also a state of being and an action: the notion of the artist as a special person, growing in the sixteenth century for reasons to which we will return, encouraged ideas of artists' spiritual endowments as well as excellent techniques.[55] The presence and the power of such grace was validated, too, by the reactions of spectators, whether or not they were well-versed in art: Vasari wrote on the impact made by the persons of Leonardo or Raphael, or by the *Moses* of Michelangelo, allegedly much admired by the Jewish inhabitants of Rome.[56]

In all the works of art mentioned here, it was not just the static perfection of the figures but also their movements, gestures, and facial expressions that created a pleasurable beauty, elements that painting and sculpture shared with dramatic or oratorical performances, thus facilitating the adaptation of many concepts into the theory of the visual arts from the poetic or rhetorical theory of Aristotle, Horace, Cicero, and Quintilian. But Renaissance art theory and conduct literature could also influence each other. In real Renaissance life, too, movement, gesture, and facial expression were both observed and controlled, as social existence came to be seen very much as a performance. The comings and goings of aristocratic personages, the enactment of religious festivities, the everyday appearance of family members or associates might be carefully noted, often with reference to the beauty or other attributes of their participants. This

meant that not only the form and the dress of the latter, but also their movements and behavior, could be taken as expressive of their honor, their *virtù*,
and, by extension, that of their families or cities: the body thus became "a
text from which good and bad character could be read."[57] The behavior literature of the sixteenth century, starting in Italy and spreading to other European
countries in translations or in similar works by non-Italian authors, developed
and promoted concepts of beautiful and appropriate movement, coining new
terms or evolving more precise or nuanced interpretations of the old and often
explicitly connecting these with moral and social status.[58] In this they built
on foundations laid in the fifteenth century. Dance literature, in particular,
has been seen as an area in which a vocabulary of beauty had developed in
relation to pleasing movement differentiated according to gender, a vigorous,
robust *gagliardo* type of movement being appropriate to the male, a gentle,
leggiadro one usually associated with the female, a usage that then influenced
art criticism.[59]

Gender is a key determinant of the well-tempered, ideal movements that
were famously analyzed in *Il Cortegiano* by Baldassare Castiglione (written
ca. 1513–1524), that prime example of the search to achieve grace in life:
a grace based on nature but shaped by the "self-fashioning" of the courtly individual into a pleasing and socially acceptable form. The male courtier, according to Count Ludovico da Canossa in the dialogue, should have a manly grace
and should be supple and agile but not languid and effeminate; his natural endowments should be fostered by physical exercise, sport, and martial activities
to achieve Castiglione's famous neologism, *sprezzatura*, most usually translated as "nonchalance."[60] This ideal of apparent effortlessness in performance
and in every action was elaborated by the author in association with *grazia*
and *facilità*, and later compared to the ease of the painter Apelles compared
to the over-diligent Protogenes, as recorded by Pliny.[61] Though he did not use
the word *sprezzatura*, a similar ideal of easy grace, in an artist's person and in
his productions, was adapted by Vasari in his exposition of *maniera* and his
comments on his exemplary artists, notably Leonardo and Raphael.[62] Later in
Castiglione's book, the beauty of the ideal court lady is described by the Mag-
nifico as to be enhanced not by "robust and manly exercises" but by restrained
music-making and dance, as well as by appropriate dress. In another reading
of the interior from the exterior, her virtue, defined as her chastity rather than
an active, male *virtù*, should be detectable in the modesty and the temperance
with which she engages in all of these.[63]

A generation later, Giovanni della Casa's *Galateo* (written 1551–1554)
gave further guidelines for the good behavior of the well-trained male (*l'uomo
costumato*), now operating not in a courtly but in a civic environment.[64] The
key principles are tempering one's instincts and avoiding giving offence to one's
associates. This means not only eliminating physical habits that are disgust-

ing (public nose picking, nail cleaning, hair combing) but also avoiding those that distort the body and thus diminish dignity (pulling faces when making jokes, puffing up the cheeks when playing a wind instrument) or that suggest a lack of control (frenzied rushing through the streets, wild gesticulation, uncontrolled staring).[65] Man must seek not only to do good deeds but to make them graceful (*leggiadre*) as well: the definition of *leggiadrìa* in this context incorporates ideas familiar from art theory, ideas of good composition, measure, and suitability combined with elegance and charm.[66] In the portraiture of the time, starting with Raphael in the 1510s and continuing through Andrea del Sarto, Pontormo, and Bronzino, one can perhaps detect these developing social ideals, spreading outward from the courts to polite society in general (Figure 6.7). Dress is clean and elegant, expensive without being vulgarly ostentatious; the stance is poised, movement is restrained, and gestures are easy and not abrupt.[67] In contrast, unrestrained movements seen as uncouth, comic, or transgressive can be found, often in scenes of carnival festivity or lower-class merriment, north of the Alps and in Italy.[68]

A more elaborate analysis of beauty as related to the female figure is found in Agnolo Firenzuola's *Delle bellezze delle donne* of 1541, which also deploys concepts from the sphere of art.[69] In sketching his ideal female beauty, a Zeuxis-like synthesis from the best features of his women listeners, Firenzuola's mouthpiece, Celso, repeats the need for just proportion in individual members and the whole figure, enhanced through dress and jewelry to maintain propriety without obscuring the basic lines of the body. This is a reworking of the Albertian formula of innate beauty completed through ornament, though it is Cicero and Aristotle who are cited specifically. Evidently thinking *bellezza* was an overly vague term, six qualities associated with feminine beauty are defined, many apparent in movement, action, speech, and other forms of social behavior, and at times having moral implications. *Leggiadrìa*, as in the *Galateo*, connotes a graceful, gentle carriage of the body conveying a lady's modesty and moderation. *Vaghezza* (derived from *vagare*, to wander, and with strongly Petrarchan associations) also denotes elegance or charm, but it suggests a rather more lively kind of movement, attracting and arousing desire. *Venustas*, etymologically connected with Venus, denotes a specifically female type of beauty that arouses reverence, translatable as loveliness. *Maestà*, stateliness or majesty, clearly associates one type of beauty with exalted social status. The interpretation of *aria*, expression or demeanor, again shows the belief that the physical reveals the moral: if a woman has "stained her conscience," this infirmity of soul will be connected with an imbalance of humors in her body, entailing an undesirable flabbiness or thinness of flesh. *Grazia*, also relating outward and inward, is an elusive, nonrational quality coming from "a mysterious proportion and a measure that is not in our books."[70] This emphasis on a more intuitively sensed harmony again exemplifies the current move away

FIGURE 6.8: Agnolo Bronzino, *Portrait of Young Man*. 1530s. The Metropolitan Museum of Art, New York, NY. © The Metropolitan Museum of Art / Art Resource, ART321313.

from an overly mathematical theory of beauty. These three writers, then, use an increasingly articulate vocabulary to evoke aspects of beauty; it is seen not as a single and absolute quality but varying in its forms of expression. Other features of Firenzuola's dialogue lead away from the expressive human body and toward two notable and interrelated aspects of the debate on beauty in Renaissance Italy and indeed Renaissance Europe: the valuation of beauty as leading to God, and the poetic rhetoric of female beauty. At the outset, Celso asserts that beauty leads the soul to contemplation and the desire for heavenly things. This connects with ideas stemming from Florentine Neoplatonism, in which, it could be said, beauty, rather than the nature extolled in Alberti or Leonardo, assumed a prime value.[71] In the modification of Platonic thought by the Florentine philosopher Marsilio Ficino, whose commentaries on Plato appeared from the 1460s onward, the earthly world of inert matter and animal existence was given value only by the presence of beauty, found solely in humankind. Beauty for Ficino was a divine principle, descending from God and the angels to animate the cosmos and generate love; it could thus be a means through which humans could know and even ascend to be united with the divine. What physical features Ficino associated with an elevating human beauty are less clear. If beauty was what raised people above the lower levels of the cosmos, of inert matter or of the brute beasts, it might be identified with nonmaterial elements: weightlessness, light, and animated movement.[72] Insofar as Ficino's philosophy was understood by artists, it might have encouraged a style downplaying mass and shadow in human bodies in favor of ethereal luminosity and animation, and keeping landscape and animals to a minimum. As has often been recognized, this is a fair description of Botticelli's later manner in *The Birth of Venus* (ca. 1480; Figure 6.3), whose whole theme is the advent of, and welcome for, celestial beauty on earth.

However this might be, if the presence of beauty was to be reverently admired, the artists who created it could be seen as privileged beings, far from the misguided copiers of shadows on a cave that Plato himself had imagined. Thus, elements of Neoplatonism could provide what could be termed, adapting Panofsky's words on Michelangelo, as a metaphysical justification for artists' works and artists' selves.[73] Even artists generally skeptical of the ideas of wordy philosophers were not unwilling at times to see themselves like Platonic magi, with privileged insight into the mind of God. "The divinity which is the science of painting transmutes the painter's mind itself into a likeness of the divine mind," Leonardo famously wrote, in creating phantoms, beautiful or otherwise, that never existed in nature, yet were utterly credible.[74] Though Platonic concepts are unlikely to have affected the imagery of Michelangelo's major commissioned projects to the extent believed some decades ago, basic notions present in his poetry, such as beauty as a signal of and a pathway to the divine were certainly shaped by his early awareness of the philosophy of Ficino.[75] A Neoplatonic hierarchy in which only mankind on earth is touched by divine beauty could have given justification to the priorities in Michelangelo's art, with its concentration on the idealized, often nude, male figure, and his reported contempt for portraiture or for the landscape and luxuriant drapery in a Netherlandish art he deemed lacking true harmony.[76] A similar denigration of art believed to be based on mere copying, not thoughtful imitation, appears, together with Neoplatonic definitions of beauty, in many later sixteenth-century treatises on art theory, notably in Gian Paolo Lomazzo.[77]

Firenzuola's dialogue, like the many treatises on love and beauty that appeared around 1540 influenced by the translation of Ficino's works from Latin into Italian, reveals how the more accessible elements of Neoplatonic ideas had become part of the common currency of literary or educated people throughout Italy.[78] Many of these shared an interest in literature and the visual arts: a prime example would be the Venetian Pietro Bembo, who like his father Bernardo owned a portrait of his beloved by a leading artist (Giovanni Bellini and Leonardo, respectively), who was made to speak eloquently on the sacred importance of beauty in Castiglione's *Courtier*, and who expanded on the subject in his popular *Gli Asolani* of 1504.[79] Themes of love and beauty pervade the lyric poetry of sixteenth-century Italy, including that of Michelangelo, especially as directed toward his masculine ideal, Tommaso Cavalieri, in whose form the poet-sculptor professed to see an elevating, heavenly beauty.[80]

However, despite the strength of the Florentine tradition of the beautiful male figure in art, from the time of Donatello and Ghiberti through Verrocchio to Michelangelo and Benvenuto Cellini, it was female beauty that came to dominate the written tradition of lauding human beauty. This was due to another aspect of literary tradition, the motif of the beautiful and beloved lady made central for Italians through Dante and Petrarch. Petrarch's "scattered" but highly emotive memories of the luminous features of Laura—her golden hair, her shining forehead, her lustrous eyes, her white hands, her radiant smile—had been partial, those of his followers became lengthy and systematic. These physical features, often with the metaphors with which they were evoked, reappeared or were amplified by such poets as Poliziano, Bembo, and Ariosto, not to mention those in other European languages, in their lengthier descriptions of beauties, fictitious or real, often as demonstrations of their literary virtuosity.[81] They also shaped both a general perception of feminine beauty, and an aspiration to attain it, in an Italy where hair-bleaching and skin-whitening were widespread.[82] Poetic descriptions of lovely women were certainly thought to be a challenge for visual artists to surpass in beauty and emotional power, testing their skill and raising their prestige. If successfully emulated with the brush, a painting of a beauty might even serve as "a synecdoche for the beauty of painting itself."[83] This and other aspects of the *paragone*, or comparison between the powers of literature and painting, became a common subject for discussion in courts and studios in Italy by the late fifteenth century. Dolce's 1557 recommendation that painters should study Ariosto's description of the beautiful magician Alcina has already been mentioned, but earlier Leonardo contributed to this debate in both words and images.[84] In a very different style, Botticelli's *Birth of Venus* also suggests the painter's response to "word-paintings" by poets such as Poliziano of the features of the beautiful beloved, in the pale skin and billowing, glowing hair of the goddess (painted with discreet use of metallic gold), though other features testify to his enjoyment of animated drapery and contrasting movements and his awareness of appropriate antique motifs.[85] Unsurprisingly in view of this wide diffusion of a literary ideal, Firenzuola combined his Neoplatonic justification of beauty with a reliance on poetic sources when analyzing its bodily particulars. He cited ancient authors like Homer or the Apuleius, whom he translated, but then he inevitably mentioned or alluded to Petrarch in relation to a "serene" high forehead and to blonde or "polished gold" hair. The skin of his ideal belle should be *candida*, defined as "shining pale," echoing a word repeated at the onset of Poliziano's evocation of Simonetta Vespucci in his *Stanze per la Giostra*.[86]

Renaissance Italy, though, was not the only site of the evolution or development of this poetic ideal. The literary rhetoric of female beauty that had originated in Latin poets like Ovid and Virgil and been elaborated in late antiquity was adapted in the vernacular romances of the Middle Ages and was thus available to non-Italian authors.[87] By the fifteenth century, aspects of it are found in French writers such as Georges Chastellain, Charles d'Orléans or François Villon. Villon's Belle Heaulnière in his *Testament* (ca. 1461), for example, recalls her youthful blonde hair, arched eyebrows, rosy lips, and shining forehead, *ce front poli*, which is found in contemporary works such as Jean Fouquet's *Melun Diptych* in the Louvre, traditionally thought to be based on Agnès Sorel, the French royal mistress, and in many other French portraits and secular paintings.[88] Encomia of the Virgin Mary or female saints could combine the poetic rhetoric with elements from the Song of Solomon (the neck like a tower of ivory) or other Biblical sources. The saintly or secular females painted by Jan van Eyck, Rogier van der Weyden, or Hans Memlinc differ from each other in many respects but do conform with the poetic ideal in their golden hair, high foreheads, and pale, smooth skin. The inscription around the enthroned Virgin in the interior of the Ghent Altarpiece by van Eyck (Figure 6.6) in a highly self-conscious fashion draws the spectator's attention to the Virgin's supreme beauty (and by implication to the artist's skill in recreating it). Using an apocryphal Old Testament text introduced on two other occasions onto van Eyck's panels, it reads: "She is more beautiful [*speciosor*—translatable alternatively as radiant or splendid] than the sun, and above all the order of stars; being compared to light, she is found before it. She is the brightness of everlasting light and the unspotted mirror of the power of God" (Wisdom of Solomon, 7:29, 26). The conception of the Virgin's beauty in terms of luminosity and celestial bodies also connects with her designation as *Stella Maris* or *Regina coeli*, or identification with the Apocalyptic Woman Clothed with the Sun, all alluded to in many fifteenth- and sixteenth-century religious paintings, Botticelli's among them.[89] Petrarch's own poem to the Virgin commences with such light imagery, as his secular verses laud Laura's sunlike hair and stellar eyes.[90]

Luminosity suited an exalted conception of beauty, whether linked with Christian or more secular females: Bembo's speech on beauty in *Il Cortegiano*, for instance, extolled the beauty of a well-formed face made more splendid when illuminated, like a jeweled vase of crystal penetrated by the rays of the sun.[91] Certain paintings by Parmigianino from around 1530 may also suggest a similar fusion of Platonic, Christian, and Petrarchan sources.[92]

However, it was the secular verse of Petrarch and his many imitators that shaped the competitive presentation of many later idealized women in Italy and Europe. By 1540 we begin to find countless blonde beauties painted by members of the school of Fontainebleau, connected with the literary *blason* tradition of Clément Marot and Maurice Scève developed from Italian poetic sources.[93] Some Elizabethan portraits, or the Italian or English portraits of Van Dyck, continue the tradition in their different ways, as do the "ladies in white satin" in the genre scenes of Gerard Terborch in the Netherlands of the 1650s.[94] These painted women might be presented very differently in costumes, environments or artistic styles, and subtle adjustments in dress, poses, or responses to the imagined spectator might help to promote different interpretations of their social status or inward characteristics, but influencing all of their artists, consciously or unconsciously, are elements ultimately derived from the poetic rhetoric, most obviously regular features, golden hair, and pale, luminous, smooth skin.

Discussion of artists from the later sixteenth century and beyond and from northern Europe prompts mention of the other major aspect of the artistic spread of the Renaissance, namely the acceptance of the idea that ancient Greece and Rome, and then High Renaissance Italy, provided criteria for excellence in art. As is well known, Vasari was crucial in promoting this notion. He saw familiarity with the larger-scale antique statues newly discovered in Rome as the main catalyst in achieving the *bella maniera* of his third stage in art, distinguished by its fuller, more assured and easy beauty.[95] Mentioned were the *Apollo Belvedere*, *Laocoon*, *Belvedere Torso* and *Ariadne* (Figure 6.9), then thought to represent Cleopatra. Guided by these and others, Vasari claimed, Michelangelo and Raphael emulated their breadth of form or grace of folds, as confirmed by the reclining Muse in the centre of the *Parnassus* (Figure 6.4),

FIGURE 6.9: Anon, Hellenistic, *Ariadne*, Vatican, Vatican Museums. Photo Credit: Mary Rogers.

for which there exists a preparatory drawing even more closely related to the *Ariadne*. Thus, the High Renaissance artists, after studying Greek and Roman figures, transformed them into new creations that, according to Vasari, could equal or even surpass the ancients' in beauty and rich inventiveness. Though exceptional artists such as Leonardo could do without such study, normally artists who failed to be guided by Raphael, Michelangelo, or classical antiquity created lesser works, as with Titian or, of course, non-Italian masters. A prescriptive quality has now come into Renaissance aesthetics: beauty in art is incomplete if uninformed by "a graceful and sweet ease" akin to that in classical antiquity. Without claiming that antiquity presented an ideal standard that artists of subsequent times could never equal, as would be the case with some aestheticians of the seventeenth or eighteenth centuries, Vasari's work does promote a canon based on the art of antiquity and of Leonardo, Raphael, and Michelangelo.

Within Europe this would foster a hegemony of Florentine–Roman High Renaissance art as demonstrating a beauty superior to other traditions, something already suggested in the *Dialogues* of Francisco de Holanda of 1547, with the author's claim that he had traveled to Rome to "steal and convey away to Portugal the excellencies and beauties of Italy."[96] Karel van Mander, in his life of Jan Scorel, repeated Vasari's claim that the discovery of antique statues had opened the eyes of artists to distinguish the beautiful from the ugly.[97] In practice, northern artists had long succumbed to the classicizing and Italianate lure, such as Dürer, as we have seen, as well as Scorel and Mabuse. Italian artists or their works found their way to France, together with casts from noted works of antiquity. These came to form a standard of ideal beauty for artists at the French court, as for instance found in the nymphs for the Fontaine des Innocents in Paris by Jean Goujon. Here is found a style of beauty obviously guided by the Hellenistic *Ariadne*, by Leonardo's painting of an ideal woman, the *Leda*, and perhaps by certain elegantly attired females by Raphael and later painters, yet which interprets them with a freedom akin to Vasari's "freedom outside the rules," achieving a distinctly French Renaissance manner in its elegant sophistication and fluid, elongated figures.

However, the High Renaissance masters and the antique in reality never formed an exclusive canon whose dominance entailed the marginalization of all other styles of art. Both in the fifteenth century and beyond, Netherlandish or German art was highly appreciated by art-buying Italians, as recent studies focusing on both Florence and Venice have made abundantly clear.[98] Even patrons and connoisseurs who greatly valued High Renaissance and antique art, such as Francis I of France or Philip II of Spain or Italian princes like Alfonso I d'Este of Ferrara, continued collecting or admiring northern, nonclassical art, such as that of Bosch, Patinir, or Dürer and others. So, too, did major Italian artists, even Leonardo, Raphael, and Titian. This is partly because of the flexibility of the aesthetic outlined earlier, which, though it could entail a conflict between perfect form and lively expressiveness, meant that different sorts of naturalism were acceptable and different sorts of ingenuity were admired. These could be justified with reference to the diverse artistic manners praised by antique writers like Pliny, who also gave room to such specialists as Peraikos, the painter of shop and market scenes or Studius, who produced rustic genre.[99] Arguably, this could eventually stimulate or sanction the genre scenes of painters such as Aertsen, Brueghel, or the early Annibale Caracci and Velasquez, not just classicizing nudes. In practice, therefore, Renaissance taste in art, like Renaissance thinking about the beautiful, was neither one-dimensionally concerned with absolute norms nor totally consistent. "What beauty is, I know not," wrote Dürer, an artist fascinated by the sheer variety of the world and all its inhabitants, and one who admired not only classical sculpture, the painting of Giovanni Bellini, and the "most precious" *Ghent Altarpiece* but also the "subtle intelligence (*ingenia*)" found in the golden artifacts made by the pre-Columbian cultures of the New World.[100]

CHAPTER SEVEN

The Marked Body as Otherness in Renaissance Italian Culture

PATRIZIA BETTELLA

Traditional views of the Renaissance as a phenomenon that glorifies the normative body as white, male, young, civilized, and contained do not render the variety and richness of representations to which the various "subalterns" or "others," have contributed to the formation of the identity of the Renaissance subject.[1] Prostitutes and courtesans, peasants, blacks and slaves, Native Americans, Muslims and Jews, old women, in short, all those perceived as being the outsiders, located on the margins, are the others that contributed to shaping, by opposition, the identity of the Renaissance elite. Social distinction of the hegemonic groups occurs by identifying the body of the subaltern classes as the nonnormative body. The representation of the other includes more or less evident marks of bodily difference: it can be made manifest as deviation from the norms of bodily conformity (nonwhite skin or disproportionate, old, ungraceful, even disgusting body), semi-nakedness, sartorial demarcation, exotic costume, or a colored badge to signal racial distinction.

PROSTITUTES/COURTESANS

The marked body exists by virtue of its opposition to the normative body, also known as the classical body. In Bakhtinian terms, the marked body displays aspects of the grotesque; it is a body that transgresses the rules of the classical. The classical body is closed, proportionate, and decorous, whereas the grotesque one is open, overflowing, disgusting, unruly, sexually excessive, and unconventional in various forms of disruption of the normative classical one.[2] It is associated with persons who are defined by their marginality, who belong to a liminal space, either on the social or geographical margins. The prostitute, and even the honest courtesan, the more refined Renaissance incarnation, occupies a marginal space in that she challenges the traditional roles available to women in early modern times. By being neither a virgin nor a wife, nor a widow, nor a nun, the prostitute (or the courtesan) had no place in a patriarchal society that viewed her as a dangerous figure whose licentiousness could corrupt and swerve men. The bodily representation of the prostitute in literary texts highlights such marginality, which leads to the depiction of a female body marked by elements of the grotesque, as defined by Bakhtin. In a letter from Verona to Luigi Guicciardini, dated December 9, 1509, Nicolò Machiavelli recounts his encounter and intercourse with a prostitute, whose head was first covered by a towel, and who later is revealed in all her physical ugliness. The body of the old prostitute is painstakingly described in its grotesqueness:

> My God! The woman was so ugly that I almost dropped dead. The first thing I noticed was a tuft of hair, half white and half black, and although the top of her head was bald, which allowed you to observe a number of lice taking a stroll, nevertheless a few hairs mingled with the whiskers that grew around her face; . . . her eyebrows were full of nits; . . . her tear ducts were full of mucus and her eyelashes plucked; . . . her nostrils full of snot and one of them was cut off; . . . her upper lip was covered with a thin but rather long moustache; . . . as she opened her mouth there came from it such stinking breath that my eyes and my nose . . . found themselves offended by this pestilence.[3]

This type of bodily description of the old and revolting prostitute, along with violent invectives, is reserved not only for common prostitutes but also for renowned courtesans, whose ambivalent role in early modern Italy elicited praise for their beauty, refinement, and intellectual skills, as well as scorn and fierce attacks. By the mid-sixteenth century in Italy there was a flourishing of literary texts attacking prostitutes and courtesans focusing specifically on their corruption, which is represented in descriptions of bodily deformity, decay, and old age. The old prostitute is depicted in a grotesque, sexually excessive body in *La Puttana Errante* (*The Wondering Whore*, dated 1530), a poem in various cantos by Venetian Lorenzo Venier, describing the immoderate lust of an insatiable whore traveling throughout Italy to experience new and excessive sexual encounters.[4] The bodily description of an old prostitute encountered by the wandering whore while in Tuscany reveals again the repulsive physical features attributed to courtesans and prostitutes:

> An old woman spoke after her,
> Who had but four teeth in her mouth;
> Her finger and toenails were long as a palm,
> Her breath reeks more than eight convents,
> She has a beard like a man, her eyes like those of a Jew,
> Her breasts are drooping like bags,
> Her hair is rare and yellowish white,
> Like the tail of a carting horse.[5]

The merciless scorn and grotesque female body Lorenzo Venier describes in the *Wandering Whore*, reappears in the poems composed by his son Maffio-Venier (1550–1586) against one of the most illustrious and admired poets and courtesans of the Italian Renaissance, Veronica Franco (1546–1591). Franco became the target of verbal aggression by Maffio-Venier who, in his violent poetic attacks, couples moral contempt for rampant sexuality with extensive description of bodily deformity, disgust, and decay. In the tailed sonnet "Veronica, ver unica puttana" ("Veronica, Verily Unique Whore"), Maffio constructs an obscene, physically repulsive, and grotesque portrait of Franco, where the beautiful courtesan is debased to the level of a common whore, accused of being a syphilitic, depraved, disfigured old procuress, a true antithesis of the Petrarchan beauty promoted in contemporary classical literature:

> Green forehead, yellow eyes,
> Rusty nose, wrinkly jaws and cheeks,
> Your ears are always laden with chilblain,
> Your mouth is full of nonsense,
> Your breath is foul, and your beautiful teeth
> Are as white as your eyelashes and hair,
> . . .
> You are so skinny and wasted
> So very scrawny, so parched,
> One who is now reduced to skin and bone,
> Who fell in ruin,
> The one whom the children in the streets often,
> Mistake for death in disguise.[6]

The main attribute accompanying the female grotesque body is old age, which renders even the most attractive woman physically repulsive. Some recurring elements in portraits of prostitutes are references to some standard marginalizing characteristics, such as masculine attributes (bearded like a man) or, like in the poem by Lorenzo Venier, resemblance to other marginal groups, such as witches and Jews. References to witches are very common, because many older women and prostitutes were accused of witchcraft, and witches were believed to be sexually excessive and to have sex with the devil.[7]

PEASANTS

On the Italian peninsula, social groups on the geographical periphery, such as peasants (*villani*), people living in the country, or even *facchini* (porters), day laborers working in the city but coming from the countryside or from the mountain regions, are represented with physical marks of difference. Their marginality vis à vis the hegemonic centers of power and culture, such as the court and the city, makes them the others, from which the ruling classes strive to distinguish themselves. Peasants, mountain dwellers, and *facchini* are the subject of comic plays and poems, where they are ridiculed for their lack of manners, and their bodies are portrayed as grotesque and disproportionate. Peasants were called *villani*, a term that defines them negatively as those living in the *villa*, or countryside village, in peripheral areas, far from the sites of affirmation of civility and good manners, that is, the court and the city. In early modern Italy there was a general sentiment of scorn toward *villani* and even more so toward *facchini* because they were viewed by the higher classes as ridiculous, lacking in language and behavioral skills. Tommaso Garzoni (1549–1589), in his book *La piazza universale di tutte le professioni del mondo* (*The Universal Square of all Professions in the World*, 1585), a sort of encyclopedic treatment of various professions existing during his time, observes that the *contadini* or *villani* used to be respected and admired in the classical era, but at his time they have become more shrewd, they are very ignorant, and they live like animals. *Facchini* or day laborers are perhaps among the most reviled workers of the early modern period in Italy (Figure 7.1).

Angelo Beolco (1502–1542), nicknamed Ruzante, was both a dramatist and a performer active in Padua at the court of landowner and patron Alvise Cornaro. Beolco developed a personal brand of rustic comedy in Paduan country dialect, which was very successful, not only locally but also in the aristocratic circles of the Venice city-state. Beolco composed various plays, mostly in Paduan dialect, about the peasants living in the rural areas between Padua and Venice. His plays were staged for the entertainment of the Cornaro and the Venetian nobility. The comic, rustic plays by Beolco place special emphasis on the rough manners and physical excess of the *villani*, who were viewed by the

FIGURE 7.1: Facchino/Porter. Day Laborer. From *Habiti antichi et moderni di tutto il mondo*, Cesare Vecellio, Venice: Gio. Bernardo Sessa, 1598. Credit: Bruce Peel Special Collections Library, University of Alberta.

Venetian patrician classes as other. Despite Beolco's sympathetic depictions of the peasants' poverty and harsh life, the *villani* are the source of comic stunts, often based on grotesque physicality. In the marriage play *Betìa* (performed in Venice in 1523) the young peasant girl Betìa is the object of the young suitor's descriptive verbal praise:

Oh sweet and beautiful girl,
Oh eyes as radiant as the sun,
Oh rosy cheeks
More pink than a salty ham,
Oh sweet lips, oh fine teeth as white as turnips
Oh mouth of honey
. . .
Oh large tits, that would please
Every cowman for their size,
Oh nice and large feet for wine harvest,
Oh big legs well rounded, oh nice gambrel,
Round, large, white and neat,
That makes a mill pond look small!
. . .
Oh white and ruddy breasts
Like a turnip in the field,
Oh glorious and holy body,
Oh arms fit for shovel and hoe,
Oh hands fit to do
Laundry more than a thousand times.[8]

The standards of perfection sanctioned for the classical female body are here completely upturned. The description of the heavy, animalized body is of special interest because the terms of comparison for the peasant girl Betìa are country products and farm animals. Betìa's body contradicts the canon of feminine beauty sanctioned in Petrarch's poetry, subverts the principles of harmony and proportion advocated in tracts like Leon Battista Alberti's *De Pictura* (*On Painting*, 1435) and books of courtly love, such as Pietro Bembo's *Asolani* (1505).

Betìa's body is disproportionate and eroticized in its abundance and sensuality—as the use of food items for parts of the female body indicates. This physical representation is pervasive in the genre of Italian rustic poetry and is also found in other regions of the Italian peninsula, for example in Tuscan poems like *Nencia da Barberino* (ca. 1470) by Lorenzo the Magnificent, and Francesco Berni's play *La Catrina* (1517), all including unconventional peasant bodies and composed to entertain high-class circles and

FIGURE 7.2: Peasant/Contadina. Contadina trivisana. Female peasant from Treviso. From *Habiti antichi et moderni di tutto il mondo*, Cesare Vecellio, Venice: Gio. Bernardo Sessa, 1598. Credit: Bruce Peel Special Collections Library, University of Alberta.

refined audiences. Hegemonic classes were amused by the representation of ill-mannered peasant others, in their ridiculous, grotesque, and highly sexualized female body. Social distinction of the elites takes place by identifying the body of the subaltern classes as nonnormative; excessive measure and distortion in the physical shape of peasants marks their nonconformity to the rules of the body (Figure 7.2).

LEONARDO DA VINCI'S GROTESQUES

Grotesque realism and bodily distortion from the norm appears also in figurative arts. Leonardo da Vinci and the Milan circle of his imitators in the mid- to late sixteenth century, produced a significant corpus of grotesque drawings and paintings. The series of drawings by Leonardo depicting grotesque bodies, especially heads, shows his interest in expressiveness and his concern with realistic features.[9] Many drawings of distorted faces not only reveal mental states and emotions of various individuals but also show the link between character and external appearance. Even though Leonardo denied scientific foundation to physiognomics, he believed the soul determined the typical form of an individual's physical features. Leonardo's theory of expression, which is discussed in his *Treatise on Painting* (1480–1516), was developed in the late Quattrocento to react against the uniformity of facial types in contemporary painting and art. Just as beauty resulted from the harmony of the body parts, so was ugliness a consequence of their discord. Leonardo believed expressiveness could be enhanced with more realism in the depicted figure. Unlike Leon Battista Alberti, who in his *De pictura* advocated the pursuit of an ideal, the most beautiful and typical human figure, Leonardo thought the painter should seek more variety of human figures:

> In the stories there must be men of different complexion, ages, incarnations, attitudes, fatness, slimness; big, skinny, small, fat, great, fierce, old, young, strong, and muscular, weak.[10]

Among the various purposes served by the grotesque, Martin Clayton finds that Leonardo used it to depict comic or evil. Although evil had a tradition of grotesque faces in European and Italian art, the comic grotesque was less common in Italian art, in contrast with the rich literary tradition of burlesque and rustic poetry, and *canti carnascialeschi* (carnival songs) that mocked the conventions of courtly love.[11] Leonardo gained exposure to such traditions of comic poetry in his early years in Tuscany, but his interest in the comic grotesque flourished when he moved to Milan at the court of the Sforza. Clayton believes the purpose of the comic grotesque drawings was to entertain the audience. It is no coincidence that Leonardo's grotesque heads date to the time he spent in

Milan (1482–1498) as court painter; comic and burlesque literature was appreciated at the court of Ludovico Sforza, whose main poet in residence was the Florentine Bernardo Bellincioni who, besides encomiastic poetry, composed many comic realistic verses that were certainly familiar to Leonardo.[12]

Leonardo's grotesques are satirical pieces, mocking the vanity of the aged and depicting bodily deformity through various forms of otherness. The wrinkled woman in ostentatious headdress shows the otherness of the grotesque old female body, where the display of withered breasts and the quasi-masculine face construct female old age as sexualized excess and vanity (Figure 7.3).

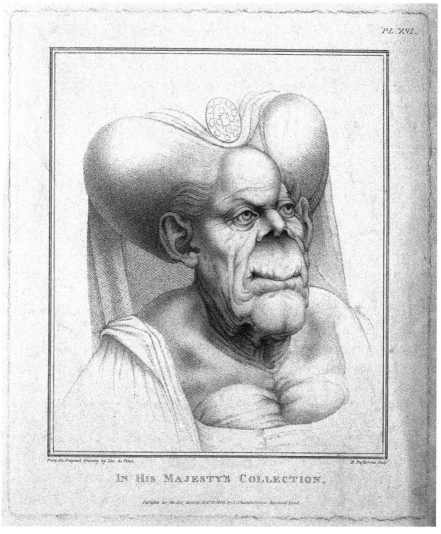

FIGURE 7.3: Old Woman. Head and shoulders of a hideous old woman wearing an escoffin over a wimple. Color Stipple engraving by B. Pastorini, 1806, after Leonardo da Vinci. Wellcome V0007456.

One grotesque drawing, *Old Man Leaning on a Stick* (ca. 1510), presents bodily otherness in the figure of the Jew. Clayton believes the old man depicted here "conforms with the caricatured Jewish type . . . with a long hooked nose, fleshy lips, avaricious eyes and a claw-like hand clutching a stick."[13]

Because Leonardo rejected physiognomics as a reference in the portrayal of different states of mind, his goal of depicting realistic human figures is achieved through the observation of different individuals. In Giovanbattista Giraldi Cintio's *Discorsi dei romanzi* (*Discourse on Novels*, ca. 1550) there is an interesting cue as to da Vinci's technique in realistic figuration:

> When he [Leonardo] wanted to paint a figure, he would consider first its quality and nature, that is whether the person had to be noble, plebeian, joyful or serious, disturbed or cheerful, old or young, enraged or tranquil, good or evil; and then . . . he would go where he knew that such people would congregate in great numbers, and he would diligently observe their faces, their manners their clothing, their bodily movements, and when he found someone who was suitable to his plan, he would sketch them in the little book that he always kept at his belt.[14]

Giraldi also relates a famous episode, included also in Vasari's *Lives,* where Leonardo expresses his difficulty in painting the head of Judas during the completion of the *Last Supper* (1498). For the face of the traitor, Leonardo had to find inspiration by going around the suburbs of Milan, in the quarters where lived all the "vile and ignoble persons and . . . evil and dreadful."[15] In the periphery of the big city Leonardo would sketch and collect realistic examples of persons that could be used in the figuration of ill-proportioned or ugly persons (Figure 7.4).[16]

According to Clayton, both Vasari and Milanese artist Giovan Paolo Lomazzo (1538–1600) contributed to the image of Leonardo as a bizarre genius, who would follow odd-looking people in the streets and visit prisoners to observe their facial expressions. The grotesque heads and distorted figures (old, wrinkled, excessive, lascivious) had a particular attraction to the Milanese and northern circle of artists (Leonardo's pupils Francesco Melzi, Giovan Paolo Lomazzo, and Aurelio Luini), who later employed these physiognomic types in many of their paintings.[17]

Leonardo's grotesque drawings may have been inspired by the observation of subaltern groups living in the city of Milan, such as peasants, *facchini,* and gypsies. Thanks to its geographical location in the Po Valley, at the feet of the mountains, and to its economic importance during and before the Renaissance, Milan attracted many poor people from the high valleys and the mountains, who were employed as daily laborers. They lived in the suburban areas and were often ridiculed for their crude language and rough manners.

FIGURE 7.4: Cretin with goiter, Or Capitan Nasotra, drawing after Leonardo da Vinci from *Le Difformes et le Malades dans l'Art* Charcot, J-M and Richter P.M.L.P Paris, Lecrosmier et Babe, 1889. Wellcome L0014188.

BODY AND MANNERS

As Georges Vigarello notes, the sixteenth century marks a cultural change regulated by the behavior of nobility, which introduces the concept of civility as a set of rules for living.[18] Texts such as Castiglione's *Libro del Cortegiano* (*Book of the Courtier*, 1528), Giovanni della Casa's *Galateo* (1558), and Stefano Guazzo's *Civile Conversazione* (*The Civil Conversation*, 1574) were considered the foundation of European culture, civility, manners, and bodily practices. Through these conduct books the higher classes defined themselves in opposition to the uncivilized other. Uncouth manners, propensity for lower bodily stratum, sexual excess, and lack of decorum were typical of the representation of lower social classes, the *contadini, villani,* or *facchini* people who lived in the *contado* or in the peripheral or mountainous regions. In a poem addressed to his friend Giovanni della Casa, comic poet Giovanni Mauro (1490–1536) pokes fun at people living in a mountain region of Latium. Mauro ridicules their boorish manners and their dirt, with particular emphasis on women: "As Nature made all of them of pure dirt / so they [mountain women] go around decorated with it from head to toe . . . / their thick hair is a forest of lice, the teeth are covered with ricotta cheese / and their breasts fall down to their knees."[19] These people of the lower classes, living in peripheral mountain regions, are isolated from the centers of refinement and decorum; hence they do not follow the codes of conduct and cleanliness established by the elite.

Castiglione's *Cortegiano* became a sort of "bible of a nobility that was introducing a detailed etiquette."[20] While setting the codes of proper behavior for the courtier, his treatise makes a much wider impact on the conduct of the higher social classes and, more generally, on anyone wishing to show civility. So any bodily practice discussed in the dialogue referring to the prince's most trusted adviser, that is, the courtier, is then viewed as a set of rules for the elite, who are attempting to define and distinguish themselves from the other, the inferior classes. Castiglione advocates a gracious behavior, a *sprezzatura* or gracefulness, which is a form of conduct where perfection is achieved without an apparent effort; includes a physical deportment devoid of any artifice and a composure that entails care for the physical body, but it also frowns upon any form of affectation and bodily manipulation, such as makeup for women, or other artifice that is considered excessive. In turn, some parts of the body can create special charm by being concealed or less visible, such as the hands, which are more attractive when covered with gloves, or the feet, which should barely appear from under a gown. Moderation and measure are emphasized in every part of the dialogue. Castiglione also takes pains at distinguishing the elite courtier from the *contadino* (peasant): "your courtier should not . . . be disgusting and dissolute in his manner of life, and act like a peasant who stinks of the soil a mile off. For a man of this sort can never hope to become a good courtier and indeed can be given no occupation other than looking after the farm animals."[21]

Giovanni della Casa's *Galateo* (1558) devotes special space to the promotion of normative bodily behavior. His book about manners became one of the most popular texts advising on correct social skills and proper and polished conduct, which would help single out members of the powerful elite. On the premise that man should always be seeking to attain beauty and proportion, della Casa, besides good speaking skills, recommends control over some parts of the body that can produce disgust or dirt. Well-mannered men, when eating, should ensure that they do not "soil their hands nearly up to their elbows, and dirty their napkins worse than their toilet towels."[22] Likewise, coughing, spitting, and sneezing are not considered appropriate when dining. Every form of secretion coming from inside the body must be carefully concealed when in public, and table manners are essential to be pleasing and to distinguish oneself from the lower classes. In the whole tract, common people and peasantry are mentioned as a measure of bad manners, low speech, and a type of behavior that needs to be avoided. Polite and graceful manners lead to measure and composure, a series of practices that excludes spitting, eating too fast, rubbing one's teeth with fingers, rinsing one's mouth with wine, or drinking to excess.[23] For della Casa, it is also important to apply a certain discipline to the body, not to reveal parts of the body normally concealed by clothing and to keep the body erect, without elbowing others when speaking. Clothing is also an important element that allows the citizen to distinguish himself according to status and age. So a gentleman should not go around the city "in plain overcoat, as if he thought he was in the country."[24] Nor should one take off his clothing in public, especially lower garments. Twisting one's mouth and eyes, puffing one's cheeks, and making unpleasant acts with the face is also not recommended. Particular mention is made of not showing one's tongue or stretching out and speaking like "a peasant waking up in a haystack."[25] The *Galateo*, then, advocates an avoidance of excess that might lead to the exposure of the grotesque body, with its fluids, secretions, and orifices that need to be contained and concealed.

PHYSIOGNOMY

During the early modern period, a conceptual tradition, available from classical times, established a link between outward appearance, that is, bodily features, and inner qualities of the individual: physiognomics. This quasi-scientific discipline that drew on pseudo-Aristotelian precepts attempted to deduce psychological and moral characteristics of an individual through features of the physical body, particularly the face. The principles illustrated in physiognomics were meant to provide a set of tools to interpret human faces and bodies. Most famous was Giambattista della Porta's (1535–1615) *De Humana Physiognomonia* (*On Physiognomics*, 1586). Drawing on Aristotelian philosophy, della Porta believed it was possible to know "particular passions of the soul from the particular shape of the body."[26] This theory of a correspondence between character and external body in one individual expresses the possibility to predict from external features all the inclinations of the human soul, a classification of humanity based on behaviors attributed to body types and mediated through the figure of the beast. It is the beast that allows for the explanation of individual features. In Book II, della Porta examines in detail all parts of the body, particularly the face, and establishes a relation to qualities of a certain animal, for example a medium-sized human face is set beside the face of a lion, which is associated with courage and virtue. A hooked nose corresponds to greed, and fleshy lips are associated with the stupidity of donkeys and monkeys. And later in chapter 4, monkeys and donkeys are listed among the lustful animals: "Lustful animals are the pig, the goat, the monkey, the donkey" (Figure 7.5).[27]

In della Porta's physiognomical system, for example, the black color of the face and of the body in general, associated with people from Ethiopia and Egypt, has negative connotation linked with the dark color: "the black color is a sign of a dreadful and deceitful mind."[28] Some features of the face that are traditionally attributed to black Africans, such as thick lips and large nose, signify stupidity. Della Porta collects a vast array of animalistic traits and uses them to describe human beings in ways that support the dominant social order.

FIGURE 7.5: Man/Donkey. From *Della fisionomia dell'huomo* by Giovambattista della Porta Collection: Rare Books. Wellcome L0027462.

This method is applied to gender difference as well.[29] The list of male attributes includes big body, big face, thin arched eyebrows, square jaw, large robust neck, and it demonstrates among animals the model of maleness that corresponds to generous, intrepid, and just, in opposition with femaleness, servility, and savagery. Femininity is a negation of all the male traits: a woman has a small head, narrow face, small eyes, hairless jaws, and delicate neck; in behavior she is thieving, deceitful, delicate, and prone to anger as well as fraudulent, stupid, unstable, and imperfect. If the lion typifies masculinity, the leopard typifies femininity as an evil, deceitful, devalued form of man.

BLACKS

The representation of visible minorities is defined by the nonnormative body in the black skin color and in the dimension of the body (too big or too small with respect to the normative body). Racial difference is defined as bodily otherness in the nonwhite skin or marked by ethnic costumes or sartorial signs of differentiation.

The Renaissance is the time when the first sustained flux of black African slaves comes into Europe. They came from sub-Saharan and West Coast Africa and were sold as slaves. There appeared to have been confusion in the naming of black and other ethnic groups in the Renaissance. In the early modern period the term *Mori* or "Moor" was used to refer to people with dark skin, including Arabs, Berbers, and Muslims (Figure 7.6).

The Moor was defined with a set of interchangeable terms—Turk, Ottomite, Saracen, Egyptian—all constructed in opposition to Christian. The Moors of North Africa were identified with Islam, and the term "Moor" was used with specific reference to people of Turkey or Morocco but more often in general terms to define the Islamic other. The Christian myth that explained the origins of dark-skinned races, including Moorish Muslims of Africa, is derived from the Old Testament story of Cham, son of Noah, who was cursed for beholding the nakedness of his father. Cham was said to be the progenitor of the black races, whose skin color was a sign of an inherited curse. At a time when the concept of race had not yet been elaborated, black skin was a defining element that separated civility from barbarism, physical beauty from ugliness.

Blackness as Ugliness, Physical Prowess, and Lust

Black skin was viewed negatively by association with deformity, with the devil, and with slavery, in opposition to whiteness, which was viewed as the norm of beauty, purity, and moral uprightness. Humanist Antonio Brucioli (1495–1566) in his *Commentary upon the Canticle of Canticles*, when expounding

FIGURE 7.6: Moro de Barbaria. Moor of Barbary. From *Habiti antichi et moderni di tutto il mondo*. Cesare Vecellio, Venice: Gio. Bernardo Sessa, 1598. Credit: Bruce Peel Special Collections Library, University of Alberta.

on the famous verses of the dark bride "I am black but comely" ("Nigra sum, sed formosa") talks about darkness as deformity, ugliness, and evil in contrast to the whiteness and fairness of all that is beautiful.[30]

Although Europeans were not capable of distinguishing the different shades of dark skin, they perceived black Africans differently from any other ethnic or religious minorities in Europe such as Jews, Muslims, or gypsies, because the skin color of the former made them more visibly recognizable. Black people living in Europe were automatically assumed to be slaves, therefore legally of inferior status. Giambattista della Porta, in his *De Humana Phisiognomonia*, considered black-skinned people evil, deceitful, and more effeminate and explained the blackness of Ethiopians and Egyptians as caused by the skin drying up in the strong sun.

The arrival of black Africans in Europe coincided with the period of European self-definition and with the notion of civilization. Therefore certain behaviors began to be viewed as civilized whereas other, non-European behaviors were considered ignorant and uncivilized. This led to the stereotyping of black Africans as other because they were non-Christian and positioned outside the culture and civilization of Europe. The concept of black African in the European mind clashed with the notion of freedom, power, wealth and civilization. Black Africans were associated with laziness, drunkenness, and lust, and as slaves they were depicted as musicians, dancers, and sometimes as bodyguards or soldiers for their physical prowess (Figure 7.7). Africans were also described as more prone to sexual transgression. As Kate Lowe notes, some body markings, scars, tattoos, or accessories such as chains or nose rings became associated with black Africans.[31] Renaissance depictions of black Africans included gold ornaments for nostrils and ears. Often these ornaments were worn by African slaves as ostentatious displays of their masters' status. Official chronicler of Venetian history, Marin Sanudo, in his *Diarii* (1466–1533) expressed his dislike for women who, like *more* (i.e., female Moors or black Africans), have pierced ears and wear earrings set in gold.[32] In many depictions of black slaves, golden ornaments, including rings, anklets, and collars, became markers of bondage. Black slaves in Europe were often forced to wear such golden chains as signs of slavery.

Already in reports of voyages to Africa, the description of black Africans contributed to stereotyping, particularly with respect to some aspects of the physical body, such as nudity and thick lips. Africans were considered savages because they did not know Christianity and because of their nudity or semi-undressed condition, which was widely reported by European travelers to Africa. Venetian Alvise Cadamosto (who was born ca. 1429 and traveled in Africa between 1454 and 1456 pursuing a commercial career sponsored by the Portuguese) wrote the first original, factual account of his two voyages to sub-Saharan Africa, an area he mistakenly called Lower Ethiopia. His

FIGURE 7.7: Black Slaves. A male and female slave playing instruments. *A collection of voyages and travels some now first printed from original manuscripts.* By Awnsham Churchill, John Churchill, John Locke and John Nieuhoff London 1744–46. Collection: Rare Books. Library reference no.: EPB 17811/D vol. 2. Wellcome L0037783.

accounts, published in the collection of *Paesi novamente retrovati* (*Newly Discovered Countries*, Vicenza, 1507), had a major influence on subsequent travel narratives. The description of many native tribes encountered on his trips included mention of seminudity and lax sexuality. For example, the non-Christian inhabitants of the Canary Islands have a warlike nature and live in a primitive state, because they constantly fight with each other; they live in caverns and "they always go naked, save some who wear goatskins, before and behind"[33] (Figure 7.8). Beside nudity, Cadamosto notes some physical deformities in the Lobi people, inhabitants of a region he calls Mali, who did not speak and had an enlarged lower lip that hung down to their breasts, exposing their gums and teeth. These people traded gold for salt. Salt mixed with water was believed to counteract the flesh putrefaction caused by the extreme power of the African sun. Cadamosto's interest in such deformity reveals that the otherness of black people in early modern times, even in a fairly balanced and accurate account such as Cadamosto's, could be viewed as a monstrosity, a legacy of classical and medieval literature on Africa and other exotic, fabulous lands.

Gender and sexuality are always included as part of the narrative of alien culture and race. Leo Africanus (ca. 1494–1544), a Moor born in Granada and brought up in Barbary, wrote in Arabic and Italian *Della descritione dell'Africa*, 1526.[34] Kim Hall thinks Leo's *Della descritione dell'Africa* gives a sense of Africa as a chaotic disordered land where hierarchies of gender disrupt the ideal order.[35] Leo gave the idea of unruly and diverse sexuality in Africa, especially the exorbitant erotic females. Leo comments on communities that do not exert enough control over their women, like the one on the Barbary Coast, and describes some women soothsayers of Fez who are dark and demonic as some type of witches who "have a damnable custome to commit unlawfull Venerie among themselves, which I cannot express in any modester termes. If faire women come unto them at any time, these abominable witches will burn in lust towards them."[36] Hall notices that the black witches are contrasted with the fair women, who are lighter in color, are beautiful, and live in accordance with the values of the patriarchal family.

Black Africans, when channeled into the slave trade and introduced into the European context, were domesticated and valued for their physical strength and prowess. They were employed to work in armies and in activities that required a sturdy body. As documented in Vittore Carpaccio's painting *Miracle of the True Cross* (1494) two black Africans in the Venetian setting were gondoliers. Black Africans were employed in this job because of their physical strength. Cadamosto, when meeting the local people of Senegal and Gambia, was impressed with the way local oarsmen rowed boats while standing, just like the Venetians. According to Marin Sanudo, the thousands of gondolas in Venice were rowed by black Muslims or other servants.

FIGURE 7.8: Man from the Canary Islands. From *Habiti antichi et moderni di tutto il mondo,* Cesare Vecellio, Venice: Gio. Bernardo Sessa, 1598. Credit: Bruce Peel Special Collections Library, University of Alberta.

Black and White Conceit

Blacks as young servants in elaborate costumes are frequently present in pictorial representations; they reflect the presence in Europe of slaves from Africa. Some Africans were bought by patrician families as exotic objects intended to showcase their prestige. In Paolo Veronese (1528–1588) we find black servants in paintings such as *Feast in the House of Levi* (1573), which includes no less then six Africans, and *Wedding of Cana* (1562/1563). Nobility enjoyed the presence of black servants to highlight their elite status. Young black servants (*moretti*, little Moors) and dwarfs were popular at many Italian Renaissance courts. Such figures functioned as exotic curiosities and were available for entertainment and to enhance the status of the elite.

As Anu Korhonen observes, the perception of black skin always "presupposed an implicit evaluation of beauty and the inherent cultural value of whiteness."[37] African appearance, skin color, and facial features were used to oppose stereotypical images of beauty and ugliness. Black skin was perceived as a spectacle produced by the opposition with white.

In Renaissance art we see the appearance of a pattern of subordination that consists of the pairing of black and white: the iconographic type of the black African attendant to a white European protagonist. Paul Kaplan sees this model of color-coded subordination (which already existed in the Middle Ages) developing in Andrea Mantegna's *Judith and Her Maidservant with the Head of Holofernes* (1492), which shows a black maidservant, believed to be the first black African servant to Judith in European art and culture (Figure 7.9).[38]

Afro-European attendants became popular components of patrician retinues as early as the Middle Ages. By the 1400s, the Aragon kings of Naples were among the most active employers of black servants. Sforza, Gonzaga, and Este, closely tied to the Aragon, contributed to the practice of the Italian Renaissance courts of keeping black slaves.[39] In the late 1400s, three fresco cycles at Ferrara and Mantua depicted Africans.[40] Mantegna's *Adoration of the Magi* (1462), which was a Gonzaga commission, shows that by 1460 black Africans were part of Mantuan cultural life. The drawing by Mantegna with the black servant is believed to include a black servant to please Isabella d'Este, whose interest and fascination with black child servants is well known.[41] Kaplan thinks the large number of black slaves in Europe in the late fifteenth century was a factor in Mantegna's decision to depict the maidservant of the *Judith* drawing as a black African.[42] Furthermore, if the drawing was executed for Isabella d'Este, it would have been particularly welcome to include a black African, because Isabella would have recognized herself in Judith and one of her *morette* in the black servant.

As is evident from her correspondence, Isabella was in the habit of securing small black children for her court and for some of her friends. From various

FIGURE 7.9: Black Servant. *Judith is putting the head of Holofernes in a bag helped by her servant,* by Johann Nepomuk Strixner, published circa 1800 after Andrea Mantegna. Collection: Iconographic Collection 573378i. Wellcome V0049939.

letters exchanged with her agent in Venice, Giorgio Brognolo, we know that in June 1491, Isabella was looking for "a *moretta* of no more than four years of age."[43] Brognolo found her one but with difficulty because he was in competition with Isabella's mother, an avid collector of black slaves, who was also in search of the same. Slaves were subject to scars and disfiguring disease that often made their physical appearance less than perfect. The letters reveal the interest in the physical appearance of the young *moretti*, with emphasis on blackness and on good shape of the body. Brognolo says he found one *moretta* who was "very black and well shaped." After welcoming her in Mantua, Isabella comments on the fact that "in blackness and shape she [*moretta*] satisfies me more than I had hoped."[44] Later Isabella was so pleased with this *moretta* that she wanted to find her a companion. When Brognolo's wife mentioned that there may be a young *moretto* available at the house of a gentleman, again Isabella expresses the importance of the physical appearance: "if he is of the highest beauty, purchase him for me . . . but if he is not black and well-proportioned, do not agree to buy him."[45] The emphasis is on the *moretti* being gracious, well-shaped, and very black, because black African slaves at the court of the Este and Gonzaga were acquired not for domestic labor but as human accessories to embellish the court and as symbols of prestige. Like buffoons, dark slaves were considered sources of amusement and objects of display.

Moretti were also objects of gift exchange among nobles: the Duke of Mantua asked his agent to purchase one to give to Lady Montpensier, and Andrea Doria gave two blacks to the Duke of Mantua as a sign of thankfulness. In 1522, Margherita Cantelmo, a dear friend of Isabella d'Este, offered her a "*mora*": "who was captured not long ago in Barbary . . . she may be sixteen or seventeen and she is beautiful . . . well-shaped . . . with a beautiful face, except that her lower lip is thick."[46] These types of comments on the physical appearance of the black servants reveal the importance of the perfect body and the critical evaluation of the thick African lips. At Renaissance courts black slaves were kept as half-pets and half-buffoons, as collector's items.[47] Ladies of high rank were keen on having dark-colored pages and slave girls at their courts, because their blackness served to highlight by contrast their light beauty. The blackness of the slave other contributed, by contrast, to the definition of the identity of the white elite.

In visual representations as a vehicle of identity construction, black servants became more important, as demonstrated by their appearance in official portraits. In Titian's *Portrait of Laura Dianti* (1524), the beautiful white woman is depicted beside a young black servant, whose presence in the canvas serves various purposes. It shows the exotic black servant as a rare commodity available only to the elite, highlights the conventional beauty of the white lady that becomes more evident when set beside the dark-skinned boy, and shows the

deference and submission of the black slave to the white.[48] French chronicler
Pierre de Bourdeille, Lord of Brantôme (1540–1614), who traveled around Europe and kept an extensive memoir of his life as a soldier and court member of
Catherine de'Medici, observed in his memoirs that "an excellent painter who,
having executed the portrait of a very beautiful and pleasant-looking lady,
places next to her . . . a moorish slave or a hideous dwarf, so that their ugliness
and blackness may give greater lustre and brilliance to her great beauty and
fairness."[49] Otherness as blackness serves to define by contrast the identity of
the white elite.

The black and white conceit was present in art and literature. The beauty
of the fair lady is opposed to the dark-skinned maid who is seen beside her.
Torquato Tasso (1544–1595) includes in a vast collection of lyrical verses some
poetry about dark women. In one *canzone* addressed to Leonora Thiene San-
vitale, countess of Scandiano, a member of the Este court in Ferrara, beside
the conventional homage to the beauty of the fair lady, Tasso extols the dark
Ancella (chambermaid) with the same formulas used in the biblical *Song of
Songs* (Solomon's *Canticle of canticles*):

> You are dark but beautiful,
> like a virgin violet; and I get so much
> contentment from your graceful appearance
> that I do not disdain the sovereignty of a maid.[50]

According to Basile, editor of Tasso's *Rime*, this maid may have been a
certain Olimpia at the service of a countess, but we do not know if she had
dark skin. The contrast between the fair beauty and the darkness of her maid
becomes more frequent in seventeenth-century poetry. In an anonymous sonnet, "Poem for a Moor Seen at the Window with Beautiful Woman," the poet
states: "Here the night around my beautiful sun / embellishes with its shadow
its splendor."[51] Baroque poet Paolo Zazzaroni also has a sonnet for a "Lady
and her Maid" where the maid is a black servant.[52]

NATIVE INDIANS

In the representation of Native Americans, otherness is defined by the darker,
olive-toned, often hairy, semi-naked body. In the accounts of travel to the new
world, there is a duality vis à vis the physical and moral representation of the
indigenous people, who are depicted as both peaceful primitives and savage
cannibals.[53] In Christopher Columbus's famous 1493 letter to Luis de Santan-
gel and Gabriel Sanchez, the Genoese admiral for the first time portrays for the
European public the image of newly discovered people of the New World. They

are described as shy, peaceful people with no religion, who show no hostility to the Christian faith and can be easily converted. Columbus (1451–1506) admits that he did not find any monstrous people here. Even though the natives all go naked, they are described "of pleasant appearance and not negro like in Guinea, they have flowing hair."[54] As such they are associated with the golden age, and with original prelapsarian bodies of Adam and Eve, and therefore good candidates for Christianization. However, some of the indigenous people that Columbus mentions are fierce cannibals, feared by the other natives who are unable to protect themselves from their attacks. Although cannibals are not viewed as monsters, they are described as uglier and with hair as long as women.

Even the most positive travel narratives on the New World attach moral judgment to behaviors such as sexual promiscuity, sodomy, and cannibalism, considered uncivilized and marked by nakedness, physical ugliness, and monstrosity. Amerigo Vespucci's *Letter of the Newly Found Islands*, sent in 1504 to Pier Soderini, reveals the European fantasy of cultural superiority. In one episode, Vespucci (1454–1512) describes savage women as cannibals who cut a young man to pieces and roasted him on a fire (Figure 7.10).[55] This incident of sexually charged violence and native savagery reveals the centrality of the native woman in the construction of otherness.

The indigenous female is often the object of special attention in European accounts of first encounters; she is the other who attracts the most interest in European travelers. Many depictions of American indigenous women in the literature of discovery and exploration are centered on erotic desire or dangerous sexuality, made visible through the dark and naked female body. Infamous is the account of Michele da Cuneo, a member of Columbus's expedition, and an avid conquistador, who viewed Amerindians as savages and beasts, criticized their sexual promiscuity, and described in cold matter-of-factness his rape of a beautiful naked young woman he received as a present from Columbus.[56]

Pietro Martire (1457–1526), the first historiographer of Columbus's voyage of discovery, describes in *De Orbe Novo* (*The Eight Decades of Peter Martyr d'Anghera*, 1511) the first encounter of Columbus and the Spaniards with the native people. Cachey notes that in this inaugural scene, Martire, rather than describing taking possession of the new land, shows the cultural encounter with the other, represented by the capture of a naked woman who becomes a figure of mediation between the hegemonic European culture and the subaltern Amerindian.[57] Although Martire presents the Amerindian woman in a positive light as cultural mediator, most depictions of American indigenous women in the literature of discovery and exploration are centered on erotic desire or dangerous sexuality, all made visible through the dark and naked body.[58]

FIGURE 7.10: Cannibalism, Native Americans. From: *Newe Welt und amerikanische Historien*. By: Bry, Theodo de. Published: Heirs of Merian Frankfort 1655. Collection: Rare Books. Wellcome L0005638.

MUSLIMS

Interest in geography and travel literature was the result of a process of self-fashioning, whereby the self is formed in relation to others and their external representation at a time when contact with the other was available. This availability was through the presence of various ethnic groups in Renaissance cities, through trade and diplomatic exchanges, and through the large accessibility of printed texts describing and accounting for different people at a very intense time of navigation and geographical exploration.

Renaissance Europe defined and measured itself in relation to the societies to its east, with their splendor and wealth. Jerry Brotton considers, for example, the significance of Gentile Bellini's painting *Saint Mark Preaching in Alexandria* (1504–1507), where the saint preaches in front of a mixture of European and Oriental figures, the latter including Egyptians, Mamelukes, North African Moors, turbaned Turks, Persians, Ethiopians, and Tartars.[59] Rather than considering these people of the east barbaric and ignorant, for

Brotton Bellini's painting reveals that the eastern cultures possessed desirable aspects that Europe did not view in mere opposition to their own, but rather as an opportunity for the exchange of ideas. The city of Venice, for example, was architecturally inspired by admiration and emulation of eastern cultures. However, the fall of Constantinople in 1453 led to the creation of the Ottoman Empire, the most powerful empire Europe had seen since Roman times. Such strength was a direct threat to Europe, particularly to the Hapsburgs, the Catholic Church, and Venice. The Church adopted a defensive and aggressive military stance against Muslims, as well as Jews, and it also tried to affirm its authority against all heretics, including Protestants and Muslims. Renaissance Europeans, particularly Venetians, had an ambivalent stance toward the Turks. Venice's territorial dominance in the Adriatic had been eroded by the Turks since the end of the fifteenth century. The opposing views of the Turks circulating in Renaissance Venice are evident in a woodcut by Nicolò Nelli *Turkish Pride* (1572), examined by Wilson,[60] depicting the face of a Turk in profile with a large turban: when the print is turned upside down the contour of the figure appears as the face of the devil. There was a dual nature in the Venetian image of the Grand Turk of danger and opportunity, threat and temptation. For the Venetians, the Turks represented what they admired and what they feared. For them and for other Europeans, the Ottomans were a defining other, foreigners whose presence sparked reflection on their identity.

Renaissance writers had a stereotypical conception of the Great Turk conjured up by images of great wealth and exotic splendor. The Mediterranean was the setting for many stories about Islamic power at sea and in the commercial ports controlled by the Ottomans. Turkish galleys were often reportedly attacking Christian merchants and presenting the Western stereotype that associates Islam with acts of violence, treachery, cruelty, and wrath. The alleged sexual excesses of Muslims or Turks were viewed as signs of evil.

Although people of the New World or Africa are depicted in their unclothed bodies (because they represent a primitive stage of becoming European and are compared with Adam and Eve), Turks and Islamic people were not viewed as easy to convert, and their representations show them in elaborate costumes. A negative perception of the Turks among Venetians derived from the different language and from stereotypes about avarice, effeminacy, and corruption. Two plays were published in Venice in 1597 and 1606 called *La Turca*. The second is a satire by Giambattista della Porta that was most likely composed in 1570. This comedy draws on contemporary stereotypes about Turks and their behavior. Turks were said to be greedy, licentious, violent, barbaric, and contradictorily lacking in masculine virtue. In Giambattista Marino's poems on the occupation of Taranto by the Turks (1594), he calls the entire Turkish race "perfido cane" (perfidious dog), where the animal comparison evokes the physiognomics of della Porta.

Venetian interest in the appearance and customs of the Turks intensified during the mid-sixteenth century, and in Venetian printing presses a wide variety of material was published, ranging from portrait books, to biographies of military leaders, to lives of Turkish sultans and their achievements. Popular historian Francesco Sansovino (1521–1586) wrote no less than seven books on the Turks, where he described his admiration for the sultans and Süleyman in particular. Sansovino, however, attacked Süleyman's son Selim for his voluptuousness, faithlessness, and dissoluteness. His most famous *Dell'Historia Universale dell'Origine et Imperio de Turchi* had seven editions between 1560 and 1654. The last quarter of the sixteenth-century was a period of ethnographic and historiographic approaches to the Islamic world through costume books that provided a visual record of the plurality of Ottoman society. As Ottoman military power in the sixteenth century threatened the Hapsburg Empire and some Italian ports, including Venice, more representations of Ottomans appeared, especially in print and costume books (nine editions published in Venice in the late 1500s). It was the costume and costume books, rather than the color of the skin, that defined the other and contributed to the formation of racial and gender stereotypes based on physical appearance.[61] Although the black person is clearly and unmistakably identifiable in his or her otherness based on the color of the skin, for other ethnic groups physical distinction may not be so obvious. Costume became an important element to classify people of different origins, especially Muslims or Turks (a term that extended to all members of the Ottoman Empire). When markers of race like black skin are not immediately visible on the body, costume can be a means to identify a foreigner. In Cesare Vecellio's costume book, *Habiti antichi et moderni di tutto il mondo* (1598), Ottomans (occupying the entire eighth book) figured prominently in comparison, for example, with the Frankish section, which is comparably small. Vecellio presented a vast array of turbaned Turks in hierarchical order, ranging from the Sultan, to the general, to the clerics and the soldier (Figure 7.11).

Sartorial, transient demarcations of nonconformity or otherness, were also in place to signal the racial distinction of Italian Jews. From the late fifteenth century throughout Italy, Jews were forced to wear a visible sign of distinction, a yellow badge or, for Jewish women, a yellow veil similar to the one prescribed for prostitutes. Such practices can be explained by the desire to control the representation of others, to dominate them or to use them to reinforce the status and identity of the hegemonic groups.[62]

Saracens, Giants, and Dwarves

Islamic others were also presented in Renaissance epics as Saracens, infidels fighting the Christian knight during eighth-century crusades. Critics have noted

SVLTAN A MVRHAT

FIGURE 7.11: Turbaned Turk. Sultan Murhat from *Habiti antichi et moderni di tutto il mondo,* Cesare Vecellio, Venice: Gio. Bernardo Sessa, 1598. Credit: Bruce Peel Special Collections Library, University of Alberta.

the persistent misrepresentation of Islam, especially in epic romance.[63] This is clear in Luigi Pulci's *Morgante* (1478–1483), Matteo Maria Boiardo's *Orlando Innamorato* (1495), Ariosto's *Orlando Furioso* (1516), and Torquato Tasso's *Gerusalemme Liberata* (1581). Renaissance writers of epic poems, such as Ariosto and Tasso, told stories of the defense of Christianity in a bygone past (the era of Charlemagne), but they often ignored the Turks' invasions of Europe that were happening at the time their great epics were composed. Although Christians saw Islam as an aggressively expanding competing form of monotheism, in Renaissance epic romance Islam is often misrepresented as paganism.[64] In chivalry romances, Muslim infidels were represented as rebellious characters, villains who came to a violent end, cursing and screaming as their souls went to hell. Saracen enemies appeared as other in physical disproportion, gigantic or dwarfish, diabolic, deformed figures. However, in comparison with medieval chivalry romances, fewer negative traits are present in the Muslim others of Renaissance epics, mainly because most Saracens who survived battles with Christians were redeemed by conversion, as in the case of the giant Morgante in Pulci's *Morgante*.

Ariosto's chivalry romance, written for the Este nobles of Ferrara (already in decline by the early sixteenth century) included scenes of defeating Saracens, but more as fantasy than as real life. In Ariosto, and in Boiardo before him, there appeared the deformed dwarfish black, Ethiopian-looking Brunello, a small person full of malice and active in theft. He has much in common with Pulci's Margutte because of the dark color of his skin. The thief *negro* of Boiardo's *Orlando Innamorato*, with short locks of black hair, in Ariosto acquires more African physical features: he has a "curly head," "his hair is black," and he has "dusky skin" and a flat nose. According to Piero Camporesi, Brunello resembles: "the *puer niger* [black boy] with Ethiopian physical features . . . linked to the devil-possession of the oldest Christian tradition."[65] Camporesi notes that in medieval culture the term "nigredo" was used not only for the devil but also for peasants and serfs, who were often represented as subhuman (hairy, with blood-shot eyes and dark skin), monstrous, in a somatic blend between devil and animal.

OLD AGE

Renaissance culture is deeply rooted in the glorification of youth and the awareness of its transience; one need only think of Lorenzo the Magnificent's famous verses "How beautiful is youth, though fleeting!"[66] Age and growing old are marked by the onset of all sorts of negative effects on the physical body. Erasmus defines old age as a time of dreadful illness and incurable disease in his elegy on old age (1506) and later reiterates the concept that "old age itself is an illness."[67] Old age is mostly defined as a period of bodily decay marked by negative effects leading to abject representations of bodily disfigurement, associated with decay and disease. Representation of the aging body was not limited to the negative valence, though that might be the prevailing one.

Gabriele Zerbi, a physician who taught medicine in Padua, Bologna, and Rome, author of *Gerontocomia* (*On the Care of the Aged*, 1489) composed the first medical book dealing specifically with the problems and treatments for old age. Zerbi identified some physical "accidents" that accompany old age: wrinkles, gray hair, baldness, lack of humidity, dryness, loss of heat. However, his book was based on the premise that, although it is impossible to prevent decay and death, old age is not an incurable illness, but a condition that can be slowed down. With appropriate measures, such as the prevention of drying out and coldness in the body, and with suitable diet and moderate exercise, people can be on the path to longevity. Even more optimistic is the famous tract by Alvise Cornaro on *Vita Sobria* (*The Temperate Life,* 1558), a classic of hygiene and a text idealizing old age. Cornaro, landowner of the Venetian mainland and patron of Angelo Beolco, stated in his book the cumulative advantages of added years and believed it was possible to increase significantly the life span in a natural way, through a good diet, a balanced and orderly life, and avoidance of physical and emotional excess. Such positions, derived from classical sources, went beyond the stoic approach of Cicero's *De Senectute* in that they stated that old age is the most beautiful period of life.

The more positive valence associated with old age fits more comfortably with the representation of male old age and is limited to the hegemonic classes (nobles and landowners like Cornaro). Furthermore, both Zerbi and Cornaro examined the condition of old age from a strictly male perspective. The positive representation of mature age for men is also in Castiglione's *Cortegiano*, but the courtesan is still a reminder that such a person is young and fit enough to fight, ride a horse, dance, and fence. There is a great fluctuation on the chronological definition of old age. A difficulty persists in the demarcation of boundaries for when old age begins. Zerbi stated that old age in men begins between the age of 30 and 40 years and extends to 50 or 60, and for Antonio Cammelli (1436–1502), a poet who lived at the Sforza and Este courts, a widow aged forty-seven described in a sonnet, carries all the signs of the aged female body, such as rotten flesh and wrinkled face.

Of importance for the marked body is the representation of female old age where grotesque realism abounds, particularly with reference to women who transgress social and class boundaries, such as prostitutes, courtesans, witches, and even widows, as shown in Cammelli's poem. The nature of feminine old age is different from that of masculine old age. Aging was believed to occur earlier in women than men because of their different physiology. If, as Zerbi stated following classical medicine, old age caused coldness and withering in the body, woman's colder, moister humors caused her to age more quickly. In the Renaissance, markers of female life stages, like puberty, marriage, and having children happened earlier in women than men. So women were perceived to age faster than men; furthermore, it is the female body in its images of youth and perfection that best exemplifies the ideals of beauty and harmony of the Italian Renaissance. Therefore, the old woman as Campbell observed, is "'Other' in both studies of representation of women and in early modern culture."[68] In comic realistic poetry, the image of the old woman is graphic and detailed, thanks to a rich tradition of rhetorical descriptive techniques. One can mention Politian's ballad, where the poetic persona mocks the old disfigured woman who engages in his courtship:

> An old woman longs for me,
> she is withered and dry to the bone;
> . . . her gums are worn out,
> from chewing too many dried figs.[69]

The grotesque portrait of the old woman includes withered breasts, droopy belly, dripping nose, and bad smell, details found also in Machiavelli's old prostitute and in poetry against courtesans. Female bodies marked by old age and physical decay were coupled with transgressive types, such as the prostitute, the courtesan, and the witch. Although the witchlike nature of Politian's old woman is inferred by some of her bodily attributes and her excessive drinking and sexuality, in Florentine burlesque poet Burchiello (1404–1449), the old woman, with disfigured body, is openly accused of being a witch and sorcerer:

> Old riotous, wicked, and evil woman,
> enemy of every food, envious
> enchantress witch and sorcerer.[70]

This old hag has dripping eyes and nose, foul breath, and facial hair, all features that contribute to the image of old woman as other and grotesque. Although comic poetry attaches moral judgment to the image of the old grotesque woman, in Giorgione's portrait of *La Vecchia* (ca. 1505–1510; possibly his mother) the painter uses toothless mouth and wrinkled body to convey the idea of the transience of female beauty and the negative effect of old age on women, as indicated by the banderole inscribed with the words "col tempo" ("with time") held in the old woman's hand. Thus, Italian Renaissance culture gendered old age in its negative connotation as predominantly female and attached to it a negative bodily and moral valence. Whereas for men, maturity and old age, though within the confines of the physical body, is represented with some positive effects.

CONCLUSION

At a time of identity formation and affirmation of the hegemonic groups in the European and Italian Renaissance, the marked body is visible in the representation of others or subalterns. As Spivak notes, subalternity should not be limited to groups at the bottom of the social scale; the subaltern has become a synonym for any marginalized or disempowered minority group, particularly on the ground of gender and ethnicity.[71] With more or less visible signs of physical difference, the marked body can be found in the grotesqueness and disproportion of outcasts and subalterns like prostitutes, peasants, *facchini*, and people of the peripheral areas, far from city and court. The marked body is evident also as deviation from the norms of conformity intended as white, young, and male: naked African blacks, slaves, Native Americans, and the old, mostly represented through the female gender. Costume and sartorial signs can also act as markers of bodily distinction for ethnic groups whose skin color or body shape is not immediately recognizable as deviating from the standard. Turbaned Turks and yellow-badged Jews serve in defining and reinforcing, by opposition and differentiation, the identity of Renaissance elites.

CHAPTER EIGHT

The Marked Body

The Witches, Lady Macbeth, and the Relics

DIANE PURKISS

To understand the marked body, we can begin by trying to understand *Macbeth*, a play filled with bodies ineradicably signed with choices made and their visible consequences. But it is also filled with bodies that bear the marks of a recently created cultural unconscious. Once, the play sometimes seems to know, there was a space for the marked body as functioning in and incorporated into a coherent culture. That culture has now been sealed over by history and can now be remembered only as a senseless and/or demonic inversion of what was once cogent and holy. The bodies of Macbeth are like an archaeological site; layers of sedimented story and layers of unreadable rubbish disclose themselves. The bodies in the play are inscribed with almost—but not quite—indecipherable signs of a past that has now been registered as the supernatural rather than as religion. What had recently been a cohesive, comprehensible, and sophisticated way of understanding and speaking with the dead had become illegitimate and extracurricular, and thus could be the subject of poetry as never before because it became a way of talking slantwise, sideways, about black and deep desires. I will be exploring four buried sets of stories about bodies and markings in the play: the story of relics, fragments, and rubbish; the story of child murder as a means to supernatural power, a story that is transferred from Jews to witches; the story of the hard body, impervious to feeling, and its power to do violence to other bodies; and the way all these stories exchange themselves between Papists, Jews, and witches, groups defined as misusers of bodies.

But to unravel at least some of that discourse implies a need to sit on exactly the historical fault line on which the play sits, which implies a willing-
ness to look backward and to look around. In this chapter I want to challenge
what have become the complacent assumptions of historicists about chronology and to argue that poetry, including drama, may contain alluvial deposits
of material and stories about material that go back much further than their
author's direct historical experience. The topic of "the body" carries a false
immediacy, but actually thinking and reading the body is a process determined
by historical forces that precede our own bodies by decades, even centuries.
The chronologies historicists now often assume—of near contemporaneity,
grounded in the question "Could Shakespeare have known this book?"—mean
we risk overlooking the way art is equipped to respond more flexibly to history
than polemic can. Historicization has come too often to ignore the long slow-
burning fuse of myth and folklore in favor of faster-moving print discourses,
even where these seem less relevant. This body of folkloric material is relevant because it continues to surface in print culture and in Shakespeare's own
reading, especially his perusal of three bodies of texts—the writings of Samuel Harsnett, which we know he read; Lucan's *Pharsalia*, with its frightening
necrophiliac night-hags, and Donne's exposition of his papist family. These
writings lay a powder trail that leads back to older and darker stories that un-
derlie the witch stories Shakespeare reinvents and sets in motion. *Macbeth* is a
play whose poetry is firmly founded on an epistemic change in thinking about
the body, a change that in part gives the play its charge.

The marker of that change is the decline and fall of the relic as object of
worship and desire. A relic is a physical object permanently saturated by the
power or personality of a saint. Because it is always already metonymic of the
saint, it becomes a way to love him or her; the relic can be kissed, fondled,
held, and journeyed for, as one might also do for love. The saint's body and, to
a lesser extent, items that have been in contact with that body bear the mark
of the saint's bodily value; they are inscribed with the saint's holiness, which
can be "caught" by the venerating believer. The ultimate relic is the eucharistic
host, which in medieval thought *is* bodily and which could convey the sacrificed Christ into the body of the believer. Like a relic, it could be devoured by
the longing gaze of love, and then held to the lips like a lover's hand. More
aggressively, of course, it was eaten—eaten even more—all the more—when
believers saw in it the body of a baby, a child, a beautiful young knight.

Macbeth is littered with a transposed discourse of the body as relic, the
body as supernatural marker, the body as lightning rod between God and
earth. The fissured response to the fragments of marked bodies evoked by
Protestant denunciations of relics becomes a way of marking off which bodies
are illegitimately supernatural and which are holy, and the play fully explores
the horror thus evoked. But for others, the relic was a physical manifestation

of love. The love of children for a father's ring or coat naturally extends to his body, said Aquinas, which allows us to venerate saints' bodies, too. And yet Jack Goody argues that the cult of relics was always "characterised not only by attraction but also by repulsion, by an attachment to the dead as well as a distancing from death, which readily becomes associated with our death."[1] Relic gathering and relic veneration required a deep transgression of normal rules for dealing with dead bodies. The constant troping of relics is, in Polycarp's words, "more precious than precious stones," and the construction of reliquaries encrusted with jewels draws attention to the problem that relics are often liable to fill the sacred space with objects likely to arouse disgust. Statues surrounding relics are a way of rendering the material immaterial.

The gold arm that surrounds the browned arm bones symbolizes their holy transfiguration, the light that suffuses them. But it also draws attention to the disparity between them, an engineered sheen and the mortal decay that overtakes the body within. The intrinsic paradox of the materiality of relics was problematic long before the Reformation. Vigilantius said, "they worship with kisses I know not what heap of dust in a mean vase surrounded with precious linen"; Jerome complained that Vigilantius had "opened his stinking mouth, casting a load of filthy rubbish before the relics of saints," but Vigilantius was anxious that the saints were being conflated with filthy rubbish.[2] Fakes were also a worry from the beginning. Martin of Tours exposed the tomb of a thief being venerated.

Just how problematic all this could become is visible in the following story. Rectors in the late Elizabethan Lake District were dismayed to discover among their congregations both magic users and papist sympathizers. There were people who kept the fast for St. Anthony or who wore beads; there was also a woman who buried "a quick newt, a dog, and a quick cock," and a woman who was a healer "for the fairies." Finally, there was also a woman named Agnes Watson, who was reported because she "kept a dead man's scalp." The interesting thing about Agnes is that we cannot be certain which list to put her in—was she keeping the scalp because it was a relic of some kind? People did keep particularly sacred items after the Henrician reformation and well into Charles's reign. Or was she keeping the scalp as a grisly trophy for use in necromancy?

The fact that we cannot know how this body part was marked is instructive, because it points toward the ideological and cultural overlap between relics on the one hand and the materials of necromancy on the other. (Necromantic use of body parts in fact antedates the cult of relics by many centuries.) In this story, the scalp is an isolated fragment in many senses; it is plainly metonymic, but we do not know anything of the whole from which it is taken, and hence we cannot know the power with which it is invested. Conversely, the fact that it is a fragment points to a link between the dismembering of the dead and iconoclasm. This linkage always troubled equations between the iconoclasts and forces of good. In some respects, relics and their powers could be understood as a licensed form of necromancy, one in which fragments of the bodies of the dead are reanimated to curative, vatic, or other miraculous purposes.

To understand more fully how that overlap worked, and how it came to seem frightening that bodies could be thus marked and used, I want to consider the idea of the fetish. The word "fetish" derives from the Portuguese word *feitica*, meaning saints' body parts. A fetish is an object specially created to carry social power. It tends to be metonymic—like the lucky rabbit's foot, which carries the luck and magic of the animal with it. The witches in *Macbeth* are fetish makers and hence relic makers, collecting and deploying fragments of personhood. The fetish allows power over the thing with which it is linked, but in a manner that is apt to collapse. The violent removal of body parts from worship—perhaps taken together with the renunciation of the doctrine of transubstantiation—may have reinforced the civilizing process that was going on around the body and its parts, creating new categories of dirt that could then be regarded with repugnance. The nausea aroused by the dead was especially strongly stimulated by relics; indeed, Catholic Robert Bellarmine remarked that "there is nothing [Protestants] shudder at so much as the veneration of relics." Calvin's *Treatise on Relics* began the revulsion:

> they not only turned from God, in order to amuse themselves with vain and corruptible things, but even went on to the execrable sacrilege of worshipping dead and insensible creatures, instead of the one living God. Now, as one evil never comes alone but is always followed by another, it thus happened that where people were seeking for relics, either of Jesus Christ or the saints, they became so blind that whatever name was imposed upon any rubbish presented to them, they received it without any examination or judgment; thus the bones of an ass or dog, which any hawker gave out to be the bones of a martyr, were devoutly received without any difficulty.[3]

The association between relics and disgust was built gradually—rather, relics' grisliness was slowly uncovered and laid bare as the license to transgress broke down. Whereas earlier denunciations of false miracles had focused on forged documents, as Langland had with his documented covered with the seals of bishops, reformers highlighted the animal origins of relics, following Chaucer and his "shoulder-bone/ which was of a holy Jew's sheep." Thomas More himself wrote of "some old rotten bone."[4] So relics are no longer metonymically linked to the divine, but to death and to animality, and later Calvin seeks to erase the marking of these bones and other bodily fragments by relabeling them as "rubbish," unmarked bones that have been picked and discarded: "even the smallest Catholic church has a heap of bones and other small *rub-*

bish [emphasis mine]"; rubbish is precisely that which has no name. Links with filth are apparent when Shaxton condemned "stinking boots, mucky combs, ragged rochets, rotten girdles, pyld purses, great bullocks' horns, and locks of hair, and filthy rags, gobbets of wood, under the name of parcels of the holy cross."[5] Pilgrims were shown linen rags, with which, they reported, "the holy man wiped the sweat from his face or neck, the dirt from his nose."[6] Lollard attacks began this trend, describing "worme-eten bonys . . . olde rages." Samuel Harsnett was similarly sickened by the bits of the English martyrs used in a Jacobean rite of exorcism, "Campion's thumb put into Fid's mouth . . . what wonders they wrought with these poor she-devils, how they made them to vomit, screech and quack like geese that had swallowed down a gag?" The association often made by reformers between relics and the female rituals of childbirth strengthened the sense that there was something messy about the whole business of the religious and powerful body part. Thomas Cromwell's 1538 proclamation explicitly outlawed "candles or tapers to images and relics, or kissing or licking the same." The phrase exposes the implicit eroticization of veneration practices that focus on bodies and their apposition.

If an image comes to signify a body, it will partake of that body's capacity to arouse and (conversely) disgust, a fetishism paradoxically increased by the dismemberment of iconoclasm itself. Elizabeth Barton's fraudulent napkin, purported to be stained by the devil's spittle, was said to be faked when Barton took soot, "and mingled it with a stinking thing, you wot what I mean."[7] Barton's spectacular dealings with the devil, which might endanger her because they could be interpreted as witchcraft, are thus reread as a sign of the dirty female body. She is "devilish" precisely because she "passeth all others in devilish devices." In contemplating this kind of figure with horror, the reformers were half consciously teaching their followers to read relics as signifiers of necromancy. Erasmus also linked saints' lives with old wives' tales, and said, "no educated or serious-minded person can read them without disgust."[8] The witches' Sabbath involved disgusting food of the kind stigmatized elsewhere as relics or rubbish—bones of discarded animals, infants whose hands had been removed "like sucking pigs," along with other animals—horsemeat, hare, buck, ravens, crows, toads, and frogs.[9] This list recalls not only the cauldron scene but also the allegations made about the true origins of fraudulent relics. Martha Nussbaum defines disgust as "a shrinking from contamination that is associated with a human desire to be non-animal . . . That desire, of course, is irrational in the sense that we know we will never succeed in fulfilling it." This is effortlessly expressed through the wish to read what was once holy as merely animal. In his history of the devil, Robert Muchembled argues that there is a point where smell becomes acutely problematic, and when bodily and animal smells became a problem, they were often linked with the evil one.[10] Smell also became an index of truth about the body and its evils; hence, disguising it with perfume was also sinful and duplicitous, associated with duplicity of femininity.[11] At the very moment when cities were reorganizing themselves to exclude tanneries and shambles, the body was also reorganized as a site where smell and impingement signified moral corruption. So the filth and garbage of relics were denounced through a kind of deconstruction that sought to break links between the relic and the saint it metonymically represented. To see how all this works in more detail, we can turn to the cauldron scene in *Macbeth*, ultimately readable as an extended commentary on the cult of relics and as a commentary on Protestant commentaries.

This is after all natural, for their critics saw relics as problematically overlapping with necromancy. "The vilest witches and sorcerers of the earth . . . are the priests that consecrate crosses and ashes, water and salt, oil and cream, boughs and bones, stocks and stones."[12] Jewel called the Agnus Dei a "conjuration." Pilkington called St. Agatha's letters "sheer sorcery," while the use of consecrated bells in a storm was "witchcraft."[13] The Lollards had spoken of "the witch of Walsingham," referring to the powerful image of the Virgin venerated there. The Feast of the Invention of the Cross was said to be hymned by "magic spells." Litanies were regarded as "nothing but an impure mass of conjuring and charming," while other prayers were also condemned as "conjuring of God." Incantatory prayers, like relics, are gestured at in *Macbeth* and also attacked by reformers. And all these issues were still urgent matters for dispute in the early seventeenth century. The Jesuit mission still proclaimed and used the power of relics, and there was some regret for their heyday due to lack of relics to cure possession.[14] Relics were kept in York Minster as late as 1695, and in countless parish churches as well.[15]

The parallels between relics and the ingredients of necromancy are made obvious in *Macbeth*'s cauldron scene. Like relics, the ingredients in the witches' cauldron of prophecy are decontextualized *bits* of animal and human bodies, detached pieces of what were once living things. The cauldron of death is an image of ruins, fragments, shards—bits of *things*, bits that suggest a whole that can never be reconstructed but can be desired persistently. We see the body parts that go into the cauldron not as part of larger wholes, but as *fragments*, in the same way that we do not see the whole beasts whose bones make up the rubbish, or the whole saint whose individual bones have become relics. And yet they are still metonyms of stories, still marked with those stories that prove more ineradicable than the reformers might have hoped.

Let us take the most striking, the "finger of birth-strangled babe, ditch-delivered by a drab." It recalls one of London's most prized relics, treasured carefully throughout the Edwardian Reformation and deployed as soon as Mary's reign was over. This was the finger of one of the Holy Innocents, returned to St. Stephen Walbrook in 1553.[16] Holy Innocents' Day was especially controversial with reformers because of its strong links with a particular form of misrule, the custom of electing a choirboy to be a boy bishop.[17] This transgressive rite was one in which children took on adult powers, so it is directly relevant to *Macbeth*'s cauldron scene, in which child apparitions take power over adult Macbeth. Henry VIII specifically banned such child carnivals.[18] But what might such a story suggest about the *cauldron* except misprision? The witches' powers seem to derive not from our *knowledge* of these stories but from our forgetting of them, even our repression of them. Rubbish is also that which we would rather forget, and when relics become rubbish the marked body resituates itself as unmarked. In this context, the witches' trafficking in relics is freighted with all the disgust reformers intended to evoke. Relics have become rubbish, silent and storyless, heaped together apparently at random with other disgust-evoking street sweepings.

But elsewhere, the play interests itself further in the notion that *any* kind of marked *body* might turn out to be a relic. Lady Macbeth, as we shall see, turns herself into a love relic of a dead child who is also her relic sacrifice to gain necromantic powers. And it is in the eroticization of the relic that we begin to see another kind of imbrication of the once sacred and the bodily. From an early stage, martyrological venerations could be troped or even practiced through the register of the erotic. Lucilla of Carthage in the fourth century c.e. is said to have kissed the bone of a martyr. Yet this kind of practice becomes acutely problematic when the licit supernatural is reimagined as illicit necromancy. The ritual kiss in witchcraft representations resembled the veneration of relics. More significantly, a number of medieval romances set out to explore the way in which love makes every dead body into a relic. In Malory's *Morte d'Arthur*, the sorceress Hallewes threatens Lancelot (Figure 8.1).

FIGURE 8.1: *Sir Launcelot and the witch Hellawes,* Aubrey Beardsley illustration. Note her extreme thinness and hungry expression. From www.artpassions.net, in public domain.

Hallewes inhabits a chapel, the Chapel Perilous, from which Lancelot must
purloin what look like healing relics, a piece of cloth and a sword. But Hallewes
threatens to turn the tables on him by making the knight himself into a relic.
She offers to kiss him, and he declines; then she explains that if he had said yes,
she would have preserved his dead body so as to be able to kiss and hold it in
her arms every day:

> And Sir Launcelot, now I tell the: I have loved the this seven yere, [but]
> there may no woman have thy love but queen Guenyver; and sytthen
> I may not rejoyse the nother thy body on lyve, I had kepte no more joy in
> this worlde but to have thy body dede. Then wolde I have bawmed it and
> sered it, and so to have kepte hit my lyve dayes; and dayly I sholde have
> clypped the and kissed the, dispyte of queen Gwenyvere.[19]

It looks as if Hallewes longs for a love relic. She plans to embalm Lancelot,
dry him out, and then venerate him. This eerily necromantic ambition is also
about the price of excessive love; a relic is actually a more manageable love
object than a living knight, though Hallewes makes it clear that it is her second
choice. As Elisabeth Bronfen remarks, a corpse is both abject and object.[20] But
her desires also mark her as a witch, and it is as such that Lancelot replies to
her. "Jesu preserve me frome youre subtyle crauftys!" replies Lancelot, which
may refer either to the "craft" of preservation or may extend more generally
to Hallewes and define her as a witch precisely because she is in the business
of using and preserving bodies. The question is, why isn't memory enough for
Hallewes? If love makes a relic of its object, then that relic should, Protestant
thought tells us, be pure because it is mental. The act of venerating or kissing
Hallewes would make Lancelot a relic.[21]

Hallewes's preservation of Lancelot as motivated by desire is expanded
upon by Malory. His source, *Perlesvaus*, also called *The High History of the
Holy Grail*, features a number of maidens already carrying knightly relics
or questing for other relics. One maiden prepares three jeweled coffins for
Gawain, Lancelot, and Percival, because they are the best knights in the world.
She plans to kill them, cut off their heads, and place their bodies in the coffins.
The otiose beheading signals the transformation of the knights from heroes to
saints; saints' relics were often decapitated so the head alone could be presented
for veneration, as was the case with Thomas à Becket. Earlier in *Perlesvaus*, we
meet another maiden bearing a head encased in a reliquary and accompanying
a cart in which lie the heads of 150 knights, some sealed in gold and silver, others in lead. Another maiden vows to love Lancelot in relic-like form:

> "Ah, Lancelot," said she, "How hard and cruel you are to me! And it
> grieves me greatly that you have the sword and that things must go so
> well for you! For if you did not have it with you, you would never part

from here of your free will, and I would have taken all my pleasure of you and had you taken back to my castle; and, powerless, you would never escape."[22]

Here the passivity of the eroticized body mimics the stillness of the relic, just as erotic exhaustion mimics death. This passivity extends to relics as well; the physical power of the puissant knight can be appropriated by the desiring woman, just as the saint's power can be taken to work magic.

The relics here show the knight becoming an emblem of the saint, a saint of romance. There is a crossover between the desires set in motion by romances and those set in motion by veneration. The knight's survival in the face of many perils also makes him like a saint, as Jacopo de Voragnie's *Legenda Aurea* often depicted saints' bodies magically resisting attempts to assault, hurt or kill.[23] By Malory's day, a woman's interaction with relics could be suspiciously erotic. Relics are characteristically venerated by kissing, so they do bring two bodies together. When that happens, the venerating or predatory woman becomes clearly marked as a sorceress.

The eroticization of both relics and images became acutely problematic for reformers. As relics cross over from being distant objects of veneration to desirability, the idea of veneration as kissing and licking becomes problematically tinged with necrophiliac sexuality. Thomas Cromwell's 1538 proclamation explicitly outlawed "kissing or licking," a phrase that makes worshippers sound very like the sorceress Hallewes. If an image comes to signify a body, it will partake of that body's capacity to arouse. This partially explains the new disgust the veneration of relics could arouse; it had become an eroticization of the dead. John Colet was repelled by kissing an arm with flesh still attached, and he refused one of the fluid-stained rags as a gift.[24] The reformers were also troubled by the feminization and even homoeroticization of the body of Christ, in the eucharistic hosts that miraculously revealed themselves to be the male body of Jesus. In *Perlesvaus*, for example, the king sees the hermit "holding in his hands a man, bleeding from his side, bleeding from his hands and feet and crowned with thorns." The *Corpus Christi Carol* similarly places the eucharistic host as a visible knight adored by a lady:

And in that hall ther was a bede,
Hit was hangid with gold so rede.
And yn that bed ther lythe a knyght,
His wowndes bledyng day and nyght.
By that bedes side ther kneleth a may,
And she wepeth both nyght and day.
And by that bedes side ther stondith a ston,
"Corpus Christi" wretyn theron.[25]

The elliptical text fails to distinguish between the erotically worshipful lady and the Virgin, the church, or the soul. The bleeding figure is doubly feminized, by its very bleeding and by being the passive object of the adoring gaze that is powerless to save or help, a gaze strongly reminiscent of the look to which Hallewes hopes to subject Lancelot. As Caroline Walker Bynum notes, the Jesus of the late Middle Ages was already a feminized figure.[26] Richard Rambuss has pointed to the sadomasochistic eroticization of Christ in Counter-Reformation poetry, but the same impulse can be unearthed in medieval texts, especially in romances where the knight's struggles can metaphorize the struggles of Christ, as they do in *Perlesvaus*.[27] Finally, Alan Stewart has shown the powerfully homophobic use of slander in the reports of the Cromwellian visitors on the monasteries.[28] The world of medieval piety was suddenly readable as erotic in a manner that especially implicated one of Catholicism's most appealing features, its influence over and connections with the body.

Hence, it became natural to read relics as perversely erotic and even as necromantic themes that reach a culmination in Donne's lyric "The Relic":

When my grave is broken up again
Some second guest to entertain
(For graves have learned that woman-head
To be to more than one a bed),
And he that digs it, spies
A bracelet of bright hair about the bone,
Will he not let'us alone,
And think that there a loving couple lies,
Who thought that this device might be some way
To make their souls, at the last busy day,
Meet at this grave, and make a little stay?[29]

Here the bone and its evocatively contrasting bright hair are love relics, just as Donne and his mistress are love's saints in "The Canonization." But here the image is grislier and more poignant. The relic is also an emblem of futility, of the inability of the dead to be connected to each other or to the living. It is a monument to an eroticism firmly declared dead but carried forward in the souls' prospective ghostly meetings. The relic is no longer redemptive in any Christian sense, and yet in its full secularization it still has the power to translate the lovers to a world beyond death.

All these discourses of desire circulated largely around two relics that were especially significant in the late medieval church: the eucharistic host and the breast milk of the Virgin. Leaving the latter to one side for one minute, moving to the eucharistic host means moving toward the discussion of flesh. The host was always miraculous, but its bodiliness was affirmed in the eucharistic miracle stories that circulated in the latter Middle Ages. Bleeding host miracles often signify an accusation: a host used in a love spell became flesh, and bled; the use of the host in this context points to a subtext of desire, and a collection of Venetian folktales records a witch who gathered eucharistic hosts for use in erotic magic. If a woman kissed a man with a host in her mouth he would always be true to her, but the ingredient in love potions could also cause abortions.[30] It is a short step from this to Alda of Siena, who tasted a drop of blood from Christ's side; once the blood fell on her girdle and she sucked it off with her mouth. The Virgin Mary too was depicted sucking on Christ's wounds while nursing him in her arms.[31] Such dangerously perverse eroticism was both blissfully seductive and worrying.

When the host became a beautiful baby, a different kind of bodily closeness was evoked. Yet was it really so different? When Hallewes imagines herself tending Lancelot's body, the act is reminiscent not only of a lover but also of a maternal, even a pietà figure. The maternalization of the cult of relics was disturbing as well, the more so because it laid bare a potential core of maternal erotics. The overlap is enacted in the host, so that "sometimes she happily accepted her Lord under the appearance of a child . . . and sometimes in the pure and gorgeously embellished marriage bed of the heart."[32] Wilburgis (d. 1289) took the host to her enclosure to help her avoid sexual temptation, and it revealed itself as a beautiful baby who spoke to her in the words of the Song of Songs.[33] Gautier of Flos saw a baby in the host, and Dorothy of Montau internalized the beautiful babe as a mystical pregnancy. It is therefore in the host itself that the bodily overlap between erotics and maternity is sketched out. Hallewes's plan for Lancelot's body accompanies the presence in the chapel of a wounded body that recalls both Christ and the Fisher King, and her treatment of it is a work of extreme mourning. The images of baby, blood, and breaking are knitted together in the eucharistic host to make a disturbing eroticism that is also detectable when the same images are traced in *Macbeth*.

Then, too, Hallewes does not long survive her own frustrated inability to venerate the dead Lancelot. The good relic revives, but the eroticized kills in its absence. The image of the witch who has dealings with a dead body that ought properly to be sacred has origins that go beyond the cult of relics themselves. The witch who embraces and deploys dead and eroticized fragments also derives from the figure of the Jew as understood and depicted in legends and folklore. Carlo Ginzburg's story about anti-heretical and anti-witchcraft stereotypes in southwest Germany epitomizes the way the problematics of how the dead could be engaged supernaturally are tied in to questions of orthodoxy and heresy. In his tale, a heretic carried ashes of a dead child; whoever ate them became at once a member of the sect.[34]

In particular, both Jews and witches steal and desecrate hosts, turning the most precious relic of all into an object with which to perform magic just like any other ritual. The Nuremberg Chronicle links witches to Jews as criminals against Christendom. The witch and the Jew became interchangeable. Legends of Jewish desecration and child murder transferred holus-bolus from Jews to women.[35] As witches replaced Jews as problematic instances of persons whom the church had failed to convert and contain, they also took over a variety of stories in which unbelievers gained power from an urge to kill and reuse Christ's body and Christians, images that lie behind Lady Macbeth's murderous desire for babies. Witches long for babies in order to turn them into a demonic equivalent of the salvific relic: "we secretly steal them from their graves and cook them in a cauldron until the whole flesh comes away from the bones and becomes a soup."[36]

Johann Weyer's catalog of popular beliefs was intended to expose practices for what they really were—the delusions of common people, but Hsia notes that the key manifestation is "the greater emphasis of ritual child murders in witchcraft discourses of the late sixteenth century," as "witches seemed to have replaced Jews as the most dangerous enemies within Christian society."[37] It included the idea that witches kill children for ritual reasons, or dig up their bodies for ritual uses. One story in particular, the story that eventually became the story of Hansel and Gretel, exchanged a Jewish child-stealing protagonist for a witch figure. Take this Serbian version:

> Then there came along some Yids, and when they saw the fire, came up to the children and asked them what they were doing there and whether there was anyone with them, and when the children had told them what and how, the Yids told them to go along with them, saying that they would have a fine time at their house. The children agreed and went with the Yids, and the Yids took them to their house. They didn't have anyone else at home, only their mother, and when they came home, they shut the boy up to get fat and made the girl a servant to their mother. One day, when the boy had been well fed and was fat, the Yids went out on some errand and told their mother to roast him, and then when they came home in the evening from their work, they would eat him.[38]

In her book on Jewish child-murder libel and the Eucharist, Miri Rubin suggests further parallels between the lost children story that eventually becomes familiar to us as Hansel and Gretel and the stories of ritual child murder. The basic story she records tells of a Jewish boy who secretly receives the eucharistic host and is thrown into an oven by his furious father. The Virgin Mary protects the boy in the oven by covering him with her cloak. The boy and his mother convert to Christianity, and the father is put in the oven himself.[39] Rubin points out that the oven represents a womb from which the child is reborn. We can detect the faint, unexorcised shadow of this story in *Macbeth* and its womblike

FIGURE 8.2: *Hansel and Gretel,* illustrated by Arthur Rackham, 1920. Note the witch's caricatured anti-Semitic appearance, with large nose and round eyeglasses. Wikimedia Commons: http://commons.wikimedia.org/wiki/File:Hansel-and-gretel-rackham.jpg.

cauldron, into which an innocent child is thrown, and out of which a supernaturally endowed child emerges. But once more this rebirth is not benign, as it is in the original story. Rather, what was once the healing Christian magic of the Virgin has become an angry and violent witchcraft. Crucial to the connection is the figure of the cauldron, the instrument of punishment in Christopher Marlowe's *The Jew of Malta.* In that play, the cauldron is the means by which the Jew is finally excised from the play. His trace is locatable in the cauldron in *Macbeth*, where we find the "liver of blaspheming Jew" marking the cauldron as a site of Jewish punishment. The admixture of the Jew's trace with parts of the baby might seem incongruous, but it makes a kind of warped sense when read in the context of anti-Semitic stories in which the Jew is specifically labeled the foe of Christian children. Usury in particular is represented as man taking baby away from its mother.[40] Similarly, a Paris Jew tests the divinity of the host in a cauldron of boiling water; the water turns the color of blood. The point here is in part that children and the host are equally vulnerable. They can be appropriated to do magic. The boy in the original story as recounted by Rubin is himself a kind of eucharistic host, baked in an oven like bread, then reborn from the Virgin. Although the contents of the cauldron in the dramas are again an inversion—unappetizing—there is a sense in which the very replication of the register of a recipe calls the possibility of consumption of the contents into action.

Another story tells of a Jew who refuses to swallow the host, keeping it for experimental purposes. He keeps it in his mouth and eventually spits it out of the cauldron of his body. It turns into an attractive little child on the palm of his hand.[41] Undeterred, he still tries to eat it, but it is too chewy. Similarly, Jews capture and murder a priest, whose heart contains "a lovely little boy."[42] Here the message is that of *Titus Andronicus*—eating people is wrong—and yet these vehemently anti-Semitic stories expose the central transgression of transubstantiation and its implications, just as relics are laid bare as decaying bones.

Ein erschröcklich geschicht Vom Tewfel
vnd einer ynhulden/ beschehen zu Schilta bey Rotzweil in der karwochen.
M.D.XXXiii Jar.

FIGURE 8.3: Execution of an arsonist witch by burning at stake. Witches' hard bodies require burning to destroy their power. From a 1533 account of the execution of a witch charged with burning the town of Schiltach in 1531. Anonymous, 1533. German woodcut. Zentralbibliothek, Zurich, Switzerland. Photo Credit: Foto Marburg / Art Resource, NY ART 31865.

The Hansel and Gretel story and the anti-Semitic legend of the blood libel have a marked resemblance, with witch substituted for Jewish man. In both cases there is a story of a beautiful child sacrificed to perverse desires, and such stories often contain the motif of a child who cannot be silenced, speaking blood, along with the Christological motif of a child made to suffer unjustly. All three are present in *Macbeth*: the sacrificed child in Lady Macbeth's baby, the children of the cauldron as the prophetic discovery of child murder, and Macduff's unjustly murdered children. And there is a further link between witches and Jews across Shakespeare. This is the trope of the hard body of the villain implicitly compared with the softness of the victim: Witchmarks had to be pricked.

Figure 8.3 illustrates that witches' hard bodies require burning to destroy their power (from an account of the execution of a witch charged with burning the town of Schiltach in 1531). Similarly, it is necessary for Shylock to actively assert his own fleshly normalcy: "If you prick us, do we not bleed?" This hard body is the antithesis of the maternal softness and succor that could also seem difficult. Neither deliquescence nor too much rigor were acceptable.

In both the blood-libel story and the story of the witches' murder of babies, the baby's body is a source of supernatural power. A third group of stories,

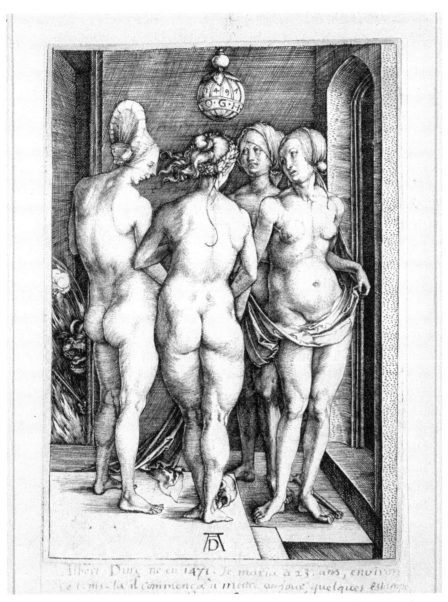

FIGURE 8.4: *Four Witches*, Albrecht Dürer, 1497. These witches are plumper and more maternal, a parody of the Graces. Engraving, Kupferstichkabinett, Staatliche Museen zu Berlin, Berlin, Germany. Photo Credit: SEF / Art Resource ART80087.

stories of women selling their babies to the fairies in exchange for supernatural sight, ties in further with these. A Scottish woman called Bessie Dunlop believed she had given up her baby in exchange for fairy powers.[43] Like a saint's relic, Bessie's baby connects her to the Otherworld to which it goes. She exchanges it for magical powers. Though Shakespeare knew nothing about Bessie Dunlop, there is an ironic and fascinating way in which her story *is* reflected in Macbeth. At a simple level, the witches' treatment of a dead baby is usually read as merely and drearily infanticidal; the birth-strangled babe who goes into the potion.

But is it? For among the spirits to come out of the potion is the spirit of a child; is this a faint trace of the *aoros* summoned by the body of the dismembered baby? And what does that child speak about? Why, about inheritance, about patrilinearity, about the rights of the father. As if this is not enough, there is also a bloodstained baby, and the baby speaks of a birth that is not of woman, a hypermasculine birth. And is there a faint analogy between Bessie's choice and Macbeth's? Bessie acquires occult powers by sacrificing not only her child but also her identity as that child's caring mother. Just so does Macbeth sacrifice the normal ties of human warmth, honor, and troops of friends for power. And when he metaphorizes these human feelings, he does so using the image of a baby, "pity, like a naked newborn babe / Striding the blast."

So it seems natural to turn back to Lady Macbeth's own metaphorization of those feelings, in which she uses the imagery of breast milk, claiming that Macbeth's nature is "too full o' th' milk of human kindness." The phrase, which has become a cliché, actually requires some explication; if Macbeth's choice resembles Bessie's, how much more does Lady Macbeth's? In the text before us, Lady Macbeth keeps alluding, not narratively, but metaphorically, to a lost or dead baby, and once, crucially, to a baby she murdered herself. Let's look very closely at her most famous speech:

> I have given suck, and know,
> How tender 'tis to love the babe that milks me.
> I would, while it was smiling in my face,
> Have plucked my nipple from his boneless gums
> And dashed the brains out, had I so sworn
> As you have done to this.

This baby is figured in extreme and sensuous materiality. Shakespeare *delays* the violence to allow us to experience an erotic and bodily closeness; Lady Macbeth's resolute "I would" is followed by an evocation of tenderness that defers the violent end and makes anticipation of it seem worse "while it was smiling in my face." Now comes the stroke of real genius; "Have plucked my nipple from his boneless gums." For a heartbeat we think *this* is the act of untender, unmaternal resolution; the refusal of food. Lady Macbeth is, for a second, merely someone who restrains her child's greed for suckling; but then comes the terrific force of "and dashed the brains out," delayed again and hence given added force by that initial "and." "Dashed" is just the right word; imagine how much "knocked" would reduce the force of the lines, losing that onomatopoeic sense of splattering that makes "dashed" so untender.

Thus, it is that the very senses that made us feel the baby so tenderly are now turned to a painful and fully bodily awareness of his destruction. The baby also becomes a kind of martyr. It is a Holy Innocent that recalls and anticipates the birth-strangled babe of the cauldron. It is also a sacrifice—or rather, it's a fantasy about the sacrifice of a baby to make something much more fixed—a kind of monument to its death. Finally, the death of the baby prefigures the transformation of Lady Macbeth herself into a kind of relic through the figures of breast milk and bloodstains.

Lady Macbeth's other long speech is also haunted by something; the ghost of a dead baby, even if he only dies in story, is a way of understanding Lady Macbeth's relations to the supernatural. For it is Lady Macbeth, not Macbeth, and not the Weird Sisters, who delivers the only authentic invocation to the powers of darkness in the play:

Come, you spirits
That tend on mortal thoughts, unsex me here
and fill me from the crown to the toe topful
Of direst cruelty. Make thick my blood
Stop up th' accents and passage to remorse,
that no compunctious visitings of nature
Shake my fell purpose, nor keep peace between
Th' effect and it. Come to my woman's breasts
And take my milk for gall, you murdering ministers,
Wherever in your sightless substances
You wait on nature's mischief.

This speech is usually read as a renunciation of the sexed body—or so it seems. But what does that mean within the context of the play? The witches, of course, are unsexed, or rather their gender is to Banquo problematically undecidable—because they have beards. But what kind of marker are beards in women? They are markers of old age, when hair begins to grow in places coded as smooth in young women. What Banquo is seeing is a body unsexed by old age, and we shall see in a moment that this is how Lady Macbeth marks her body, too.

Old age brings the functions of the female body to a halt. Lady Macbeth is making, in effect, the same choice as Bessie Dunlop, but she is making it much more comprehensively. She is wishing for early menopause, and this is why she asks that her blood be made thick. A witch's blood was thought to be so thick with old age, so lacking in fire that it was impossible to extract it, and it was this idea that lay behind the notion that a witch's body could not be pierced by shot or by a pin. Like the Jew's body, the witch's body was hard and insensible. Such hardness is inimical to the soft body of the mother. Yet this is the body Lady Macbeth desires for *herself*—a body that is dried and preserved, just what Hallewes wants to do with Lancelot, a body that is a dead end, not capable of multiplying. With her reproductivity denied, her hard body seems inimical to time, not unlike the body of a virgin. Her body now seems static, caught forever at the instant of her crime, like that of an inverse martyr. The witch's body is thus like the body of the saint in being the way by which supernatural power transmits itself to other parts of the material world.

(It's also interesting how often and how derogatorily the word "old" is used in the critique of relics; they are "old bones," as though being old means they are especially repugnant.) When witches are depicted, there is sometimes a strident contrast between the sexual poses they adopt and their withered bodies (Figure 8.6).

Similarly, Lady Macbeth offers to substitute gall for her breast milk. To early modern medicine, breast milk was impure blood from the womb that was made white and pure by the burning fires of maternal love, which also drew it upward through the body until it reached the breasts.[44] By contrast, the gall Lady Macbeth substitutes for milk is a signifier that her heart has failed in maternal love and of the poison in which witches are thought to deal. In an era when babies who were not breast-fed were far more likely to die, she imagines herself murdering her child, via the trope of a refusal to feed it. Lady Macbeth's double refusal of breast milk marks her as a witch, too, because witches were beings who stole the milk of other animals and mothers, substituting unnourishing blood for it. And she also imagines herself choosing not to feed the child, but to feed something else, to feed the familiar spirits she summons. Worse still, these spectral images of infanticide are haunted by the image of the female demons whose breast milk acted as poison to babies, especially the Jewish mother-demon Lilith.

FIGURE 8.5: *Witches Riding*, Francisco de Goya, ca. 1798. Wikimedia Commons: http://commons.wiki media.org/wiki/File:Linda_maestra.jpg

FIGURE 8.6: *The Witch*, Albrecht Dürer, ca. 1500. Hanging breasts and loose skin mark the body as asexual, while the pose suggests sexuality. Engraving, 11.4 × 7.0 cm. Inv. 4536-1877 (B.67). Photo: Joerg P. Anders. Kupferstichkabinett, Staatliche Museen zu Berlin, Berlin, Germany. Photo Credit: Bildarchiv Preussischer Kulturbesitz / Art Resource, ART182485.

This concern with the protocols of maternal love and their physical expression is also tied into concerns with relics and their truth or value. So too it was when breast milk relics were also exposed as hard, perhaps as unfeeling. Fake breast milk, and especially the fraudulent breast milk of the Virgin Mary, was one of the most popular targets for reformers keen to hunt out fake relics. Erasmus began the trend by his account of his visit to Walsingham. The shrine's premier relic was a crystal vial of this milk. In his colloquies, Erasmus complained that it was remarkable that a woman with only one child should have produced so many milky relics.[45] The East Anglian visitors noted at least seven specimens of the Virgin's milk. In St. Paul's was a miraculous vessel said to hold the milk, but which proved to contain a piece of chalk. The fraudulent milk disconnected the relic from the body from which it supposedly came and the body on which it was supposed to act magically. That metonymy was based on an assumed metaphoricity, or likeness. Breast milk forms a kind of stream of bodily marking. But if it is just chalk, then it no longer has the power to evoke milk. It deprives the believer of nourishment, just as Lady Macbeth herself does. Breast milk revealed to be hard rather than liquid, indigestible rock rather than nourishing food. The revelation that it is hard rock mimics Lady Macbeth's murderous request.

Through her child sacrifice, her wish to be prematurely ancient and withered, her summoning of familiars, and her story of sensuous child sacrifice, Lady Macbeth is the play's only true witch. She also becomes a relic, or perhaps more accurately an anti-relic, of her own wicked deeds. In the play's final act, she is glimpsed again as the victim of a marked body whose marks cannot ever be erased, despite her neurotic efforts. Her shocked repetitive washings in the sleepwalking scene uncannily and chillingly replicate the many rituals of relic devotion, but these ritual cleansings are ineffective because nothing can wipe out the stain of murder from her sight, nor its smell.

Here all the themes discussed in this chapter come together. First, blood spots are simple signs of witchcraft. In England, a witch's mark was a demon's suckling place, but it could also be wounds or bruises on waking that the witches cannot explain. The devil left a secret mark on those who made a pact with him. Usually the mark took the form of some kind of sign of sexuality with the devil, but in England it was more commonly a sign of perverse nurturance, a misplaced teat from which devils rather than children were fed. The marks on Lady Macbeth's hands tie her acts with her will to murder children through half-remembered stories of host theft and child murder. Just as Lady Macbeth denied the supernatural power of Duncan's royal body, so power manifests itself forcibly in her eyes by the reappearance of his blood, just as it did for the child-murdering Jews, witches, and infanticidal mothers of popular story and legend. The relic-like power of the blood Lady Macbeth sees and smells on herself has the ability to transcend death and to bring the living into close—

FIGURE 8.7: Reliquary for the arm of St. Luke with the coat-of-arms of Sancia of Mallorca, ca. 1337–1338. Produced in Naples. Silver, crystal, gold, enamel. Louvre, Paris, France. Height: 0.480 meters. From the Treasury of Medina del Campo. Executed for the queen, Sancia of Mallorca (1309–1343), wife of the king of Naples, Robert d'Anjou. Photo: Daniel Arnaudet. Photo Credit: Réunion des Musées Nationaux / Art Resource, ART346778.

too close—apposition with the dead. The appearance of blood upon hands is not only a common criminal sign, but it is also especially associated with the Jewish blood libel. And yet the bloodstained bodies of Jews in these medieval legends are themselves hard, like the body for which Lady Macbeth longs. Finally, like Duncan, the host is always stabbed repeatedly in these legends, and the miraculous abundance of the host's blood is correlated with this multiple stabbing. "A Jew has mutilated the / Host of the holy sacrament / By striking ten blows or more / And making it bleed abundantly."[46]

At the same time, the finger that entered the witch's cauldron as perhaps their most elaborated ingredient is replicated here in Lady Macbeth's telltale hand. Both fingers and hands were relics especially fraught with power.[47] The finger relics of Holy Innocents were echoed in the innocent baby slain by his guilty mother in *Macbeth*. In the same way, arm and hand relics of the great saints are among the most common of the Middle Ages, especially relics of the right hand.

Healing miracles were celebrated with models of healed limbs, which went with display of bones of ordinary people in ossuaries—crucial distinction between the recent dead and clean bones.[48]

As the marked body was progressively displaced by the very jewelers' arts that were supposed to represent it, the body's own resonances of mortality and disgust were displaced too, and relics ceased to be a visible way of loving the dead; they became the kinds of objects of disgust associated with necromancy. As that process occurred, visceral loathing was transferred to the body of the witch created by necromancy, which had all too much in common with the relics themselves, being fixed, static, infertile, hard, duplicitous, dead, and disgusting because it was the body of an old and infertile woman. In *Macbeth*, this marked body, the body as relic, is set alongside and in some respects set against the fragmentary and inscrutable scraps of leftover personhood that constituted relics. Ironically, just as stories of Jewish child murder, host theft, and witches' murder of infants coalesce in the single figure of Lady Macbeth, so the same stories are scattered throughout the witches' readings of the cauldron and its ingredients. The play is thus readable as an interrogation of the ways in which heaven and earth touch at the point of the dead body, marking its substance forever with the will of the supernatural.

CHAPTER NINE

Fashioning Civil Bodies and "Others"

Cultural Representations

MARGARET HEALY

It is seemly and fyttynge that a man be well fasshyoned in soule / in body / in gesture / in apparel; and in especyall it besemeth children all maner of temperaunce / and in especyall in this behalfe noble mennes sonnes. All are to be taken for noble / which exercise their mynde in the lyberall science. Lette other men paynte in their shyldes Lyons / Egles / Bulles / and Leopardes: yet they have more of verye nobylyte / whiche for their badge may paynte so many ymages / as they have lerned sure the lyberall seyence. Then that the myne of a chylde well burnysshed may upon all sides evident apere / for it apereth moste clere in the visage or countenaunce.[1]

In the early sixteenth century, the leading northern humanist, Erasmus of Rotterdam, penned a fascinating little volume in Latin that instructed young boys how to "fasshyon" [fashion] their souls, bodies, gestures, and apparel in order to "be taken" for nobility. In his estimation painted shields—the time-honored symbols of lineage and chivalry—were passé, while "verye" or true "nobylyte" [nobility] could now be evidenced through a display of one's knowledge in the liberal sciences but also through a skilled performance of civility. Erasmus's conduct book was rapidly translated into the European vernaculars, and it is hard to overestimate its importance in widely disseminating this optimistic, liberating humanist message: "No man can chose to hym selfe father and mother or his countrey / but condycion wyt / and maners any man maye countrfet" (sig. D3r).

Erasmian writings, generally, served to destabilize essential notions of the self by foregrounding the socially constructed nature of class and gender categories, but *De civilitate morum puerilium* [*A lytel boke of good maners for children*] effectively authorized and taught schoolboys how to "countrfet" [counterfeit or simulate] the behaviors suitable to a new, expanding class of "humane" and civil men.

In his groundbreaking study, *The Civilizing Process*, Norbert Elias asked the crucial question why a preoccupation with "uniform good behaviour," associated with the concept of "civilite," became so acute in the sixteenth century.[2] He concluded that the transformation in manners—the increased embarrassment, shame, and closer and more private attention to social bodily rituals suggested by Renaissance conduct books—could be attributed almost entirely to changes in the structure of European societies, which in turn necessitated alterations in social behavior. Elias designated the sixteenth century a "fruitful transitional period" between two great epochs characterized by "more inflexible social hierarchies," a phase in which "the social circulation of ascending and descending groups and individuals speeds up."[3] For the first time men could be fashioned into gentlemen—made not born.

MANNERS AND CONDUCT

Beginning with the eyes—the windows of the soul—and the face, Erasmus's manual instructed its young readers in a choreographed performance of somatic movements together with a semantics of exterior fabrication that would lead the onlooker to concede the inner nobility of the body observed ("to represent to the eyes of men, these laudable gyftes of the soule" sig. A2v) or its opposites—sluttishness and folly, or worse, madness and sin. Indeed, negative examples form the greater part of this book's wisdom. The eyes must, for example, be stable and look honest, and must not frown, "hang" down, or roll (sigs. A3r-v). Frowning is a "signe" of cruelty, rolling eyes of madness, "hanging downe" eyes of folly, and winking with one eye is "unfyttynge" (sigs. A3r-v). Twitching brow movements speak of mental instability, and a nose "full of snyvell" is the giveaway sign of a "sluttysshe persone" (sigs. A3v, A4r). Much attention is given to nasal excrement: it must not be dried with one's "cappe" or "cote" or "daube[d] on thy clothes"; these gestures are the "propertie of fysshmonger" (sigs. A4r-v). The most civil course is to use "thy handkercher" [handkerchief]—a novel French invention (sig. A4r). Visible bodily secretions and other spontaneous emissions (sneezing, spitting, and breaking wind) are the height of incivility and require special management. Thus, after sneezing, divine assistance must be invoked: one must "bless the mouth with the sign of a cross," and when another person sneezes one must utter the words, "christe helpe" (sig. A5r). The privy members should always be well hidden and any emergent wind must be expelled under the cover of a cough: "Let him close the fert [fart] under colour of a coughe" (sig. B1v). Clearly, the body's emissions must be restrained or camouflaged—the civil body is contained and nonleaky. Among other examples of indiscreet behaviors, biting the lip is a "syne of malyce," licking the lips is a sign of folly, and poking the tongue out indicates "knaves scoffyyng" (sig. A5r). Wearing brightly and diversely colored clothing is the giveaway sign of "ydiots and apes" (sig. B3r).

The manual is attentive to the customs of various countries, indicating, for example, that "Some make curtresye [curtsey] with bothe knees bowed (as yrisshmen)," and it advocates the judicious modification of one's manners to suit the place (sig. B1v). How the head, shoulders, and hands are held and move are indicative of character and mood, and the government of one's body at mealtimes provides a veritable lexicon of one's inner worth. Good table manners involve inspecting the condition of one's nails beforehand, placing both hands on the table, not touching your "bely" or leaning with the elbows, and knowing exactly what to do with one's napkin, glass, and knife (sig. B6r). All facets of dining correctly are dealt with at length, and moderation of the appetite is crucial. Deportment in church is of special importance, too. Kneeling must be conducted in a very particular fashion, for example, so as not to give the wrong impression: "To touche grounde with the one knee and the other standing up / upon which the lyfte elbowe doth leane / is the gesture of the wicked jewes and gentyles" (sig. B5r). Even bodily conduct in the bedchamber and the gaming house is subject to scrutiny and governed by rules (sigs. D2r-D3r).

In fact, Erasmus's conduct book leaves the modern reader with the strong impression that being an effective "counterfeiter" of civility was an extremely unrelaxed, exacting affair: prying eyes were busily observing your every move at all times; one slip in the sign system, one wrong gesture or facial expression could expose you to ridicule, or worse, relegation to the sluttish, dodgy, or downright evil camps. The manual also suggests that this was an intensely theatrical culture in the making: playing a part well comes across in the end as being more important than any other factor in establishing one's social credentials. A striking passage in Erasmus's *The Praise of Folly* serves to reinforce this perception:

> If anyone tries to take the masks off the actors when they're playing a scene on the stage and show their natural faces to the audience, he'll certainly spoil the whole play and deserve to be stoned and thrown out of the theatre for a maniac. For a new situation will suddenly arise in which a woman on the stage turns into a man, a youth is now old, and the king

of a moment ago is suddenly Dama, the slave, while a god is shown up as a common little man. To destroy the illusion is to ruin the whole play, for it's really the characterization and make-up which holds the audience's eye. Now, what else is the whole life of man but a sort of play? Actors come on wearing their different masks and all play their parts . . . Now he plays a king in purple and now a humble slave in rags. It's all a sort of pretence.[4]

The whole life of man is a play, and we are all performing roles: in the topsy-turvy carnival world of *A Praise of Folly* (1511), kings can become slaves with a quick change of costume and makeup—fabrication and performance are all. This is the voice of Folly, however, and elsewhere, as here in *The Education of a Christian Prince* (1516), Erasmus is clear that a real king should ideally be far more than an actor: "Do you want to know what distinguishes a real king from the actor? It is the spirit that is right for a prince: being like a father to the state."[5] Erasmus's dispiriting conclusion was, however, that most of Europe's princes were failing to live up to this standard—most were simply acting a part and doing that badly.

Humanism insisted that roles should be played well, and throughout the sixteenth century conduct books poured off the European presses instructing men and women how to behave and dress appropriately in a range of situations from boyhood, to marriage, to death. For example, Juan Luis Vives's book of instruction for well-to-do Christian women encourages would-be readers on its title page that "diligent" study of its contents would lead to the knowledge of many things, but "specially women shall take great co[m]modyte and frute towarde the [in]creace of virtue [and] good maners."[6] A performance of good manners here, as in Erasmus's book for little boys, is an outward display designed to convince others of one's inner worth. Further, many conduct books directed at a male audience urge the seminal importance of speaking eloquently and decorously. In his important book, *Renaissance Self-Fashioning*, the new historicist critic Stephen Greenblatt argued that "the chief intellectual and linguistic tool" in Europe's transitional cultures was rhetoric:

> Which held the central place in the humanist education to which most gentlemen were at least exposed . . . Rhetoric served to theatricalize culture, or rather, it was the instrument of a society which was already deeply theatrical.[7]

Building on Jacob Burckhardt's famous nineteenth-century thesis (in *The Civilization of the Renaissance in Italy*, 1860) that the political upheavals in Italy in the later Middle Ages brought about a radical change in consciousness and that "princes . . . secretaries . . . ministers, poets [etc.]" were "cut off from established forms of identity and forced by their relation to power to fashion a new sense of themselves and their world," Greenblatt insisted that it was above all the art of eloquence—rhetoric—that enabled men to fashion "the self and the state as works of art."[8] It would seem that the Renaissance, like the cusp of the twenty-first century, had a very pronounced conception of the person as a construct or artifice, as the product of social intervention and cultural shaping.

"REGIMES" OF THE BODY

Self-fashioning was not, however, restricted to the body's grooming, coverings, vocalizations, and expressive performances: its material composition was imagined to be malleable and in need of intervention, too; indeed, one's general well-being depended on the observance of a strict regimen of self-management in order to maintain the body's humors in optimum balance for health. Again, developments in printing played a crucial role in circulating accessible knowledge in this area throughout Europe. The medical regimens that poured off the presses in the early sixteenth century were humanist-inspired, pocket-size self-help guides to managing the body in health and sickness, and as such, they are rich repositories of information about how people construed their bodies and minds. Written in the vernaculars to reach a wide audience, they were penned by physicians, clerics, lawyers, grammar school teachers, statesmen, and "men of letters": professional categories that were much more fluid and hybrid than they are today, and authors of regimens often worked, in true humanist fashion, across several of these fields.[9]

As might be expected in this period, techniques of self-government were inspired by Greek and Roman models, and Michel Foucault's study of ancient theories of regimen suggests the crucial importance of these in the classical world:

> "Diet" itself—regimen—was a fundamental category through which human behaviour could be conceptualized. It characterized the way in which one managed one's existence, and it enabled a set of rules to be affixed to conduct . . . Regimen was a whole art of living.[10]

Concerned with "a whole art of living," discourses of proper regimen were not confined to the ancient physicians' writings but occur widely in treatises of moral and political philosophy, notably Plato's *Republic* and Aristotle's *Politics*. As will become clear in my discussion of sixteenth- and seventeenth-century regimens, health and disease constructs are shaped by—and in turn affect—other cultural phenomena, and early modern discourses of proper regimen are particularly sensitive indicators of social and intellectual change.

Thomas Paynell's free translation and adaptation of the Salerno physician Joannes de Mediolano's Latin verse, *Regimen sanitatis Salerni* (1528) and Sir Thomas Elyot's synopsis of "the chiefe" ancient authors of Physyke (notably Galen and Hippocrates), *The Castel of Helth* (1534)—were flagship sixteenth-century English regimens. Both were highly esteemed and popular: Paynell's went through nine editions between 1528 and 1634, and Elyot's claimed seventeen editions between 1534 and 1610.[11] Paynell's book opens with a dedication to no less than the "hyghe Chamberlayne of Englande": proper regimen was an important English matter. Making the commonplace observation that people lived longer in times gone past than "nowe adayes," Paynell proceeds to offer two possible explanations (sig. A2r). The cause is either "our myslyvynge and filthy synne," or "our mys diete" (sig. A2v). Significantly, especially given that Paynell was a cleric, he favors and stresses the natural cause (as opposed to divine intervention):

> Surfet and diversities of meates and drynkes letting and corrupting the digestion febleth man . . . Yll diete (as me thynketh) is chief cause of all dangerous and intolerable diseases: and of the shortenes of man's life. (sig. A2v)

In this early sixteenth-century regimen, medicine is a pragmatic discipline concerning the corporal body, and disease is construed as entirely the result of "putrified" humors caused initially through poor diet and habits and ungoverned emotions. Three prime rules for health are first, to live joyfully; second, to maintain a tranquil mind; and last but far from least, to eat a "moderate diete" (sig. B2r-v). The regimen proceeds in cheerful mode to discuss aspects of daily hygiene—washing, sleeping, and eating—and to detail at length the digestive qualities of various foods and drinks before considering the most suitable times to bleed and purge the body.

Elyot's *Castel* is a very humanistic regimen informed by Linacre's new translation of Galen. It methodically describes the body's composition and details the "signes" of various "complexions" or humoral types. The melancholic individual, for example, is lean; has hard skin and plain thin dark hair; is watchful; has fearful dreams; is stiff in opinions, timorous, and fearful; is prone to anger; seldom laughs; and has slow digestion, weak pulse, and watery urine (f.3r). The reader's task is to ponder the signs of his own body and work out his particular "complexion" (determined by a predominance of black bile, yellow bile, blood, or phlegm): is he melancholic, choleric, sanguine, or phlegmatic? This established, he must proceed to learn about "Thynges not naturall" ("ayre, meate and drynke, slepe & watche, mevyng & rest, emptynesse & repletions," f.11v)—habits and rules to observe to remain in prime condition—subsequently applying the program of bodily care specific to his humoral type. The *Castel* instructs the choleric man, for example, in what he should eat and drink, and when he should wash, sleep, engage in "Venus" (sexual intercourse), have his blood let, and be purged, in order to maintain his bodily vessel in an optimum balance. Compared with later sixteenth-century regimens, morals and God rarely enter Elyot's book overtly, but he does lament the "contynuall gourmandyse . . . the spirit of gluttony" that, in his view, is tormenting the realm with sicknesses (f.45r). Observing a moderate diet, and thus maintaining the body in optimal humoral condition (in homeostasis), is the key to health. In the humoral medical schema, only an unregimented, imbalanced body will succumb to contagious diseases like plague, which is associated with "miasmic" (poisoned) air. The nation's diet—its pattern of consumption—establishes the nation's health. Disease is never a private affair; it always has a social dimension, and in humoral medicine, drinking, eating, and doing anything excessively is inevitably troublesome.

Let us examine this crucial understanding further. In Erasmus's conduct book for boys, temperance and bodily control (especially around eating practices) are key themes, as they are in the books of regimen. The same regard for containment and moderation emerges in Renaissance discourses of philosophy and aesthetics, too; consider, for example, Leonardo de Vinci's famous illustration of the Vitruvian Man (Figure 9.1): in this image, the ordered, well-tempered male body fits with extended hands and feet into the most perfect geometrical shapes, the circle and the square.

The circle had associations with God; the geometry of circumscribed Vitruvian Man is, therefore, divine as well as beautiful, and this is why many Renaissance churches and cathedrals were modeled on this perfect form. As the historian Rudolph Wittkower has foregrounded, the sketchbooks of Renaissance artists and architects are littered with inscriptions of this mystical geometry that was felt to embody the harmonies of the universe; in his view it had a hold on the imagination that cannot be overstated.[12] In fact, this ideal of disciplined control, equilibrium, and harmony informed all Renaissance artworks, including the artfully fashioned body of man.

Significantly, however, as the sixteenth century progressed, the medical regimens bear witness to an increasing, urgent emphasis on self-government: attending to one's diet, sleeping and waking habits, including exercise and the frequency and timing of sexual intercourse, become ever more exacting concerns with far-reaching implications. The physician William Bullein's regimen, *Bullein's Bulwarke of defence* (1562), rails, for example:

> There are many idle people in citees, and in noble houses, dooe thinke the chiefe felicitie onely, to be from bedde to bellie . . . to bedde again: none other lives thei wil use, then Cardes, Dice, or pratlying title tatle excepted . . . slepyng, eating and laughing.[13]

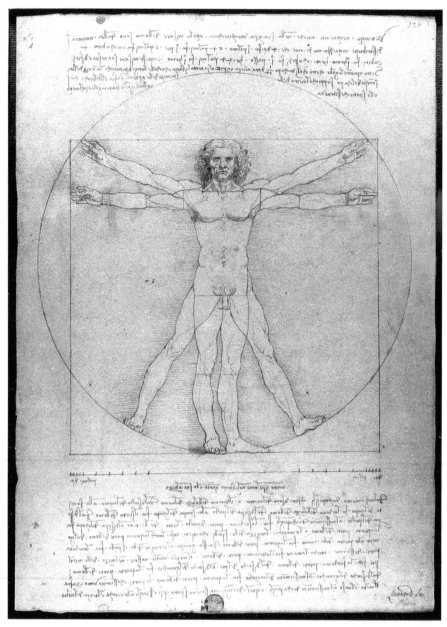

FIGURE 9.1: *The Proportions of the Human Figure* (Vitruvian Man), Leonardo da Vinci, 1490. Pen, ink, and watercolor over metalpoint. Wikimedia Commons. Photo Credit: Luc Viatour, www.lucnix.be http://commons.wikimedia.org/wiki/Image:Leonardo_da_Vinci-_Vitruvian_Man.JPG.

This manner of "idle" living, the reader is instructed, quickly makes a noble person "a deformed monstrous man" and reduces him to beggary (f.lxvij.r). Dramatic interludes from the mid-century obsessively stage this message, too. Similarly, the lawyer, Thomas Newton's *The Touchstone of Complexions* (1576) urges:

> There be many excellent witts and very towardly natures, which . . . do degenerate from their good inclination of nature, and become altogether rebellious, wilfull, lewde and barbarous.[14]

The reader is instructed that his "chiefest care and whole diligence" (f.2v) should be "perfectly" to know:

> The exacte state, habite, disposition, and constitution, of his owne Body outwardly: as also the inclinations, affections, motions, & desires of his mynde inwardly, so that he can regulate them accordingly.

The consequences of not knowing one's self—in a very literal sense—and of subsequent inadequate regimen are dire:

> If the bodye do abounde and be full of ill humours, if the Spirites bee unpure, and the brayne stuffed full of thicke fumes proceedinge of humours, the bodye and Soule consequentlye cannot but suffer hurte, and bee thereby likewise damnifyed. (f.19v)

The ungoverned body is now a terrible danger to the soul: the Protestant Reformation is clearly exerting a powerful effect on discourses of bodily regimen. "We are what we eat" is the ever-present maxim of this text, and through a mere change of diet the body can be "refourmed into better" (f.3v). The consequences of immoderate eating, "surphet, and dronkennesse" are "lewd affections, and unbridled motions," the diminishment of reason and "venerous luste":

> For when the body is bombasted with drincke, and bellycheere, the privities and secrete partes do swell, and have a marveylous desire to carnal couture. (f.10v)

Morals, health of the soul, behavior, and the well-being of others are at stake in the crucial business of proper regimen.

The Manchester grammar school teacher Thomas Cogan took this even further in the *Haven of Health* (1584), warning his schoolboy readers that maintaining a "meane and temperate dyet, in the feare of God" is a spiritual matter with wide-ranging consequences.[15] In "the haven of health"—Protestant England—personal bodily self-restraint and management, especially in the spheres of diet and sexual conduct, is the linchpin to the nation's "health"—physical, spiritual, moral, and social. Perfect knowledge of one's own complexion, as construed through the reading of "signes," is the prime key to salvation of the individual soul and the godly nation.

In regimens of the second half of the sixteenth century, therefore, discourses of bodily management increasingly intersect with those of religion and the state, and the emergent "castel" is a charged political site, and not just in England. The new temperance movement was far from restricted to Protestant countries; the Counter-Reformation produced a similar impulse toward heightened self-control and curbing the body's excesses. This was accentuated, too, by European humanism's increasing flirtation with the philosophical discourses of Platonism and Stoicism. In *The Republic of Plato*, the inability to rein in the body's passions is notably represented as synonymous with despotism, violence, and criminality:

> When a master passion is enthroned in absolute dominion over every part of the soul, feasting and reveling with courtesans and all such delights will become the order of the day . . . Soon he will be borrowing and trenching on his capital; and when all resources fail, the lusty brood of appetites will crowd about him clamouring. Goaded on to frenzy by them . . . he will look out for any man of property whom he can rob by fraud or violence.[16]

Such lack of self-mastery makes men totally unfit for husbandry and rule. Similarly, Stoicism, which underwent a considerable resurgence in the late sixteenth century, emphasized the crucial importance of self-discipline. The widely disseminated texts of the philosopher and political theorist Justus Lipsius instructed European readers in the 1580s and 1590s, for example, that they should construct "fortes and bulwarks" to withstand the furious assaults of "lustes": too many delights—notably "feasts and banquets"—lead to excess and effeminizing "luxuriousnesse."[17]

We can conclude, therefore, that by 1600 the ideal, civil body throughout Europe was imagined to be disciplined, ordered, restrained, temperate, and thus subject to reason. This message was widely disseminated via an expanding print culture and reiterated by literary, philosophical, religious, and staged texts as well as by conduct books and regimens; indeed, it is often difficult to separate these works into clearly demarked generic categories. For example, Edmund Spenser declared of his epic poem, *The Faerie Queene*: "the generall end . . . of all the booke is to fashion a gentleman or noble person in vertuous and gentle discipline."[18] He devotes the whole of his epic's second book to "fashioning Temperaunce" in his gentleman and noble readers because:

> In a body, which freely yield
> His partes to reasons rule obedient,
> And letteth her [the soul] that ought the scepter weeld,
> All happy peace and goodly government
> Is settled there in sure establishment. (II. xi. 2. ll.1–5)

"Sober government" renders the "body" (of man, the state, and religion) the most "faire and excellent" of all God's works; however, its antithesis, the body, "fowle and indecent, / Distempred through misrule and passions bace," gives rise to 'Monster[s]'" (II. ix.1.ll.4,2,5–6,7).

MONSTROUS AND SAVAGE "OTHERS"

> The frame therof seemd partly circulare,
> And part triangulare, o worke divine;
> Those two the first and last proportions are,
> The one imperfect, mortall, feminine;
> Th'other immortall, perfect, masculine,
> And twixt them both a quadrate was the base.
> —*The Faerie Queene*, II. ix. 22. ll.1–6

Before his detailed poetic rendering of dietary regimen in Book II of *The Faerie Queene*, Spenser notably evokes an image of Vitruvian Man that foregrounds the imperfect nature of the feminine in relation to the Castle of Temperance. In the humoral bodily model, male bodies were hotter and dryer, stronger, and more reactive than sluggish female bodies as this passage from a popular medical book explains:

> Now herein he [Galen] putteth the difference betwixt her and the male, that in males the parts of generation are without the body, in females they lie within because of the weakness of the heat which is not able to thrust them forth. And therefore he saith that the neck of the womb is nothing else but the virile member turned inward.[19]

As colder, moister, imperfect versions of men, women were rendered inferior by their physiology; their defective, uncontained "leaky" bodies were imagined as more prone to unreasonable, wavering, and libidinous behavior— the sins of mother Eve could not be easily cast off.[20] As this homily on matrimony puts it:

The woman is a weak creature, not endued with like strength and con-
stancy of mind, therefore, they be sooner disquieted, and they be the more
prone to all weak affections and dispositions of mind, more than men
be, and lighter they be, and more vain in their fantasies and opinions.[21]

In this construction of sexed and gendered bodies, God and nature appeared
to dictate that women were simply unable to govern themselves adequately; it
made perfect sense, therefore, for them to be subjugated by men for their own
well-being. All women needed a "head"; that is, male governance in the form
of a father, husband, or brother. Once married, they were entirely subject to the
"desires" and "law" of their spouse:

Women have no voice in parliament. They make no laws, consent to
none, they abrogate none. All of them are understood either married or
to be married, and their desires are subject to their husband: I know no
remedy, though some women can shift it well enough. The common law
here shaketh hand with divinity.[22]

We can see, therefore, that the humoral bodily schema, together with its close
partner, proper regimen, were, in fact, powerful props of patriarchal power relations and an obsessive late sixteenth-century representational system of civil,
temperate bodies versus "others" designated as undisciplined and uncontained
worked to underpin dominant and emerging social hierarchies.

This understanding is highly illuminating of the phenomenon that the literary critic Louis Montrose has described as "the work of gender in the discourse
of discovery."[23] If we look closely at one of the most widely distributed early
modern visual representations of the New World—Theodor Galle's engraving
of America depicting Amerigo Vespucci's discovery (Figure 9.2)—it is not difficult to apprehend what "the work of gender" might mean in this context. In
the foreground, there is a prominent, semi-recumbent, buxom nude wearing a
feather headdress, raising herself from her hammock, and apparently beckon-
ing to Amerigo Vespucci who has just stepped foot on the shore of the New
World.

The title of the engraving is "America," and the well-formed naked wench
is a personification of the New World welcoming her discoverer. To her right,
clinging to the trunk of a tree, is a sloth—a symbol of her torpor (she has obviously been asleep)—and in the background, close to the vanishing point, we
can just about make out a severed human leg and haunch cooking on a spit
over a fire—a cannibal feast is in progress. The clear message is that the man-
eating inhabitants of this land are savages. The personification of America and
her cannibals contrasts markedly to the visual information to the left of the
image. Vespucci's upright, armored and heavily clothed figure meets the gaze

Americen Americus retexit, & ... AMERICA. ... Semel vocauit inde femper excitam ...

FIGURE 9.2: *America*, an engraving from *Nova Reperta*, by Theodor Galle, after Jan van der Straet, ca. 1580. Print Collection, Miriam and Ira D. Wallach Division of Art, Prints and Photographs, The New York Public Library, Astor, Lenox and Tilden Foundations.

of naked America directly. In his right hand he is holding a cruciform staff with a banner bearing a cross (he is manifestly a representative of Christianity) and in his left he displays an astrolabe: a symbol of the advanced fifteenth-century knowledge and technology that made his discovery possible. Behind him, his impressive ship is anchored just off the coast.

Galle's engraving is thus a representation of a civilized European male about to possess and reform female, savage America—there will be no more cannibal dinners after his intervention and conquest. The motto accompanying the engraving declares, "Americen Americus retexit, & Semel vocavit inde simper excitam," which translates as "Americus rediscovers America; he called her once and thenceforth she was always awake."[24] Sloth-like, savage America thus welcomes her new master: his masculine wisdom, rule and government will shake her out of her unproductive torpor and reform her barbarous behavior. This is, as Montrose points out, a "crude and anxious misogynistic fantasy, a powerful conjunction of the savage and the feminine"—and it serves to foreground the raw ingredients of an emerging colonialist ideology.[25] America's past was sleep and barbarity; taken by Vespucci, she (the New World) is interpellated within European history—America has no meaningful history of her own. Effectively the ideology of gender hierarchy explored here sanctioned the conquest and subjugation of the indigenous peoples of the New World.

But there were critics of the powerfully justifying Eurocentric discourse pre-
mised on the savage versus civil body dichotomy; notably Michel de Montaigne
in his astonishingly astute essay, "Of the Cannibals," which effectively decon-
structs the proto-colonialist binary. Beginning by reflecting on how the ancient
Greeks called all "strange [foreign] nations . . . barbarous," even the Romans,
the essay proceeds to argue that "the rule of reason" should be the judge of
such matters not "common report."[26] Describing in detail the behavior and
rituals of the indigenous people of "Antartike France," and contrasting these
with the practices of European invaders, Montaigne gradually and skillfully
unravels how that which is termed "barbarous" is actually a matter of custom, perception, and self-interest. Pointing out that cannibalism is not so much
a form of nourishment as honorable ritualistic revenge performed on a dead
body, he proceeds to contrast this with the way in which Europeans torture
people by burying them up to the middle, shooting arrows at them, and hanging them up while still alive—the latter, he suggests, is "more smartfull, and
cruell."[27] He concludes: "I am not sorie we note the barbarous horror of such an
action [cannibalism], but grieved, that prying so narrowly into their faults we
are so blinded in ours."[28] To push his point further home, Montaigne records
the reactions of three of this so-called barbarous nation when they visit France.
They are horrified by European practices such as grown men paying homage to
a child king and exhibiting disregard and disrespect for other human beings:

> They had perceived, that there were men amongst us full gorged with all
> sortes of commodities, and others which hunger-starved, and bare with
> need and povertie, begged at their gates.[29]

The exposure of such unreasonable, greedy behavior effectively turns the
tables on "Reason"—that great stalwart of old world patriarchy—along with
civility and its close associate, temperance, demonstrating that these qualities
and their antitheses are relative and often defined by custom in the service
of power relations. Indeed, through the course of this tract European bodies
are refashioned as intemperate and diseased, as barbarous outsiders intruding
upon and corrupting an idyllic world: European smugness and hypocrisy are
exposed.

A famous English explorer's narrative of his discovery provides a blatant
example of the type of self-interested Eurocentric discourse that Montaigne's
rhetoric was working to undermine. Sir Walter Raleigh's *Discoverie of Guiana*
is a fascinating text that particularly exploits the hierarchy-enforcing gender
differentials of Renaissance culture to advance its author's colonizing ambitions. This is not, however, without some considerable irony because, in common with other Englishmen, Raleigh was ultimately subjugated to a female

head—Elizabeth I. As one of the queen's favorites in the 1580s he had led an exploration to the New World and founded a colony, which he named Virginia after the Virgin Queen. He subsequently got into enormous trouble by daring to seduce and secretly marry one of the queen's virginal ladies-in-waiting. Thrown into prison, and finally released to undertake a second voyage, Raleigh's account of *The Discoverie of the Large, Rich and Bewtiful Empyre of Guiana* (1596), reads rather like a penitential narrative.[30] In this revealing passage (which plays on the early modern trope of "Countrey" as female sexual parts) he asks for the queen's permission—this time—to take a "Maydenhead":

> To conclude, Guiana is a Countrey that hath yet her Maydenhead, never sackt, turned, nor wrought, the face of the earth hath not beene torne, nor the vertue and salt of the soyle spent by manurance, the graves have not beene opened for gold It hath never been entred by any armie of strength . . . This Empire is made knowen to her Majesty by her own vassal, and by him that oweth to her more duty than an ordinary subject . . . The countrey is already discovered, many nations won to her Majesties love and obedience . . . Her majesty may in this enterprise employ all those souldiers and gentlemen that are yonger brethren, and all captains and Chieftains that want employment.

With the queen's go-ahead, Englishmen will effectively rape and possess Guiana: plunder her for gold and—less convincingly—win love for her majesty.

There is a detectable and understandable tension in Raleigh's sexualized narrative of violent entry and exploitation addressed, as it is, to a female governor. Indeed, it is built upon the very patriarchal gender constructions that made Elizabeth I a puzzling and dangerous contradiction—a female ruler. As we have seen, according to humoral physiology women could not rule themselves, let alone govern others. On this matter the Calvinist divine John Knox railed in *The first blast of the trumpet against the monstrous regiment of women*: "I affirm the empire of a woman to be a thing repugnant to nature . . . For their sight in civil regiment is but a blindness . . . their counsel foolishness: and judgment frenzy."[31] In this scheme of things, Mary Queen of Scots, Mary I, and Elizabeth I, were all cold, moist, inferior men—monstrous others in terms of patriarchy and yet, paradoxically, monarchs. No wonder then that, given this logic, Mary Queen of Scots was construed as unable to contain her lust and thus not only committed adultery but was also implicated in murder—both excessive, barbarous acts. The problem of their gender was accentuated for the two English queens by their father's insistence on making much of his phallic power as the basis of his rule. The copy of a painting by Holbein the Younger, which once adorned a wall of Whitehall Palace (The Whitehall Mural, 1537), gives vivid expression to this. The painting was a celebration of the Tudor dynasty, and it depicted Henry VII, a Lancastrian, and Margaret of York in the background—their marriage had ended the Wars of the Roses by uniting the two great, factious houses. Standing larger than life in front of his parents, engaging the viewer's gaze directly and confidently, is the bulky, muscular figure of Henry VIII, and across from him stands the diminutive figure of Jane Seymour, the mother of his only surviving male heir. As several commentators have stressed, Henry's prominent codpiece and strategically positioned pointer stick serve to emphasize his virility; the propaganda message implicit in this symbolic image and underlined by the Latin verses on a tablet that Henry is demonstrating, is that his manly virility has produced a son who will carry the Tudor line forward, guaranteeing peace and stability for England.[32]

This was an impossible act for his daughters to follow. Mary I dealt with her contradictory status by marrying and effectively being governed by Philip of Spain, her spouse. Elizabeth I was determined to rule England herself and, in the early days of her reign, she formulated shrewd, body-centric ways of dealing with the apparently insoluble cultural dissonance of a female governor. She created two master narratives, both associated with her body: the first was that she had special status as a chaste Protestant virgin and mother to her country and the second that she had two bodies—her body natural and her body politic—and she governed her kingdom through the latter. Elizabeth's virginity became the central feature in the web of mythologies woven about her body and fashioned by those writers and artists who were engaged in producing the texts, icons of majesty, and performances in which the queen was variously represented to her people, her court, and foreign powers. *The Armada Portrait* (1588) (Figure 9.3) is one of these highly symbolic representations. In this image, the queen's generous costume lends masculine bulk and a circular form to a virgin body that is emphasized by a large bow and pearl in the groin region.

Elizabeth's right hand covers the globe of the world, suggesting English colonial ambitions and rights, and behind her a window vignette depicts the Spanish Armada crashing on the rocks, defeated by the superior English fleet and the weather, over which Elizabeth's majesty has influence: Renaissance monarchs were imagined to be invested with divine powers (the divine right of monarchs) giving them some control over the elements and the ability to heal certain diseases, such as scrofula, through touch. It is significant that Elizabeth's body appears completely enveloped and sealed by her voluminous costume: her mystical, inviolable, contained virgin body (there is no leakiness here) with its almost circular (and thus divine) form is Protestant England that, under Elizabeth's rule, is impregnable and thus secure from invasion by exterior hostile forces such as Roman Catholic, Antichrist Spain.

The Armada Portrait is, in fact, a very graphic illustration of the way in which, to cite Mary Douglas's famous assertion, "the body is an image which

FIGURE 9.3: Elizabeth I, *The Armada Portrait*, by George Gower. Oil on panel, ca.1588. Wikimedia Commons: http://commons.wikimedia.org/wiki/File:Elizabeth_1_(Armada_Portrait).jpg.

can stand for any bounded system":[33] Elizabeth's body represents England and the Protestant church; her mystical virginity and intactness guarantee and signify the protection of both under her rule. This is similar to the way in which the virgin Una (the One) in Spenser's *Faerie Queene,* is a type of Elizabeth I and the true Church, which is under continuous attack from the hostile forces of Antichrist in this poem. As metaphor theorists such as Mark Johnson have demonstrated, symbolic bodies and their metaphors facilitate particularly rapid and easy transitions across domains—psychic (soul), physical (medical), and social (religious, political, economic). He sees metaphor (and metonyms) as "one of the chief cognitive structures" through which we make sense of the world: "through metaphor, we make use of patterns that obtain in our physical experience to organize our more abstract experience."[34] In this respect it is noteworthy that the English medical regimens of the latter part of the sixteenth century demonstrate considerable fear about the body's vulnerable boundaries and its proneness to attack from outside. As we have seen, this was accompanied by heightened exhortations to maintain English bodies in a disciplined, temperate, and vigilant state in order to protect the godly nation. Two contagious infections were particularly associated with this heightened medical anxiety—the pox or syphilis and plague. Syphilis is especially interesting in this context because it was linked both to European exploration—Columbus's sailors were thought to have caught the infection from lusty native women and imported it into Europe—and, in the English Protestant imagination, to papal Catholicism, especially its forces in Spain. In fact an entirely new model of the body emerged in the sixteenth century in medical tracts about the cure of this new disease and an old one too, the alleged corruptions of Roman Catholicism.

Theophrastus von Hohenheim, Paracelsus as he was commonly known, established his medical reputation by publishing two treatises on the fearful new sexually transmitted infection and introducing the famous mercury treatment. As Charles Webster has described, for this enigmatic and audacious physician, medical and religious reform were intimately intertwined, and both required radical renewal. Paracelsus declared Galenic medicine heathen and obsolete, like Roman Catholicism—its practitioners greedy like Catholic priests; and its treatments ineffective like Catholic absolution.[35] His rival model of bodily functioning was intensely spiritual and informed by the mystical approaches of alchemy and Neoplatonism. A regimen written by one of his English disciples, R. Bostocke, *Auncient and Later Phisicke* (1585) reveals the strong Platonic underpinning of Paracelsian theories: "For all thinges good or bad, be derived and doe flowe from Anima . . . into the body and to every parte of man."[36] Although the Galenic universe was remarkably secular and devoid of mystical powers and magic, the essence of Paracelsus's religion of medicine is an enchanted cosmos inhabited with spirits, in which stones, roots, plants, and seeds all have "powers" accessible via the practice of chemistry, which can only be channeled into the service of medicine by the true, reformed Christian—in this bodily schema religion and medicine are inseparable.[37]

The new disease was a ready target for appropriation in the service of reform because its modes of transmission—associated with lust and excess—were rapidly established and documented. The Valencian physician Pintor warned in 1500 of the extreme infectiousness of the disease and added that coitus with a contaminated female was the main cause. The rich nobleman, astronomer, and doctor Juan Almenar of Valencia declared in 1502 that the French sickness was usually transmitted by coitus, but also by kissing, breast-feeding, and, much less commonly, by corrupt air. It was a stigmatizing infection that traveled rapidly round the globe, and all nations sought to project it onto others: thus, depending on your perspective, it was the Spanish, French, Italian, Neopolitan pox, *il morbo gallico*, or syphilis.[38] Albrecht Dürer's representation of a Pox victim (1496) (Figure 9.4) is illuminating of the early understanding of the disease.

In this woodcut image a foppish man sporting a large feathered hat and fashionable wide-toed, slashed shoes is displaying his syphilitic sores as though they are stigmata. His isolated victim is a suffering Christ type then, but Dürer's representation also works to suggest that he is an undisciplined dandy and

FIGURE 9.4: Albrecht Dürer, image of a syphilitic accompanying a broadside of Theodoricus Ulsenius, 1496. Wellcome L0014503EB.

the five planets in the sign of the scorpion—ruling the genitalia—implicate the stars in his fate. As the century progressed, the pox became known as the disease of Venus, closely associated with unruly prostitutes and, tellingly, in an engraving by Jan Sadeler I (ca. 1590; Figure 9.5), Venus is represented as a statue in a fountain contaminating the stream of water flowing from it with her polluted breast milk, which an unsuspecting youth is thirstily and greedily imbibing.[39] The shepherd Syphilus—a mythical character who features in Fracastoro of Verona's Latin epic poem, *Syphilis* (1530)—and the hunter Ilceus are being warned against yielding to the temptations of Venus.[40]

But lecherous dandies and wanton women were not the only stereotypes of the disease. While Paracelsus was busy tying the corruptions of Galenic medicine and papal Catholicism into a tight knot linked to the new venereal contagion, others, such as Erasmus and John Bale in England, were spreading the message via popular print culture and the stage that Catholic clerics spread "the Romish Pox." Erasmus wrote four of his humorous didactic dialogues—the Colloquies—about the horrors of the disease ("the new leprosie") and how to avoid it and in one, set in a brothel, a Romish priest client reveals his "bleare

FIGURE 9.5: Warning against syphilis. Engraving by Jan Sadeler I after Christoph Schwartz, 1588/1595. Wellcome L0031325.

eyed, palsey shaken, and crooked" pox disfigurement caused by his lechery.[41] Shortly after this dialogue was published, Simon Fish's polemical tract, *A Supplicacyon for the Beggars*, deployed the motif of the pox-ridden priest to further the process of ecclesiastical reform in Britain. The force of his attack lay in the representation of a class of men who should be the spiritual elite, as even more reprehensible than harlots:

> These be they that corrupt the hole generation of mankind yn your realme; that catche the pokes of one woman, and bere theym to an other . . . ye, some one of theym shall bost emong his felawes, that he hath medled with an hundredth wymen.[42]

A Supplicacyon's clear message is that the priests' idleness, luxuriousness, and lust have spread the pox round England. The cumulative effects of such rhetorical tracts undoubtedly assisted Thomas Cromwell's task of closing down monastical institutions. Indeed, in many polemical tracts, which included plays for public staging, the pox is conflated with papal Catholicism itself: both are venereal corruption—the quintessential "disease" of failure of regimen and intemperance, of ungoverned excess.

CODA

This has been a strange and tangled story of bodies fashioned temperate and civil and their binary others in Renaissance Europe. It has illustrated how the body was both a site on which social messages were inscribed and a sensual/ practical actant that shaped and produced culture. England has been put under the spotlight, and we have seen how fashioning the English nation in the post-feudal era—where a collapse in the old systems and ways of doing things necessitated major cultural shifts and social re-formations—involved continuous recourse to the bodily nexus. Although such habitual thinking in terms of somatic correspondences was undoubtedly encouraged by Neoplatonic, Pauline, and Parcelsian analogies of the body and society in the Renaissance—by pronounced macrocosm, microcosm links—cultural theory of the body suggests that this way of making sense of experience was far more than simply an early modern habit encouraged by these philosophical paradigms. Indeed, the early modern philosopher Giovanni Battista Vico detected something of a more heuristic nature underpinning the widespread tendency to analogize from the body:

> It is noteworthy that in all languages the greater part of the expressions relating to inanimate things are formed by metaphor from the human body and its parts and from the human senses and passions. Thus head

for top or beginning; the brow and shoulders of a hill . . . all of which is a consequence of our axiom that man in his ignorance makes himself the rule of the universe.[43]

Vico's astute theorizing sounds precociously modern and not centuries apart from that of the phenomenologist Maurice Merleau-Ponty in *Phenomenology of Perception* (1945), "one's own body . . . puts forth beyond itself meanings capable of providing a framework for a whole series of thoughts and experiences";[44] or that of cognitive philosophers like Lakoff and Johnson who argue for the embodied nature of all understanding.[45] The remarkable insight of Vico's *New Science* was that people imagine and make sense of their world through their gendered bodies. In fact, as the cumulative reflections of the anthropologist Michael Lambek foreground, human embodiment has complex ramifications that extend well beyond individual cognition:

> Bodies serve as icons, indices, and symbols of society and also of individuals and of the relationship between them. At the most basic level, in posture, adornment . . . in touching, in relative positioning, in gathering together, in coitus, in feeding and being fed, in working cooperatively and in consuming together and again in refusal of engagement or consumption, in the maintenance of taboos, in all these situated practices— persons and ongoing social relationships are not simply signified but actively constituted . . . If sociality is embodied, conversely human bodily activities are culturally mediated, hence infused with mind.[46]

Certainly, the cultural representations explored here reveal a complex interplay between the body and society: fashioning and performing the civil body in the Renaissance context was inextricable from fashioning and enacting civil society.

CHAPTER TEN

Renaissance Selves, Renaissance Bodies

MARGARET L. KING

Just like our own, the "selves" that Renaissance people inhabited, and acted through, and acted upon, were in some way—once known to theologians and now debated by cognitive scientists—bound to bodies. This chapter examines four pairs of figures whose bodies figure prominently in their experience as persons.

The first pair, the young and the old, are selves whose biological age preconditions their social existence: the young are young and the old are old because of their bodily states. The second pair, prostitutes and Jews, are figures whose social roles or cultural commitments—as sellers of sex and adherents of Torah and Talmud, respectively—caused their bodies to be caricatured, signed, and abused by the many who found them dangerous.

The third pair, women who embraced the different destinies of mother and saint, performed their roles through the vehicle of their bodies, alternately the vehicles of fertility and birth in the one case, and of sacrifice and annihilation in the other. In the same way, the fourth pair, men who waged war and pleased princes, warriors, and courtiers, performed their roles by the discipline of their bodies, as commanded by their acute intelligence. The warrior fought and triumphed; the courtier shed his armor to primp and sparkle as dictated by the civilizing force of the lady whose demands groomed him to serve the prince.

THE YOUNG AND THE OLD

In 1983, Leo Steinberg pointed out to a surprised audience what should have been perfectly obvious: depictions of the infant Christ in Renaissance art not only did not hide, but bared and celebrated his genitalia.[1] The purpose of these representations, he argued, was theological: the focus on the male member asserted and reaffirmed the humanity of Christ, the incarnate God. The society that produced these representations, we could also argue, was looking at babies and finding them interesting.

Looking at babies was not that easy to do: as soon as infants were born, they were swaddled tip to toe with long bands of cloth and displayed to public view like a little mummy. Only mothers and nurses, as they unwrapped and renewed the bands, viewed the infant body. Men in general and artists in particular would have had to intrude upon the business of female caretakers to admire the fleshly presence of an infant.

And so they did. Not only do nude infants figure in innumerable Renaissance paintings of the nativity, but others prance about as mischievous Cupids (Cupid, or Eros, was the god of love), and as charming putti, decorative rows of plump and playful babies incorporated into sculptural and architectural programs. These representations of the infant body were new: the image does not figure in medieval art, and it is rare in classical Greek art, despite its celebration of nudity in general. Only Hellenistic art, imitated by the Romans, featured nude infants, although they are not so central a motif as in the Renaissance.

In the same centuries (from the fifteenth to the seventeenth) in which the nude infant blooms forth in the visual arts, infant birth emerged as a matter of concern for physicians. Well, naturally, we would say. But in Christian Europe, the birthing of babies was a matter left to midwives, who managed the job rather well. Without formal methods of training, they mastered their profession empirically and transmitted that knowledge across the generations to their younger assistants, sometimes immediately to their daughters.

In the fifteenth century, university-trained physicians, beneficiaries of the medical learning of ancient Greece and medieval Islam, began to intrude upon that zone of midwife expertise. Around 1460, Michele Savonarola wrote a treatise of practical gynecology and pediatrics addressed to the midwives and women of his city of Ferrara, written in the vernacular so that women could read it.[2] About fifty years later the German Eucharius Rösslin, then town physician of Worms (modern Germany), wrote a best-selling guide for midwives: the *Rosengarten*, or "rose garden for pregnant women and midwives," later translated and plagiarized in several European languages; it became the bible of obstetrical knowledge for the next two centuries.[3] His goal, Rösslin explains in his singsong verse preface, was to save the lives and souls of infants by correcting the incompetence of midwives:

Who know so little of their call
That through neglect and oversight
They destroy children far and wide . . .
And earn for this a handsome fee.[4]

This concern for infant life expressed in obstetrical literature responds to
the single greatest fact about children's existence in the premodern centuries
around the globe: the tremendous toll of early death. Overall, some 25 to 50 percent of all children died, most in the first year or two of life: a harsh statistic
that was not altered until the twentieth century, when sanitation secured a safe
water supply, substitutes were developed for breast milk, and antibiotics and
immunization quelled infantile disease. So frequent was child death that few
families escaped it, and few women saw all the children they had borne live to
adulthood. Many lost all their children, as did Aphra Behn, the English playwright and poet who, around 1685, lamented the loss of her youngest infant,
the seventh of those she had lost before:

The softest pratler that e'er found a Tongue,
His Voice was Musick and his Words a Song.[5]

The vulnerability of Renaissance infants to disease and death is a central
reality of the era. In his famous *Centuries of Childhood,* as this fulcral work
is known in its 1962 English translation, French historian Philippe Ariès suggested that the frequency of infant death might have encouraged premodern
parents not to invest affection in their offspring.[6] His suggestion gave rise to a
historiographical contest waged by advocates and critics of the Ariès theory.
Among them, British historian Lawrence Stone viewed Tudor–Stuart England
as virtually an infanticidal society, and the American Steven Ozment detailed
the close affectionate bonds over several generations among elite Nuremberg
families.[7] This scholarly duel, encouraging the close study of a range of re-
gional, social, and religious milieus, has documented the complex and not yet
wholly resolved issue of parent–child relations.

Among the poor, children certainly enjoyed less protection and affection
from parents than among the wealthy. The Renaissance launches an era of
infant abandonment and, in some settings, infanticide. The first home of the
Renaissance, Italy, was also the pioneer of the foundling home: a refuge specifically dedicated to found and abandoned children. Florence's *Ospedale degli
Innocenti* (Hospital of the Innocents), its main structure designed by no less
an architect than Filippo Brunelleschi, creator of the famed cathedral dome,
admitted infants at a rate of about one hundred per year in the fifteenth century, assigned them to wet nurses, and trained them for service, marriage, or

employment.[8] In similar hospitals in Padua and Bologna, the rate of admission was some tens per year.[9] The institution spread throughout Europe, reaching England and St. Petersburg in the eighteenth century.

In hindsight, the foundling home appears to be as much an institution for the destruction of children as for their nurture. When, as late as the nineteenth century, the mortality rate at the Infants' Hospital on Randall's Island in New York City was about 65 percent, the observer might wonder why an establishment so profoundly unsuccessful in pursuing its mission was permitted to continue.[10] Outside of the foundling home scenes of child death abounded as well. Catholic and Protestant societies alike stepped up their prosecution of maternal infanticide.[11] In innumerable "baby farms," the prototypes of Charles Dickens's unfeeling caretaker Mrs. Mann starved, drugged, and abused unwanted or superfluous children, with the inevitable outcome.[12]

Poor children who survived their earliest, most dangerous years often rotated out of their own families to work in others.[13] As early as age eight or nine, boys and girls might enter service, their parents formally signing a contract with the employer who pledged to feed and house the youngster and send him off, after a term of years, with a small sum—in the case of a girl, her dowry.[14] Less formally, many other children became unofficial servants, becoming attached to an adult male in exchange for sustenance and protection. The figure of the picaro was thus born, the little boy whose wits are sharpened by his early adventures, like Lazarillo de Tormes, in the anonymous, eponymous sixteenth-century Spanish tale, who escaped death so many times he might truly be said to have been, like the biblical Lazarus, raised from the dead.[15]

Among the artisan and merchant families of the burgeoning European cities, children (mostly male, but also female) might be sent to other households not as servants but as apprentices.[16] Apprenticeship was a formal arrangement where the employer, or master, pledged not only to feed and house but also to train his young charge in certain skills that would suit him, in time, to seek an independent livelihood. Apprenticeship might begin as early as age twelve and continue into late adolescence or early adulthood, depending on the craft. A floating crowd of pubertal adolescents, prone to drunkenness, brawling, and rape, were the primary cause of social disorder in the cities of Europe.

The wealthy, although they might still neglect and abuse their children, were less likely to abandon them than the poor. They kept their daughters closely confined, reserved for their destiny of marriage—or, in Catholic societies, the convent. In either case, they departed their father's home with a dowry, a portion of patrimonial wealth assigned for their lifelong maintenance, the disbursement of which removed the daughter from her natal household and any claim on the inheritance.

Sons were prized, for they sustained familial honor and inherited the patrimony. The Italian humanist Leon Battista Alberti, himself born illegitimate, wrote not merely one but two substantial dialogues, one in Latin and one in the vernacular, about the patriarchal household, determined by the need to bear, rear, promote, and endow sons.[17] In high noble and royal families, of course, the value of a male heir was paramount: upon him depended not only the welfare of the family but also of the state. The succession of the marriages of English king Henry VIII, marked by the divorce, annulment, and execution of infertile wives, is the emblematization of the hunger for sons among European elites.[18]

Though they were prized, the sons of the wealthy were not, initially, well educated—for no model of schooling existed for those who would become the rulers of society.[19] For most of earlier European history, only those destined to ecclesiastical positions required schooling, and that schooling was provided within the walls of the monastery or cathedral. Merchants also required literacy, and so schools to train their sons in basic arithmetic and vernacular literacy sprang up in the busier cities.

The Renaissance humanists, the intellectual leaders of the era, were the first to devise an education for elite males.[20] Based on the study of the classics, its aim was not vocational, but rather to train the mind and, it was hoped, in consequence, the character of those who were to succeed to leadership positions. That humanist curriculum remained the model for secondary education into the nineteenth century. The Dutch humanist Desiderius Erasmus explained in his *On Education for Children* why fathers must tend to the proper education of their sons:

> There is no beast more savage and dangerous than a human being who is swept along by the passions of ambition, greed, anger, envy, extravagance, and sensuality. Therefore a father who does not arrange for his son to receive the best education at the earliest age is neither a man himself nor has any fellowship with human nature.[21]

Like most other humanist authors of pedagogical treatises, Erasmus opposed the corporal punishment of children. This was commonplace; indeed, no one knew, before the Renaissance, how to teach children without beating them: the standard image of the teacher was of a stern tyrant holding a rod. These beatings (along with the oafish ignorance of his teachers) had made his own schooling a horror for Erasmus. In his premier work, *The Praise of Folly,* he excoriated the schoolmasters who were the despots of the classroom, or rather, their "treadmill and torture chamber": there "they terrify the trembling crowd with threatening voice and looks, thrashing their wretched pupils with cane, birch and strap."[22]

The routine flogging of pupils was only one aspect of the physical abuse inflicted on children in earlier times—and not only on children but on all who were, like children, social subordinates by definition: slaves, servants, and wives. The power of the old over the young was expressed, and impressed, on the bodies of their subjects.

Renaissance art provides innumerable images of infants, but very few of young children. Figure 10.1 showing the haunting image by Domenico Ghirlandaio of an old man with a child of some six years—perhaps a grandson—is a notable exception. Highlighted here is the contrast of ages: the elder's grizzled face, lank gray hair, wart-studded nose; the boy's limpid skin, fair flowing curls, small tender hand. One portrait cannot tell the story of the relations between young and old, but it does tell us that in at least this one case, such a relationship could be warm and complete.

Few Renaissance people reached old age because they died as infants, children, or adolescents—or, as young adults, in battle or childbirth. Those who survived could live into great old age, much as we do. Doge (*dux*, or ruler) of Venice Enrico Dandolo was at least eighty-four years old (and blind) when he led the crusader forces that seized Constantinople in 1204; he died the following year, remarkable but not unique among the gerontocrats who ruled that city.[23] Cassandra Fedele, a female humanist of middling social rank, delivered her last public oration in 1556, aged ninety-one, two years before her death.[24] The painter Titian (Tiziano Vecellio) fell victim to the plague at an uncertain age, but certainly an advanced one, between eighty-six and ninety-nine. Queen Elizabeth I of England was a sturdy seventy at her death in 1603; King Louis XIV of France was a few days short of seventy-seven at his, leaving his throne to his great-grandson as all his other male heirs had predeceased him.

Though they might live long, these Renaissance seniors felt the pains and defects of advancing age without the remedies modern medicine provides. The Italian poet and humanist Petrarch (Francesco Petrarca) wrote of the dependence on spectacles that old age had wrought, along with an "array of discomforts."[25] Erasmus whined about his ailments in letters to his friends, and likened the condition of doltish old men to that of infants: both were bald and short, pablum-consuming, toothless prattlers.[26] The French creator of the essay, Michel de Montaigne, was more philosophical. "We must meekly suffer the laws of our condition," he wrote in 1587/1588, shortly before his death, in "Of Experience"; "We are born to grow old, to grow weak, to be sick, in spite of all medicine."[27] What is the point of turning to doctors? The kidney stones from which he mightily suffered were the proper preparation for death: "Consider how artfully and gently the stone weans you from life," the author's mind speaks to him, "and detaches you from the world."[28]

Another illness affected the minds of old men, or so contemporaries thought: that of avarice. Lust is the vice of the young, writes Italian humanist Poggio Bracciolini in his dialogue *On Avarice*, ambition of the mature, but avarice of

FIGURE 10.1: Domenico Ghirlandaio, *Old Man with Young Boy* (Großvater und Enkel, 1488) (Source: The Yorck Project: *10.000 Meisterwerke der Malerei*. DVD-ROM, 2002). Wikimedia Commons: http://commons.wikimedia.org/wiki/Image:Domenico_ Ghirlandaio_003.jpg.

the old, a portrait of whom is offered: "He will suffer the continual thirst for riches; no day, no hour will pass free from this preoccupation. He will seethe and writhe, anxious to satisfy his ever-growing hunger for gold."[29] Old men holding on to patrimonial wealth deprived young men of the means to launch

FIGURE 10.2: Three children of Henry VII and Elizabeth of York, 1748. Wellcome L0021667.

and pursue careers. Like old Volpone in English playwright Ben Jonson's comedy of that name, their miserliness turned kin and friends into grasping haters. What if they were to give away their wealth to a younger generation yearning to take charge? Then they were fools, like Shakespeare's King Lear: fools recognized as such by fools and, driven insane by the treachery of their offspring.

If old men were fools when they gave away their wealth, and misers if they retained it, old women were positioned to worse labels: crone, whore, witch. These denunciations applied mostly to women alone; elderly women under the protection of kinsmen, convent, or shelter were not intended. But widowed women were prime targets, viewed as hideous, lustful, and malicious. The witch charge could be fatal. The profile of a typical victim of a witch charge was a sole, elderly woman with a sharp tongue. In some cases, a loyal son might appear to defend his mother. Such was the good fortune of the mother of German mathematician and astronomer Johannes Kepler.[30] Kepler defended his mother over the months of her trial in person, and he intervened with learned men to vouch for her veracity and so save her from torture. He won her acquittal in the end, although she soon died, mortally weakened from her long imprisonment.

FIGURE 10.3: Leonardo da Vinci. *Self-portrait as an Old Man,* ca. 1510–1513. Wikimedia Commons: http://com mons.wikimedia.org/wiki/File:Leonardo_self.jpg.

Even widows who escaped the ferocity of the witch charge were vulnerable to poverty. Historian David Herlihy tracks in compelling statistical tables the decline in wealth experienced by Florentine women with each increasing category of age, pronouncing that the story the numbers tell is one of the immiseration of widows.[31] In northern European societies, where women were more likely to have the lifetime use of a portion of their husband's wealth, the problem of contempt remained, if that of poverty was less. Erasmus, uniquely, defended the widow who wished to remain alone, the mistress of her house. She could pledge herself to chastity, the proper memorial to her love for her husband, and engage in charitable work—especially, urges this intellectual whose student days had been bleak, the relief of young students. "Do not despair, widows, because you appear in this world to be powerless and oppressed, for you have . . . consolation and security in Christ."[32]

PROSTITUTES AND JEWS

Like the witch, Renaissance societies targeted prostitutes and Jews for hatred. Like the witch's body, most often perceived as that of an elderly woman, the bodies of these outcasts were imagined as deformed. A prostitute, though outwardly beautiful, was inwardly hideous—or indeed, once syphilis had been introduced in the late fifteenth century, was conspicuously disfigured. A Jew, whatever his or her real appearance, was in the European imagination constructed as a loathsome figure, with pallid skin and protuberant nose. In both cases, to match an imagined existential deformity, Renaissance societies often imposed an identifying sign to be worn prominently on the body of the vilified person. In modern times, the hooker's revealing getup and the yellow star with which the Nazis labeled the Jews they would attempt to annihilate are the descendants of these earlier hostile badges.

Prostitutes were necessary, yet reviled members of Renaissance cities. Often their brothels were cautiously located outside the walls. Elsewhere, prostitution was openly acknowledged and even fostered. In the German lands, Frankfurt led the way in 1396, Nuremberg by 1400, Munich by 1433, Memmingen (in Bavaria) by 1454, and Strasbourg by 1469. In France, Dijon opened its "Great House" in 1385, and Toulouse operated a municipal brothel from 1363 or 1372. In Italy, public brothels were opened in Venice in 1360, in Florence in 1403, and in Siena in 1421. Venice, notoriously, was bursting with prostitutes. A decree of 1460 ordered all the whores of Rialto, the principal island of that island city, be confined to one house and kept under guard.[33] Another decree of 1543 lamented that "there are now excessive numbers of whores in this our city," who "have put aside all modesty and shame" and mix with "our noble and citizen women."[34]

The effort to contain prostitution mounted across Europe in the sixteenth century, fueled by the Reformations, both Protestant and Catholic. Fear of the plague, venereal disease, and crime, as well as concern for the souls of male exploiters and female enablers, transformed the prostitute from an urban worker to despised species. Officials shut down the public brothels in most regions of Europe. They survived in the shadows as private, shameful, and expensive facilities. Brothel keeping was criminalized; in France, the rape of a prostitute was no longer deemed a crime.[35]

All these regulations concerned ordinary prostitutes. In Italy a new category of whoredom arrived with the courtesan: *cortigiana*, in Italian, linguistically the feminine form of *cortigiano*, or courtier, yet very different—the former groomed to amuse, the latter a male aristocrat resident at a noble or royal court. These were women of elevated social skills—they might be witty conversationalists or fine musicians—who served such upscale populations as the presumably celibate clerics of Rome and those nobles in Venice who chose not

FIGURE 10.4: Jews. Wellcome V0015823.

to marry to prevent the dispersion of patrimonial wealth. The homes of the
most famous courtesans were salons where men resorted for witty conversation and to see and be seen.

As French historian Jacques Rossiaud describes her, the courtesan "was
richly attired and she lived in respectable street; she . . . did not even keep a
private bordello but received her admirers and paid 'visits' to important personages . . . Far from being 'common to all,' she was mistress or concubine to
only a few, confounding the comfortable categories of traditional typology."[36]
She lived in lavishly furnished apartments, like Julia Lombardo, whose residence included three large bedrooms, a receiving room, a maid's room, and

FIGURE 10.5: Prostitute, 1524. Wellcome V0042062.

overflowing storerooms with chests full of linens, rugs, clothing, shoes, gloves, stockings, purses, and sixty-four fine white *camicie* (shirts).[37]

Among the 215 women listed in the 1570 *Catalogue of All the Principal and Most Honored Courtesans of Venice*, one was Veronica Franco, one of the foremost poets of Italy.[38] Franco's poems candidly discuss her own experience of love. Yet she did not recommend prostitution as a way of life. In the 1570s, when a courtesan acquaintance proposed to introduce her own daughter to the career of prostitution (as Franco had herself been introduced to the profession), Franco discouraged her:

> You know how many times I have begged and warned you to protect her virginity; and since this world is so dangerous and frail and the homes of poor mothers are not immune from the temptations of lusty youth, I showed you how you can shelter her from danger and assist her in settling her life decently and in such a way that you will be able to marry her honestly.[39]

The life of a prostitute, she argued, was the worst possible one for a woman: "To become the prey of so many, at the risk of being despoiled, robbed, killed . . . exposed to so many other dangers of receiving injuries and dreadful contagious diseases . . . What riches, what comforts, what delights can possibly outweigh all this? Believe me, of all the world's misfortunes, this is the worst."[40]

If the prostitute's body was thus at risk at the hands of lustful men, the bodies and prospects of the Jews of Europe were safe at no time in its history—perhaps not even now. Entering Europe in late antiquity as merchants, regarded as strangers wherever they settled, Jews were the ethnic minority most widely encountered in European cities. Gathering in the shadow of the town synagogue, they maintained their own social and religious customs while sharing wherever possible in the surrounding cultural world.

They were not always permitted to do so. As Christian Europe developed its own merchants to carry on functions for which Jewish traders were once valued, resentment swelled against perceived outsiders. Christians blamed the Jews in their midst for accidents or disasters, or held them guilty of imaginary and horrible crimes. If a child died or disappeared under mysterious circumstances, people blamed the Jews. Such ritual murder charges became "blood libels" when Christians accused Jews of mixing the blood of these allegedly kidnapped and murdered children with the unleavened bread, or matzoh, used during the Jewish holy days of Passover. Jews were also vulnerable in times of plague—not only to the disease but also to rioting neighbors who tried to purge the contagion by killing Jews.

Although churchmen sometimes participated in anti-Jewish pogroms, the official position of the Church forbade the persecution of this still despised minority. In 1272, citing the precedents of earlier pontiffs, Pope Gregory X (r. 1271–1276) reprimanded those who raised the blood libel: "And most falsely do these Christians claim that the Jews have secretly and furtively carried away these children and killed them, and that the Jews offer sacrifice from the heart and blood of these children." He added, "Nor should Christians attempt the forced conversion of Jews" nor "seize, imprison, wound, torture, mutilate, kill or inflict violence on them."[41] Gregory's successor, Pope Clement VI (r. 1342–1352), reigning during the time of the Black Death, similarly rebuked those who blamed the Jews for the plague: "Recently, . . . it has been brought to our attention . . . that numerous Christians are blaming the plague . . . on poisonings carried out by the Jews at the instigation of the devil." None, he ordered, should "dare to capture, strike, wound or kill any Jews or expel them from service on these grounds."[42]

Even as the Church forbade physical attacks against Jews and deplored the false accusations against them, anti-Semitism burgeoned in the communities of medieval Europe. Jews were expelled from England and France (in 1290 and 1306, respectively), from many cities in the German lands (Nuremberg, 1349; Cologne, 1424; Speyer, 1435), and from Spain and Portugal (1492 and 1496, respectively), where the Inquisition notoriously persecuted those Jews who had converted for returning to the observance of Jewish rituals.

Jews remained in Italy, where they constituted a tiny minority of Italians during the Renaissance: about 20,000 in the late sixteenth century, most living in Rome and the Papal States where they had resided since antiquity.[43] The different cities of Italy had different policies with regard to the Jews. In general, they were welcome when their supplies of capital made them useful as sources of loans, or to provide pawnbroking services for the populace. When they were no longer needed for these purposes, they might be expelled. At times, anti-Semitism spiked, as at Trent in 1475, when a blood libel charge fomented a cruel pogrom.[44]

This rich Jewish cultural life in Italy flourished despite the harsh regulations imposed on Jewish residents. A Florentine regulation of 1463 required that all Jews, "male or female above the age of twelve, . . . wear a sign of [a yellow] O . . . on the left breast, over the clothing in a visible place; it shall be at least one foot in circumference and as wide as the thickness of a finger." In addition, the city allowed no more than seventy Jews "of both sexes, masculine and feminine, large and small" to practice the trade of moneylending. Nevertheless, all religious practices were permitted in this decree, and religious and scholarly books might be "possessed, read, studied, and copied."[45]

In Venice in 1516, the Jews were required to live in an area of the city called the "Ghetto"—the first such place to which European Jews were confined.[46]

"Given the urgent needs of the present times," the degree of founding reads, "the . . . Jews have been permitted to come and live in Venice . . . but no god-fearing subject of our state would have wished them . . . to disperse throughout the city, sharing houses with Christians and going wherever they choose by day and night, perpetrating all those misdemeanours and detestable and abominable acts which are generally known and shameful to describe."[47]

The Venetian dialect word "ghetto" (derived from gettare, "to cast") originally designated an iron foundry. Ghetto residents (there were as many as 5,000 at its height) had frequent and sometimes cordial relations with Christian merchants, scholars, and patricians. Within their walls, they freely followed their own rites (requiring the importation and preparation of special foods in order to observe the dietary laws). Yet those walls were also a prison: the gates were locked at nights and during certain Christian holidays, purportedly for the security of ghetto residents. Over later centuries, ghettoes were established in other cities, eventually becoming a commonplace of Jewish social life within European Christendom.

Amid these many restrictions, the Jews of Italy persevered in their ancient traditions while creating new works of philosophy, art, and literature. They contributed, indeed, to the Renaissance culture of their neighbors, which was, in its Christian assumptions, in many ways antithetical to the Jewish way of life. Humanist students of Hebrew necessarily acquired that language from Jewish teachers. Jewish philosophers, in turn, had an affinity for the Platonic and Neoplatonic ideas that infused Renaissance humanist thought. The *Dialoghi dell 'amore* (*Dialogues on Love*, 1535) by the Sephardic Jew Leone Ebreo (Judah Abravanel), a principal work of sixteenth-century Platonism, influenced noted philosophers of the next century, including Giordano Bruno and Baruch Spinoza.

Jewish authors also wrote in the key humanist genres of poetic and epistolary composition. Sara Copio Sullam (ca. 1592–1641), a Jewish woman thinker of the seventeenth century and resident of the Venetian ghetto, was a poet and a *salonière*, in whose home the learned gathered to talk about the latest ideas. Her most famous work is a defense of her belief in the immortality of the soul, against a Christian accuser. The inscription on her tomb in the Jewish cemetery of San Nicolò del Lido, composed by the scholar Leon da Modena, her teacher, reads in part:

A lady of great intelligence
Wise among women
Supporting those in need
The poor found a companion and friend in her . . .
May her soul enjoy eternal beatitude.[48]

MOTHERS AND SAINTS

Prostitutes and Jews were marginal figures in Renaissance society, but mothers and saints—two alternative careers for women—were central. The bodies of these two idealized groups of women were essential to their identity, and in quite opposite ways. The fertile body of the mother was celebrated in representations as diverse as Piero della Francesca's painting of a pregnant Madonna proudly displaying her swollen belly and the abundant, crude woodcuts of a woman on a birthing stool (supported by a servant and attended by a midwife) that adorned the first modern obstetrical manuals. The starved body of the saint, especially the female saint, in contrast, was depicted on altarpieces and celebrated in the hagiographies that circulated widely among the faithful. Both ideals came together in the omnipresent figure of the Virgin Mary, the most frequently depicted subject of the great art of the Renaissance, who was both mother and virgin.

Representations from the thirteenth through sixteenth centuries of the virgin mother Mary holding the infant Christ testify to an intensified awareness of the maternal in that era.[49] At first depicted as a stiff figure holding an equally rigid, elderly Jesus, Mary was transformed in the Gothic figure into a blooming young woman with a soft, beaming, naked child wrapped in her arms—the latter an image "wholly unknown" before 1300.[50]

This mother–child image became a principal subject of Italian Renaissance painting and sculpture in renditions by Donatello, Michelozzo, Luca della Robbia, Fra Angelico, Giovanni Bellini, among others, as even the casual tourist tripping through the corridors of European museums becomes aware. The painter Raphael, a specialist in the type, alone produced more than thirty such representations of the Madonna and child, cloying to some modern tastes but in high demand in their time.

It would be naive to assume that these lovely depictions of mother–child attachment factually document contemporary familial relations. It could as well be, as one historian suggested, that they record the frustrations of artists deprived of maternal warmth in infancy![51] At a minimum, though, they testify to an interest on the part of artists and audience in birth, fertility, and nurture. These maternal images were often mounted in private homes, and indeed in bedrooms, where they served not only as objects of religious devotion but also as spurs to affection and reminders of lineage responsibilities.[52]

Mary was not only a pure mother, impregnated not by man but by the Holy Ghost, but a pure mother: her whole being was motherhood. She was the New Testament reiteration of Hannah, whose prayer to God for a child resulted in the birth of Samuel, offered to the Lord as a servant of the high priest in the temple as soon as he had been weaned (1 Samuel). Both existed to give birth to remarkable sons, much as women everywhere in the premodern world were valued primarily for their capacity to give birth.

FIGURE 10.6: Childbirth, Midwives, ca. 1583. Wellcome V0014910.

Just as mothers gave birth to infants in a flux of blood, they nurtured them with a flow of milk—or so, trumpeted a host of male experts, they should.[53] Driving this stampede of breast-feeding advocacy was a belief, rooted in biological concepts virtually universal in premodern times, of the identity between blood and milk—and spirit. Blood was seen as a liquid life force. In the female body, during pregnancy, it nurtured the fetus; after birth, it became transformed into life-giving milk, poured forth from the breasts.[54] The milk was of the same substance as the blood, and both embodied the spiritual essence of the mother. The nurse—maternal or alien—delivered not only sustenance, but soul, character, life itself, to the nursing.

The identification of blood and milk, and of both with spirit, meant that mother's milk was to be preferred to any other, not just as a matter of health or sentiment, but because of the innermost resemblance of that milk to the blood that had engendered the nursing baby. So Cicero wrote, as the Romans generally believed, that an infant sucks in virtues with his mother's milk.[55] That notion is embraced by Renaissance humanists for whom the shaping of human character was a preeminent concern.

Accordingly, in his 1415 treatise *On Marriage*, the Venetian humanist Francesco Barbaro (1390–1454) explains at length how maternal character was transmitted through breast milk, making even a highly qualified servant a poor substitute.[56] The Spanish humanist Juan Luis Vives, in his 1523 treatise *The Education of a Christian Woman*, sees the act of breast-feeding as an expression of maternal commitment to the child.[57] In his colloquy "The New Mother," Erasmus depicts the harshness of sending the baby away from his mother: "[I]sn't it a kind of exposure to hand over the tender infant, still red from its mother, drawing breath from its mother, crying out for its mother's care—a sound said to move even wild beasts—to a woman who perhaps has neither good health nor good morals and who, finally, may be much more concerned about a bit of money than about your whole baby?"[58]

In Tudor–Stuart England and its colonial offspring, instructions to mothers in handbooks, sermons, and other literature become abundant. Protestant spokesmen joined Catholics in exhorting women to meet their responsibilities to their offspring, especially that of nursing, as did William Gouge in his many sermons and his treatise *On Domesticall Duties*.[59] New England churchmen preached the same message in sermons and treatises, like Cotton Mather, who denounced women who did not nurse their own infants as "dead while they live."[60]

Yet, in Christian Europe, defying the flood of learned opinion in support of maternal breast-feeding, women of the high aristocracy and urban elites—or rather their husbands—hired wet nurses to care for infants until the age of weaning: often at eighteen months for girls, and twenty-four for boys.[61] Various theories are offered for the Western preference for wet-nursing: the mother's laziness or desire to retain her figure; the husband's desire for sexual access to his wife (forbidden during lactation), or his desire not only for sexual contact, but in the interest of his lineage, for the consequent new pregnancy and birth.

Advocates of lactation, therefore, had to overcome the resistance of elite mothers reluctant to nurse and elite husbands concerned with their sexual success. It was mothers who tipped the balance. In seventeenth-century England, for the first time, elite women began to encourage breast-feeding to each other and to their daughters.[62] From the eighteenth century, generally in the advanced societies of the West, mothers embraced the role of nurse, as well as birth-giver, to their children. In 1758, the naturalist Carl Linnaeus identified the maternal breast as the "categorizing organ" of the order of animals he named, after it, *mammalia* (mammals).[63]

FIGURE 10.7: Birth of St. Edmund, Wet Nursed. Wellcome V0014976.

Mothers were defined by their breasts and wombs, those organs essential for childbirth and nurture. The denial of breasts, wombs, and the body wholesale, defined the holy women and female saints whose explosive presence characterizes the era, inexplicably coincident with its intellectual and scientific innovations.

Female sanctity during the Renaissance was above all penitential and ascetic: holiness was constituted by constant prayer and genuflection, sleeplessness and torment, self-mutilation and self-starvation. Beneath her hair shirt, Colomba da Rieti wore chains studded with sharp points around her hips and her breasts.[64] Francesca Bussa de' Ponziani thrashed herself with a cord studded with nails, and dripped hot candle wax on her genitals so that her wounded organs could not feel pleasure.[65] Margery Kempe wept, and Angela of Foligno screamed uncontrollably.[66] Dorothy of Montau devoted her nights to spiritual exercises, imitating Christ: "shuffling about on her knees, crawling, arching her body in the air with her forehead and feet on the floor, joining her hands in front of herself in the form of a cross, falling on her face with her hands behind her back as if they were bound."[67]

FIGURE 10.8: Donatello, *Mary Magdalene*, Maddalena, sculpture by Donatello in Opera del Duomo Museum Museo in Florence, Italy ca. 1453–1455. Wikimedia Commons: http://commons.wikimedia.org/wiki/Image:Maddalena_di_donatello.jpg.

Two Renaissance saints noted for their asceticism were also noted for the impressive visions with which they were graced, the reward of their self-abnegation. The austerities of the fourteenth-century Italian saint Catherine of Siena were so great—she fasted so strictly that her stomach could retain "not even a bean"[68]—that she died of starvation at age thirty-three. Her brief lifetime, however, was long enough for heroic charitable endeavors and public action: she famously summoned Pope Gregory XI to return from Avignon to Rome and end what was known as the Babylonian Captivity of the church. It was also graced by divine visions, as when Christ took her as his bride:

> with his own right hand [he] took [her] right, and having a golden ring adorned with four most precious pearls and a very rich diamond, put it on [her] ring finger, saying these words: "Behold I betrothe you to me in a faith which will endure from this moment forward forever immutable, until the glorious wedding in heaven, in perfect conjunction with me, in eternal marriage, when face to face you shall be allowed to see me and enjoy me entirely."[69]

In sixteenth-century Spain, a land dominated by the Inquisition and the demands for conformity asserted by the Council of Trent, Teresa of Avila (1515–1582) founded a reformed order of Carmelite nuns, pledged to a maximum standard of austerity—sufficient to allay the suspicions of sharp-eyed inquisitors. Her followers adopted vows of absolute poverty, ceremonial flagellation, and discalceation, or barefootedness, that is, the wearing of sandals rather than shoes. Like Catherine, Teresa's piety was enriched by visions, including a famous one later represented in sculpture by Gian Lorenzo Bernini, now in the church of Santa Maria della Vittoria in Rome. Teresa saw an angel from God in whose hands was

> a great golden spear, and at the iron tip there appeared to be a point of fire. This he plunged into my heart several times so that it penetrated to my entrails. When he pulled it out, I felt that he took them with it, and left me utterly consumed by the great love of God. The pain was so severe that it made me utter several moans. The sweetness caused by this intense pain is so extreme that one cannot possibly wish it to cease, nor is one's soul then content with anything but God.[70]

The holy women of the Renaissance raised themselves beyond the body by their austerities. They received the grace of God, however, impressed upon their bodies, which experienced union with the divine in the flesh as in the spirit.

WARRIORS AND COURTIERS

As the bodies of women performed the tasks of mothers and saints, so the bodies of men, and not only their bodies, performed the tasks of warrior and courtier, skilled on the battlefield and in the salon, expert in arms and in letters.

In the rough centuries that followed the collapse of Roman authority in Europe, war was the province of the nobleman and it was the whole of what

he did. Violence was everywhere and all the time. When there was no assembly
for battle, there was preparation for battle: games in the tiltyard and hunts in
the forest were also military endeavors. The knight was a battle-ready soldier
welcomed into that circle by a noble superior as a reward, theoretically, for
valor displayed in combat. Central to his identity was his horse, as the terms
for knight in the European languages attest: *cavaliere, caballero, chevalier, Rit-
ter*, all are horsemen.

In Italy, populated by many knights and not too many dukes or kings, a
knight could make himself a lord—a *signore*. Born to a family recently come
to power in Verona, Can Grande ("Big Dog") della Scala (1291–1329) was
knighted as an infant, and assumed sole rule at twenty. A champion of the
Ghibellines against the Guelphs in the endless factional struggles of the era, in
less than twenty years he cobbled together a northern Italian empire, adding to
Verona, at least for his lifetime, the important cities of Vicenza, Treviso, and
Padua, plus smaller localities. At his death in 1329—by poison, most likely—
Can Grande was buried in Verona in a grand late Gothic sepulcher topped by
an equestrian statue of the invincible knight in full dress—one of the first such
monuments since ancient times.

For all his ferocity, Can Grande was also a lover of the arts, better known
for his patronage of the poet Dante Alighieri than for his conquests. And in
Paradise, Dante's guide Beatrice pointed out the "great Lombard" whose generosity would win him eternal fame: "His generosity is yet to be so notable that
even enemies will never hope to treat it silently."[71]

A century later, military power in Italy lay in the hands, as political philosopher Niccolò Machiavelli so expertly observed, of either petty princes, who
sought to protect and expand their territories, or men who had no state: who
"being bred to arms from their infancy, were acquainted with no other art, and
pursued war for emolument, or to confer honor upon themselves."[72] These
were the *condottieri*, military captains who negotiated a *condotta*, or contract,
with a city or ruler who required a mercenary force—and then, once the contract had expired or even before, insouciantly signed another with someone
else, and fought for the other side. True military experts, networks of condottieri fought the wars of northern Italy that raged from the 1420s to the Peace of
Lodi of 1454, some loosely gathered in the groups of *sforzeschi* and *bracceschi*
after their primal leaders of the Sforza and Braccio clans.

Prominent among these were two who fought much of their time for Venice,
who lived to see their seventieth birthdays: Gattamelata and Colleoni, "Tabby
Cat" and "Big Balls." Gattamelata, born Erasmo da Narni, son of a miller,
joined the Venetian forces as captain general in the early 1430s.[73] In 1439,
aged nearly seventy, he suffered an apoplectic stroke. He retired to Padua supplied with a generous pension from the Venetian government, which also gave
him a splendid funeral at his death in 1443, attended by the doge and adorned

by humanist eulogies. His family commissioned Donatello to erect a bronze equestrian statue of the hero in front of the cathedral of San Antonio in Padua (1447–1450), and they built a splendid tomb within the cathedral.

Meanwhile, Bartolommeo Colleoni, the orphaned son of a minor nobleman, had been trained in the military profession and sought preferment in the Venetian military, winning at last, in 1454, the coveted title of captain general.[74] The rewards of his Venetian employment permitted him to live like a great lord, surrounded by objets d'art, at his castle in Malpaga. His will bequeathed 300,000 ducats to Venice to pay for a great bronze equestrian statue to be erected in the Piazza San Marco. Upon his death, Venice took the money and commissioned the statue—but erected the mammoth statue instead on the smaller square in front of the church of SS. Giovanni e Paolo, burial site of many of the doges.

FIGURE 10.9: Donatello, statue of condottiere Gattamelata, 1453. Wikimedia Commons: http://en.wikipedia.org/wiki/Image:Gatta melata.jpg.

The principal condottiere of the age, however, was Francesco Sforza, the last, triumphant leader of the *sforzeschi*.[75] Illegitimate son of Muzio Sforza, Francesco reached the zenith of achievement for the Renaissance condottieri: a state of his own, the duchy of Milan, that he ruled and bequeathed to his descendants—who held it for a while before they lost all that he had gained. Arrived from earlier conflicts in Lombardy, he married, in 1441, the illegitimate daughter (and sole direct heir) of Filippo Maria Visconti, ruler of Milan. In 1447, when Visconti died intestate, Sforza made his claim for that inheritance, as did other powerful contenders. But the Milanese revolted, declared a republic, and, imprudently, hired Sforza as their captain general. Having turned on his employers and starved the city to submission, Sforza triumphed in 1450.

Sforza, too, was a patron, a builder of a vast castle complex, an admirer of learning, who kept historians on his payroll, famously rewarding his court humanist Francesco Filelfo with a jeweled cup of immense value.[76] Like Can Grande, like Gattamelata, like Colleoni, he, too, was to have a memorial horse. Leonardo da Vinci—also in his circle of clients—had modeled a prototype when the French invaded Milan, and destroyed it.[77]

On the battlefield, the condottiere sat athwart his living horse of flesh and bone and sinew, and in the public square, his imagined horse of marble or bronze, the only pedestals mighty enough to sustain the weight of his power. Horse and man, fused, rode as one.

Muscular men, men on horses, these warriors were yet men of intellect and discernment. They pioneered a type of elite male who would exhibit a still greater refinement, the type that would define the aristocracy of early modern Europe and beyond, until the mud of the trenches of the Western Front consumed it at last: the courtier.

The courtier knew how to fight. But he knew, too, to shed the soil of battle before he entered a lady's drawing room, where the skills that were required of him were grace, wit, and that attitude of insouciant excellence summed up in the word *sprezzatura*, the untranslatable term central to the multiple themes of *The Book of the Courtier* by Baldassare Castiglione. The powerful horseman came to court, where he was expected to dance properly and comport himself elegantly. By the end of the Renaissance, a new ideal had emerged, the fusion of rough soldier and delicate savant: the courageous and honorable gentleman and scholar who could with equal ease wield sword and pen.

In the hilltop palace at Urbino built by of one of these warrior princes, under the rule of one of his weaker descendents, high-born ladies and important men—diplomats, authors, rulers—gathered in the evenings for games and conversation. Castiglione memorialized in his *Book of the Courtier* one such session, a series of four evenings that had occurred in 1507. Composed during the years 1513 to 1518, the *Courtier* was not to be published until 1528, one year before the author's death. No mere document of those long-ago conversations, Castiglione's book is a reinvention of the events he describes and a meditation on the courtly society of which they are the epitome.

Two things are important: first, the setting, and second, what was said. The setting is a grand hall where the participants, both men and women, sit in a circle. The duke was not well, and he retired after dinner. Presiding over the guests is the duchess, Elisabetta Gonzaga, together with her companion and aide, the vivacious lady Emilia Pia. The two women guide the evening sessions of music, dancing, games, and conversation. The participants are "countless . . . very noble gentlemen,"[78] lords and prelates, learned and powerful men.

The prototype for this gathering is the court of love, the product of French medieval society, where a lady was elevated as the object of desire of all the knights gathered at her court, and ruled over them, transmuting their violence and crass sexuality into poetry and song. Its descendent is the Enlightenment salon, where *philosophes* gathered in the drawing rooms of cultivated ladies to engage in learned repartee and imagine new worlds. Castiglione sketches the Renaissance fulcrum between these two moments of European culture, where the games of old still linger and the discourse of philosophy is fresh and new.

What is said over the four nights of that conversation in Urbino concerns the construction of a new masculine ideal: not warrior, not priest, not sage, not fop, but a man capable of action and reflection, a cultivated and effective leader. The courtier should be of noble birth (there are rare exceptions) and attractive in appearance: "endowed by nature not only with talent and with beauty of countenance and person, but with that certain grace which we call an 'air,' which shall make him at first sight pleasing and lovable to all who see him."[79] He must be accomplished in arms, as well as in all kinds of related skills—swimming, jumping, running, throwing stones. But he should not be a sullen warrior unequipped for the life of the court, who might be reproved by a lady thus: "I should think it a good thing, now that you are not away at war or engaged in fighting, for you to have yourself greased all over and stowed away in a closet along with all your battle harness, so that you won't grow any rustier than you already are."[80]

Above all, the courtier is to wear his physical attainments lightly, as well as all his others, cultivating a kind of nonchalance that seems, in pretending that the task is easy, to make its accomplishment all the more impressive: "to practice in all things a certain *sprezzatura*, so as to conceal all art and make whatever is done or said appear to be without effort and almost without any thought about it."[81] The term *sprezzatura*, as famous as it is untranslatable, and which the speaker himself introduces conscious that it is a neologism, implies an attitude of easy superiority, an elegant disdain for the difficulty of the task.

These are by no means the courtier's only accomplishments or concerns, which occupy the wandering discourses of the next three nights. Among these is a discussion of women: admired because, reflecting contemporary discussions, it awards a status to women higher than the norm; critiqued because it still fails to envision a gender-neutral society. Whereas the courtier is to possess "a certain solid and sturdy manliness," the lady must display "a soft and delicate tenderness, with an air of womanly sweetness in every movement, which, in her going and staying, and in whatever she says, shall always make her appear the woman without any resemblance to a man."[82] This definition of the difference between courtier and lady is stark. It would characterize the assumptions of European society for the next three centuries.

So even as the courtier is presented as a new ideal, the court lady is not freed. But that is not the point. More importantly, Castiglione presents women as the tamers of men. In their company, the warrior puts aside his armor, sharpens his wit, restrains his tongue, subdues his instincts, and disciplines his body. Here there is no advance for women, but there is a giant transformation of men.

Fragile youths and fading dotards, despised prostitutes and outsider Jews, prolific mothers and anorexic saints, tested warriors and agile courtiers, the people of the Renaissance in their diversity of social rank and cultural attainment are bodies and selves intertwined in the mysterious and capricious patterns that our own experts have yet to decipher, that are perhaps indecipherable—as may be our own.

NOTES

Introduction

1. Elizabeth Eisenstein, *The Printing Press as an Agent of Change* 2 vols. (Cambridge: Cambridge University Press, 1979); Adrian Johns, *The Nature of the Book: Print and Knowledge in the Making* (Chicago: University of Chicago Press, 1998). Because the essays in this volume have such full bibliographical apparatus, I have kept the notes to this introduction to a minimum.
2. Jonathan Sawday, *The Body Emblazoned: Dissection and the Human Body in Renaissance Culture* (London: Routledge, 1995); K. B. Roberts and J.D.W. Tomlinson, *The Fabric of the Human Body: European Traditions of Anatomical Illustration* (Oxford: Clarendon Press, 1992); Andrea Carlino, *Books of the Body: Anatomical Ritual and Renaissance Learning* (Chicago: University of Chicago Press, 1999).
3. Roger Chartier, "The Practical Impact of Writing," in *A History of Private Life: Passions of the Renaissance*, ed. Roger Chartier (Cambridge, MA: Belknap Press, 1989). Many of the other chapters in this volume touch on themes that this present volume addresses.
4. A recent study, sympathetic to the Catholic Church, but by two leading Galileo scholars, is William R. Shea and Mariano Artigas, *Galileo in Rome: The Rise and Fall of a Troublesome Genius* (Oxford: Oxford University Press, 2003).
5. The classic study is still Frances A. Yates, *Giordano Bruno and the Hermetic Tradition* (Chicago: University of Chicago Press, 1964).
6. A good example of the genre is Anthony Pagden, *European Encounters with the New World* (New Haven, CT: Yale University Press, 1993).
7. The classic monograph is Margaret T. Hodgen, *Early Anthropology in the Sixteenth and Seventeenth Centuries* (Philadelphia: University of Pennsylvania Press, 1964); a good sample of primary literature is in J. S. Slotkin, ed., *Readings in Early Anthropology* (London: Methuen, 1965).
8. A good introduction to this worldview is Frances A. Yates, *The Rosicrucian Enlightenment* (London: Routledge and Kegan Paul, 1972).

9. Steven Shapin, *The Scientific Revolution* (Chicago: University of Chicago Press, 1996); Stephen Gaukroger, *The Emergence of a Scientific Culture: Science and the Shaping of Modernity* (Oxford: Clarendon Press, 2006).

10. Owsei Temkin, *Galenism: Rise and Fall of a Medical Philosophy* (Ithaca, NY: Cornell University Press, 1973); Owsei Temkin, *Hippocrates in an Age of Pagans and Christians* (Baltimore: Johns Hopkins University Press, 1991); David Cantor, ed., *Reinventing Hippocrates* (Aldershot, UK: Ashgate, 2002).

11. Harvey's debt to Aristotle is most systematically elucidated in Walter Pagel, *William Harvey's Biological Ideas* (Basel, Switzerland: Karger, 1967); for Descartes, see Stephen Gaukroger, *Descartes: An Intellectual Biography* (Oxford: Oxford University Press, 1995), and, as a wonderfully accessible introduction, A. C. Grayling, *Descartes: The Life of René Descartes and Its Place in His Times* (London: Free Press, 2005).

12. The Eliot quotation comes from *Sweeney Agonistes*.

13. For an archivally rich study of midwifery, see Doreen Evenden, *The Midwives of Seventeenth-Century London* (Cambridge: Cambridge University Press, 2000).

14. A recent well-informed survey is John Kelly, *The Great Mortality: An Intimate History of the Black Death* (London: Fourth Estate, 2005).

15. A wide-ranging history of pain and its relief is Thomas Dormandy, *The Worst of Evils: The Fight against Pain* (New Haven, CT: Yale University Press, 2006); Roselyne Rey, *History of Pain* (Paris: La Découverte, 1993) is also illuminating.

16. For the early history of syphilis, see Jon Arrizabalaga, John Henderson and Roger French, *The Great Pox: The French Disease in Renaissance Europe* (New Haven, CT: Yale University Press, 1997).

Chapter 1

1. Bernard McGinn, *Antichrist: Two Thousand Years of the Human Fascination with Evil* (New York: Columbia University Press, 2000).

2. See, for example, Helen King, "As If None Understood the Art That Cannot Understand Greek: The Education of Midwives in Seventeenth-Century England," in *The History of Medical Education in Britain*, ed. Vivian Nutton and Roy Porter (Amsterdam: Rodopi, 1995), 184–98; Lisa Forman Cody, *Birthing the Nation: Sex, Science, and the Conception of Eighteenth-Century Britons* (Oxford: Oxford University Press, 2005); and Laura Gowing, *Common Bodies: Women, Touch and Power in Seventeenth-Century England* (New Haven, CT: Yale University Press, 2003).

3. See, for example, Philippe Ariès, *Images of Man and Death*, trans. Janet Lloyd (Cambridge, MA: Harvard University Press, 1985); Edelgard E. DuBruck and Barbara I. Gusick, eds., *Death and Dying in the Middle Ages* (New York: Peter Lang, 1999); and Bruce Gordon and Peter Marshall, eds., *The Place of the Dead: Death and Remembrance in Late Medieval and Early Modern Europe* (Cambridge: Cambridge University Press, 2000).

4. David Cressy, *Birth, Marriage, and Death: Ritual, Religion, and the Life-Cycle in Tudor and Stuart England* (Oxford: Oxford University Press, 1997).

5. I refer to the "unborn child" because it was the term used most commonly during the early modern period. My decision is strictly historical and is not informed by the current abortion debate.

6. Jacqueline Marie Musacchio, *The Art and Ritual of Childbirth in Renaissance Italy* (New Haven, CT: Yale University Press, 1999), 25, uses evidence supplied by David Herlihy and Christiane Klapisch-Zuber, *Tuscans and Their Families: A Study of the Florentine Catasto of 1427* (New Haven, CT: Yale University Press, 1985), 277.

7. Doreen Evenden, *The Midwives of Seventeenth-Century London* (Cambridge: Cambridge University Press, 2000); Hilary Marland, trans. and ed., *'Mother and Child Were Saved'. The Memoirs (1693–1740) of the Frisian Midwife Catharina Schrader* (Amsterdam: Rodopi, 1987); Hilary Marland, "Stately and Dignified, Kindly and God-fearing: Midwives, Age and Status in the Netherlands in the Eighteenth Century," in *The Task of Healing: Medicine, Religion and Gender in England and the Netherlands 1450–1800*, ed. Hilary Marland and Margaret Pelling (Rotterdam, Netherlands: Erasmus Publishing, 1996), 271–305; and Nina Rattner Gelbart, *The King's Midwife: A History and Mystery of Madame du Coudray* (Berkeley: University of California Press, 1998).

8. Irvine Loudon, *Death in Childbirth: An International Study of Maternal Care and Maternal Mortality 1800–1950* (Oxford: Clarendon, 1992); Roger Schofield, "Did the Mothers Really Die? Three Centuries of Maternal Mortality in the World We Have Lost," in *The World We Have Gained*, ed. Lloyd Bonfield, R. M. Smith, K. Wrightson, and P. Laslett (Oxford: Blackwell, 1986), 231–60. See also B. M. Willmott Dobbie, "An Attempt to Estimate the True Rate of Maternal Mortality, Sixteenth to Eighteenth Centuries," *Medical History* 26 (1982): 79–90.

9. Adrian Wilson, *The Making of Man-Midwifery: Childbirth in England 1660–1770* (London: UCL Press, 1995), 15–19.

10. Lisa Forman Cody, "Living and Dying in Georgian London's Lying-In Hospitals," *Bulletin of the History of Medicine* 78 (2004): 309–48. Her work refutes the negative depiction of lying-in hospitals in Roy Porter, *The Greatest Benefit to Mankind: A Medical History of Humanity* (New York: Norton, 1977), 266, and Irvine Loudon, *The Tragedy of Childbed Fever* (New York: Oxford University Press, 2000), 59.

11. The fear of dying in childbed pervades, for example, Elizabeth Jocelin's *The Mothers Legacie to Her Unborn Childe,* 2nd ed. (London: John Haviland, 1624). In this case, Jocelin's prediction proved correct. See also Adrian Wilson, "The Perils of Early Modern Procreation: Childbirth with or without Fear?" *British Journal for Eighteenth-Century Studies* 16 (Spring 1993): 1–19.

12. Jacques Gélis, *La sage-femme ou le médecin: une nouvelle conception de la vie* (Paris: Fayard, 1988), 330–32, notes that the rate of publication increased during the seventeenth and eighteenth centuries.

13. Lianne McTavish, *Childbirth and the Display of Authority in Early Modern France* (Aldershot, UK: Ashgate, 2005).

14. From the second half of the eighteenth century, most obstetrical treatises focused on the mechanism of normal birth and included fewer case studies. For an early example of this change, see André Levret, *L'art des accouchemens* (1753; repr. Paris: Alexandre Le Prieur, 1761).

15. François Mauriceau, *Observations sur la grossesse et l'accouchement des femmes* (Paris: Chez l'Auteur, 1695), 21.

16. Mauriceau, *Observations sur la grossesse*, 19.

17. François Mauriceau, *Des maladies des femmes grosses et accouchées* (Paris: Jean Henault, 1668), 273–76.

18. Wendy Arons, trans., *Eucharius Rösslin: When Midwifery Became the Male Physician's Province* (Jefferson, NC: McFarland, 1994), 87–88.

19. Adrian Wilson, "The Ceremony of Childbirth and Its Interpretation," in *Women as Mothers in Pre-Industrial England,* ed. Valerie Fildes (London: Routledge, 1990), 68–107.

20. Mauriceau, *Observations sur la grossesse,* 10, notes, for example, that he once expelled an uncooperative husband from the lying-in room.

21. Lianne McTavish, "Blame and Vindication in the Early Modern Birthing Room," *Medical History* 50 (2006): 447–64.

22. For an analysis of this debate see McTavish, *Childbirth and the Display of Authority,* 151–53.

23. Wilson, *The Making of Man-Midwifery,* 97.

24. Scipione Girolamo Mercurio, *La Commare o riccoglitrice* (Venice: C. Cioti, 1596).

25. Nadia Maria Filippini, "The Church, the State and Childbirth: The Midwife in Italy during the Eighteenth Century," in *The Art of Midwifery: Early Modern Midwives in Europe,* ed. Hilary Marland (London: Routledge, 1993), 152–75.

26. François Rousset, *Traitté nouveau de l'hysterotomotokie, ou enfantement caesarien* (Paris: D. Duval, 1581), 16–30.

27. Jacques Guillemeau, *De l'heureux accouchement des femmes* (Paris: Nicolas Buon, 1609), 329–34.

28. Mauriceau, *Des maladies des femmes grosses et accouchées,* 356–67; Cosme Viardel, *Observations sur la pratique des accouchemens naturels, contre nature & monstrueux* (Paris: Edme Couterot, 1671), 172–80; Philippe Peu, *La pratique des accouchemens* (Paris: Jean Boudot, 1694), 322–36; Pierre Dionis, *Traité général des accouchemens* (Paris: Charles-Maurice d'Houry, 1714), 310–18; and Percivall Willughby, *Observations on Midwifery,* ed. Henry Blenkinsop (Wakefield, UK: S. R. Publishers, 1972), 340.

29. Roy Porter, *Medicine: A History of Healing* (New York: Marlowe and Company, 1997), 16.

30. Renate Blumenfeld-Kosinski, *Not of Woman Born: Representations of Caesarean Birth in Medieval and Renaissance Culture* (Ithaca, NY: Cornell University Press, 1990), 26.

31. Blumenfeld-Kosinski, *Not of Woman Born,* 104; and Mireille Laget, "La césarienne ou la tentation de l'impossible: XVIIe et XVIIIe siècle," *Annales de bretagne et des pays de l'ouest* 86 (1979), 177–89.

32. McTavish, "Blame and Vindication in the Early Modern Birthing Room," 447–64.

33. Arons, *When Midwifery Became the Male Physician's Province,* 92–93 (original punctuation).

34. Pierre Franco, *Chirurgie,* ed. E. Nicaise (Geneva, Switzerland: Slatkine Reprints, 1975), 239–42. For the original edition see Pierre Franco, *Traité des hernies contenant une ample déclaration de toutes leurs espèces, et autres excellentes parties de la chirurgie* (Lyon, France: Thibaud Payan, 1561).

35. Dionis, *Traité général des accouchemens,* 315.

36. William Smellie, *A Treatise on the Theory and Practice of Midwifery* (London: D. Wilson, 1752), 379–84.

37. Jean-Louis Flandrin, *Families in Former Times: Kinship, Household and Sexuality,* trans. Richard Southern (Cambridge: Cambridge University Press, 1979), 200.

38. Arthur E. Imhof, *Lost Worlds: How Our European Ancestors Coped with Everyday Life and Why Life Is So Hard Today,* trans. Thomas Robisheaux (Charlottesville: University Press of Virginia, 1996), 164–66.

39. Philippe Ariès, *Centuries of Childhood: A Social History of Family Life*, trans. Robert Baldick (New York: Knopf, 1962). See also Patrick H. Hutton, *Philippe Ariès and the Politics of French Cultural History* (Amherst: University of Massachusetts Press, 2004), 92–112.

40. Stephen Ozment, *Ancestors: The Loving Family in Old Europe* (Cambridge, MA: Harvard University Press, 2001); and Andrea Immel and Michael Witmore, eds., *Childhood and Children's Books in Early Modern Europe, 1550–1800* (New York: Routledge, 2006).

41. Jacques Gélis, *Les enfants des limbes: mort-nés et parents dans l'Europe chrétienne* (Paris: Louis Audibert, 2006), 18, 26.

42. Gélis, *Les enfants des limbes*, 75.

43. Gélis, *Les enfants des limbes*, 97.

44. For baptisms by women see Cressy, *Birth, Marriage, and Death*, 17–23. For an early modern example of conditional baptism see Dionis, *Traité général des accouchemens*, 317.

45. Peu, *La pratique des accouchemens*, 179.

46. Cressy, *Birth, Marriage, and Death*, 100.

47. Katharine Park, "The Criminal and the Saintly Body: Autopsy and Dissection in Renaissance Italy," *Renaissance Quarterly* 47 (1994): 1–33.

48. Jonathan Sawday, *The Body Emblazoned: Dissection and the Human Body in Renaissance Culture* (New York: Routledge, 1995), 54–84.

49. Charles Donald O'Malley, *The Illustrations from the Works of Andreas Vesalius of Brussels* (New York: Dover, 1973), 14.

50. Katharine Park, *Secrets of Women: Gender, Generation, and the Origins of Human Dissection* (Brooklyn, NY: Zone Books, 2006).

51. Sawday, *The Body Emblazoned*, 118–19, 123.

52. Anatomical dissections were not always designed to produce knowledge about the body. Roger French, *Dissection and Vivisection in the European Renaissance* (Aldershot, UK: Ashgate, 1999), 78–81, argues that some displays were meant to position anatomy as the basis of medical authority. See also K. B. Roberts and J.D.W. Tomlinson, *The Fabric of the Body: European Traditions of Anatomical Illustrations* (Oxford: Clarendon, 1992); Andrew Cunningham, *The Anatomical Renaissance: The Resurrection of the Anatomical Projects of the Ancients* (Aldershot, UK: Ashgate, 1997); and Andrea Carlino, *Books of the Body: Anatomical Ritual and Renaissance Learning* (Chicago: University of Chicago Press, 1999).

53. French, *Dissection and Vivisection in the European Renaissance*, 193.

54. Cited in French, *Dissection and Vivisection in the European Renaissance*, 237.

55. Ambroise Paré, *Les oeuvres de M. Ambroise Paré* (Paris: Gabriel Buon, 1575), 784.

56. John Stow, *The annales of England faithfully collected*, 2nd ed. (London: Ralfe Newbery, 1592), 1261.

57. Robert Bartlett, *The Hanged Man: A Story of Miracle, Memory, and Colonialism in the Middle Ages* (Princeton, NJ: Princeton University Press, 2004), 6.

58. See, for example, Paré, *Les oeuvres de M. Ambroise Paré*, 783.

59. Donald F. Duclow, "Dying Well: The *Ars moriendi* and the Dormition of the Virgin," in *Death and Dying in the Middle Ages*, ed. DuBruck and Gusick (New York: Peter Lang, 1999), 386.

60. Peter Marshall, *Beliefs and the Dead in Reformation England* (Oxford: Oxford University Press, 2002), 12–18; and Nancy Caciola, "Spirits Seeking Bodies: Death,

Possession and Communal Memory in the Middle Ages," in *The Place of the Dead: Death and Remembrance in Late Medieval and Early Modern Europe*, ed. Bruce Gordon and Peter Marshall (Cambridge: Cambridge University Press, 2000), 66.

61. Jacques Le Goff, *The Birth of Purgatory*, trans. Arthur Goldhammer (Chicago: University of Chicago Press, 1984), 4–11.

62. Dante Alighieri, *The Divine Comedy of Dante Alighieri*, trans. John Aitken Carlyle and Philip H. Wicksteed (New York: Modern Library, 1932).

63. Le Goff, *The Birth of Purgatory*, 11–12, and Penny Roberts, "Contesting Sacred Space: Burial Disputes in Sixteenth-Century France," in *The Place of the Dead*, ed. Gordon and Marshall, 132–33.

64. Marshall, *Beliefs and the Dead in Reformation England*, 48.

65. Ibid., 148–56.

66. Ibid., 265–308.

67. Willughby, *Observations on Midwifery*, 271–73.

68. Charles Kite, *An Essay for the Recovery of the Apparently Dead* (London: C. Dilly, 1788).

69. Sonja Boon, "Last Rites, Last Rights: Corporeal Abjection as Autobiographical Performance in Suzanne Curchod Nicker's Des inhumations précipitées (1790)," *Eighteenth-Century Fiction* 21 (2008): 89–107. I would like to thank Sonja Boon for allowing me to consult and cite her publication before it was published.

70. William Tebb, *Premature Burial and How It May Be Prevented* (London: Swan Sonnenschein, 1905).

71. Duclow, "Dying Well," 379.

72. Boon, "Last Rites, Last Rights"; Philippe Ariès, *Western Attitudes toward Death: From the Middle Ages to the Present*, trans. Patricia M. Ranum (Baltimore: Johns Hopkins University Press, 1974), argues that a concern with preparing for one's death increased during the fifteenth century, in keeping with a growing emphasis on individual rather than collective judgment.

73. Duclow, "Dying Well," 383.

74. Richard Wunderli and Gerald Broce, "The Final Moment before Death in Early Modern England," *Sixteenth Century Journal* 20, no. 2 (1989): 264.

75. Wunderli and Broce, "The Final Moment before Death," 265.

76. Lucinda Becker, "The Absent Body: Representations of Dying Early Modern Women in a Selection of Seventeenth-Century Diaries," *Women's Writing* 8, no. 2 (2001): 253. See also Mary Sidney Herbert, Countess of Pembroke, "From *A Discourse of Life and Death*," in *Women's Writing of the Early Modern Period 1588–1688: An Anthology*, ed. Stephanie Hodgson-Wright (New York: Columbia University Press, 2002), 7–14.

77. Becker, "The Absent Body," 261.

78. The best-known example of this gesture is found in Giotto di Bondone's *Raising of Lazarus*, 1304–1306, fresco, in the Cappella Scrovegni (Arena Chapel), Padua.

79. Vanessa Harding, "Whose Body?: A Study of Attitudes towards the Dead Body in Early Modern Paris," in *The Place of the Dead*, ed. Gordon and Marshall, 170–87.

80. Imhof, *Lost Worlds*, 180. See also Philippe Ariès, *The Hour of Our Death*, trans. Helen Weaver (New York: Knopf, 1981), 491–94.

81. C. Echinger-Maurach, "Michelangelo's Monument for Julius II in 1534," *The Burlington Magazine* 145 (May 2003): 336–44; and Charles de Tolnay, *The Tomb of Julius II* (Princeton, NJ: Princeton University Press, 1970).

82. Sergio Bertelli, *The King's Body: Sacred Rituals of Power in Medieval and Early Modern Europe*, trans. R. Burr Litchfield (University Park: Pennsylvania State University Press, 2001), 31–32.

83. Caroline Walker Bynum, *The Resurrection of the Body in Western Christianity, 200–1336* (New York: Columbia University Press, 1995).

84. Marshall, *Beliefs and the Dead in Reformation England*, 112.

85. Nigel Llewellyn, "Honour in Life, Death and in the Memory: Funeral Monuments in Early Modern England," *Transactions of the Royal Historical Society*, 6th series, 6 (1996), 191.

86. See www.bodyworlds.com/en.html.

Chapter 2

1. Francis Bacon, *The Essayes or Counsels Civill & Morall* (London: J. M. Dent, 1900), 98. All citations are to this edition.

2. Bacon, *The Essayes*, 98–9. For more information about medical regimens in this period, see Margaret Healy, "Fashioning Civil Bodies and 'Others': Cultural Representations," Chapter 9 in this volume (*A Cultural History of the Human Body in the Renaissance*, edited by Linda Kalof and William Bynum, Berg Publishers, 2010).

3. Sir Thomas Elyot, *Castel of Helth* (London, 1534) f.1r.

4. On Paracelsian medicine, see Healy, "Fashioning Civil Bodies and 'Others'."

5. For more on early modern syphilis and its representations, see Healy, "Fashioning Civil Bodies and 'Others'."

6. Arien Mack, ed., *In Time of Plague: The History and Social Consequences of Lethal Epidemic Disease* (New York: New York University Press, 1991), 1–3.

7. See P. Wright and A. Treacher, eds., *The Problem of Medical Knowledge* (Edinburgh: Edinburgh University Press, 1982), 8–15.

8. Sander Gilman, *Disease and Representation: Images of Illness from Madness to AIDS* (Ithaca, NY: Cornell University Press, 1988), 3.

9. Thomas Dekker, *London Looke Backe*, (London, 1630), sig. A4v. All citations are to this edition.

10. For an extended discussion of plague writing in early modern England, see Margaret Healy, "Defoe's Journal and the English Plague Writing Tradition," *Literature and Medicine* 22 (2003): 25–44.

11. Thomas Paynell, *A Moche Profitable Treatise* (London, 1534), sig A2r.

12. See, for example, Thomas Dekker, *The Wonderfull Yeare* (London, 1603), sigs. D1v, D3v, D4r (all citations are to this edition); Daniel Defoe, *A Journal of the Plague Year*, ed. L. Landa (Oxford: Oxford University Press, 1990), 76 (all citations are to Landa's edition).

13. See Meyer Fortes, "Foreword," in *Social Anthropology and Medicine*, ed. J. B. Loudon (London: Academic Press, 1979), xix.

14. For a survey of plague deployments in early literature, see Margaret Healy, *Fictions of Disease in Early Modern England: Bodies, Plagues and Politics* (New York: Palgrave, 2001), 54–64.

15. Anon., "A Warning to Be Ware" in *The Minor Poems of the Vernon MS*, Part II, ed. F. J. Furnivall (London: Kegan Paul, 1901), verse 8, 719.

16. Thomas Lodge, *A Treatise of the Plague* (London, 1603), sig. C3r.

17. Thomas Nashe, *Christes Teares over Jerusalem* (London, 1594), sig. Y2r.

18. On various signs and symptoms of the plague in 1665 according to H. F., see Defoe, *Journal*, 200. On the modern microbiology of plague, see Charles T. Gregg, *Plague: An Ancient Disease in the Twentieth Century* (Albuquerque: University of New Mexico Press, 1985), 113–28.

19. Defoe, *Journal*, 75, 203.

20. William Bullein, *A newe booke Entituled the Governement of Healthe* (London, 1558), f. xliir-v.

21. Defoe, *Journal*, 20, 22.

22. Thomas Dekker, *Newes from Graves-end* (London, 1604), sig. C4v.

23. On Helmontian medicine and the plague, see, Ole Peter Grell, "Plague, Prayer and Physic: Helmontian Medicine in Restoration England," in *Religio Medici: Medicine and Religion in Seventeenth-Century England*, ed. Ole Peter Grell and Andrew Cunningham (Aldershot, UK: Scholar Press, 1996), 204–27.

24. See Paul Slack, *The Impact of Plague in Tudor and Stuart England*. (Oxford: Oxford University Press, 1991), 248–49.

25. Ibid., 329–31.

26. Epistle 296 to Servatius, cited in R. J. Schoeck, *Erasmus of Europe: The Prince of Humanists 1501–1536* (Edinburgh: Edinburgh University Press, 1993), 92–93.

27. Desiderius Erasmus, "Festina lente," in *Erasmus on His Times: A Shortened Version of the Adages of Erasmus*, trans. Margaret Mann Phillips (Cambridge: Cambridge University Press, 1967), 14.

28. See Harold J. Cook, *Trials of an Ordinary Doctor: Joannes Groenevelt in Seventeenth-Century London* (Baltimore: John Hopkins University Press, 1985), 83.

29. Michel de Montaigne, *The Complete Works*, trans. Donald M. Frame (London: Hamish Hamilton, 1957), 836. All citations from the essays are to this edition. For an extended discussion of Montaigne's approach to his illness, see Margaret Healy, "Journeying with the 'Stone': Montaigne's Healing Travel Journal," *Literature and Medicine* 24 (2005): 231–49.

30. Roy Porter and G. S. Rousseau, *Gout: The Patrician Malady* (New Haven, CT: Yale University Press, 1998), especially 24, 25, 32.

31. Available for modern ears in the collection, *Le Labyrinthe & autres histories, with Paolo Pandolfo playing basse de viole* (Spain: Glossa, 1999).

32. Parveen Kumar and Michael Clark, eds., *Clinical Medicine* (London: Saunders, 2002), 627; R. Weiss, N. George, and P. O'Reilly, eds., *Comprehensive Urology* (London: Mosby, 2001), 329.

33. Elaine Scarry, *The Body in Pain* (Oxford: Oxford University Press, 1995); Chris Shilling, *The Body and Social Theory* (London: Sage, 2003), 184.

34. Epistle 2343, cited in Schoeck, 341–42. On Erasmus's medical acquaintances, see Peter Kravitsky, "Erasmus' Medical Milieu," *Bulletin of the History of Medicine*, xlvii (1973): 113–54.

35. Cited in Johan Van Beverwijck, *Treatise on the Stone* (Amsterdam, 1652), trans. into modern English in *Opuscula Selecta Neerlandicorum De Arte Medica* XVII (1943), 103.

36. Cited in Leon E. Halkin, *Erasmus: A Critical Biography*, trans. John Tonkin (Oxford, UK: Blackwell, 1993), 221–22. Erasmus's metaphorization of disease provides a compelling example of the process Susan Sontag described so powerfully in *Illness as Metaphor* (New York: Farrar, Straus, and Giroux, 1977).

37. Michel de Montaigne, *Travel Journal*, trans. and with an introduction by Donald M. Frame (San Francisco: North Point Press, 1983) 165. All citations are to this edition. See James Spottiswoode Taylor, *Montaigne and Medicine* (New York:

Paul B. Hoeber, 1921); Charles R. Mack, "Montaigne in Italy: Of Kidney Stones and Thermal Spas," *Renaissance Papers* 1991: 105–24; J. Starobinski, *Montaigne in Motion*, trans. A. Goldhammer (Chicago: University of Chicago Press, 1985); Anne Jacobson Schutte, "Suffering from the Stone: The Accounts of Michel de Montaigne and Cecilia Ferrazzi," *Bibliotheque d'Humanisme et Renaissance* 64 (2002): 21–36.

38. Robert Latham and William Matthews, *The Diary of Samuel Pepys: A New and Complete Transcription*, vol. 2 (London: G. Bell and Sons, 1970–1972), 194; all citations are to this edition.

39. Latham and Matthews, *The Diary of Samuel Pepys*, vol. 3, 153.

40. Latham and Matthews, *The Diary of Samuel Pepys*, vol. 5, 150.

41. Beverwijck, *Treatise*, 65, 61. See also Cook, *Trials of an Ordinary Doctor*, 86.

42. Beverwijck, *Treatise*, 61, 71.

43. Beverwijck, *Treatise*, 69.

44. See, for example, *Clinical Medicine*, 625–27 and *Comprehensive Urology*, 325–29. Most stones are composed of calcium oxalate and phosphate. Hypercalciuria has a familial tendency (fifty percent of first-degree relatives). Stones can also be composed of uric acid and cystine. Management consists in prescribing analgesia, a very high fluid intake, and increased physical activity. Calcium-rich dairy products should be avoided. In the past two decades the management of urinary calculus disease has been revolutionized by technological advances, including shock wave lithotripsy: *Comprehensive Urology*, p. 329.

45. Johannes Baptista Van Helmont, *Oriatrike or Physick Refined*, trans. J. C. of Oxon (London, 1662), 544. This influential Flemish iatrochemist wrote a popular treatise devoted to the stone in 1644, see Cook, 86.

46. George Thomson, *The Direct Method of Curing Chymically* (London, 1675), 17–18.

47. Beverwijck, *Treatise*, 67.

48. Ibid., 69.

49. Ibid., 69.

50. Ibid., 107.

51. Ibid., 105.

52. Ibid., 71.

53. Ibid., 65, 71.

54. *Epistle 1759 to John Francis*, English translation in *Opuscula selecta Neerlandicorum de Arte Medica* XVII, 5, 7. All citations are to the *Opuscula* translation, pp. 7–9.

55. Ibid., 7.

56. *Epistle to Theophrastus Paracelsus*, translated in Johan Huizinga, *Erasmus and the Age of Reformation* (London: Phoenix Press, 2002), 242–43.

57. Beverwijck, *Treatise*, 105.

58. Ibid., 111.

59. Ibid., 105.

60. Ibid., 109.

61. Ibid., 111.

62. See Schoeck, 120; *Epistle 1735 to William Cop* in *Opuscula*, 3, and *Epistle 1759 to John Francis* in *Opuscula*, 5.

63. Scaliger's charge occurred in an *Oratio pro m. Tullio Cicerone contra des. Erasmum Roterodamum* (September 1531); see Craig R. Thompson's comments in his introduction to "Penny Pinching" in *The Colloquies of Erasmus*, trans. Craig R.

Thompson (Chicago: University of Chicago Press, 1965), 489. All citations from the *Colloquies* are to this edition.

64. Thompson, *Colloquies*, 350.

65. Ibid., 350.

66. *Epistle 1759 to John Francis*, 7.

67. See, for example, *Journal*, 153, 142: "at the time the migraine first seized me again I had drunk a great quantity of Trebbiano ['sweet, heady white wine'], when heated by travelling and by the season, and when its sweetness did not quench my thirst."

68. Montaigne, *Journal*, trans. Frame, 8. All citations are to this edition.

69. Ibid., *Journal*, 152.

70. Latham and Matthews, *The Diary of Samuel Pepys*, vol. 5, 359 and vol. 6, 67.

71. Ibid, vol. 4, 346.

72. E. S. De Beer, ed., *The Diary of John Evelyn*, vol. 3 (Oxford: Clarendon Press, 1955), 3–4.

73. See Pepys's biography by Claire Tomalin, *Samuel Pepys: The Unequalled Self* (Penguin Viking: 2002), 61–65.

74. Tomalin, *Samual Pepys*, 529.

75. See Beverwijck in *Opuscula*, 117; Cook, *Trials*, 89–91; Lisa Jardine, *Ingenious Pursuits: Building on the Scientific Revolution* (London: Abacus, 2000), 288.

76. Jardine, *Ingenious Pursuits*, 288.

77. Ibid., 296–98.

78. Antoni Van Leeuwenhoek, "A Part of the 69th Letter of January 4th 1692 to the Royal Society in London," trans. into modern English in *Opuscula Selecta Neerlandicorum de Arte Medica* XVII, 177–83.

79. Leeuwenhoek, "A Part of the 69th Letter," 182–83.

80. C. Solingen, *Manuale Operatien Der Chirurgie* (Amsterdam, 1684), trans. into modern English in *Opuscula* XVII, 130–75, 137.

81. Kumar and Clark, *Clinical Medicine*, 627; Weiss and O'Reilly, *Comprehensive Urology*, 329.

82. Modern medicine is vague about why stone disease declined in Europe from the eighteenth century, suggesting that changes in diet might be responsible: see Kumar and Clark, *Clinical Medicine*, 627. Children with diets low in animal protein (as in large parts of the developing world today) undergo phosphate depletion, which encourages stone formation. Our three literary sufferers ate notably protein-rich diets; this was unlikely, therefore, to have been a contributing factor in their disease.

83. Cystine stones can be made smaller and even to disappear completely simply by drinking large quantities of water, see C.A.C. Charlton, *The Urological System* (Harmondsworth, UK: Penguin, 1973), 101.

84. Erasmus died on July 12, 1536, the immediate cause was dysentery, see Schoeck, 358. Montaigne died on September 13, 1592 from "throat inflammation . . . stone and other ailments," see Frame, Introduction, *Complete Works*, xii. Pepys died on Wednesday, May 26, 1703; at autopsy his left kidney was "a mass adhering to his back" and full of stones, while his old wound, bladder and gut were all septic, see Tomalin, *Samuel Pepys*, 378.

Chapter 3

1. Jacques Guillemeau, *The Happy Deliverie of Women* (London: Hatfield, 1612), 7.

2. Michel Foucault, *History of Sexuality*, vol. 1, *An Introduction*, trans. Robert Hurley (New York: Vintage, 1978).

3. Critics of Foucault are legion. One way to put them into perspective is to consider Foucault's intentions in describing the emergence of sexual identity as a product of discourses. See David M. Halperin, "Forgetting Foucault: Acts, Identities, and the History of Sexuality," *Representations* 63 (1998): 93–120.

4. Clement of Alexandria, "Stromata," in *The Ante-Nicene Fathers. Translations of the Writings of the Fathers down to A. D. 325*, ed. Alexander Roberts and James Donaldson, 9 vols. (New York: Charles Scribner's sons, 1896–1926), I, xv, 315; V, iv, 449–50; V, xii, 463.

5. Eusebius of Caesarea, *Eusebium Pamphili De evangelica praep[ar]atione Latiunu[m] ex Graeco beatissime pater iussu tuo effeci* (Venice: Leonhardus, 1473); Anicius Manlius Severinus Boethius, *The Consolation of Philosophy*, trans., ed., and intro. William Anderson (Carbondale: Southern Illinois University Press, 1963); Fulgentius Placiades, C. *Iulii Hygini Augusti Liberti, Fabularum Liber, Ad Omnium poëtarum lectionem mire necessarius & ante hac nunquam excusus* (Basel, Switzerland: Joan Hervagium, 1535); Lucius Caelius Firmianus, *Lucii Coelii Lactantii Firmiani, Opera quae extant, Ad fidem MSS. Recognita et Commentariis illustrata* (Oxford: Theatro Sheldoniano, 1684).

6. On the levels of reading in the Middle Ages, including changes in their conceptualization over time, see, for instance, Jane Chance, *Medieval Mythography from Roman North Africa to the School of Chartres, A.D. 433–1177* (Gainesville: University Press of Florida, 1994), 36–42.

7. Arnulf of Orléans, *Allegoriae*, in *Arnolfo d'Orléans, un cultore di Ovidio nel secolo XII*, ed. Fausto Ghisalberti, *Memorie del Reale Istituto lombardo di scienze e lettere 24* (Milan: Libraio del R. Istituto Lombardo di scienze e lettere, 1932), 157–234. Bibliothèque Nationale de France MSS lat. 7996, 8001, 8010, 14135; Reims, Bibliothèque municipale MS 1262, fols. 2–25; St. Omer, Bibliothèque municipale MSS 670, 678, fols. 104r–111r.

8. Originally written in the early fourteenth century, the *Ovide moralisé* circulated extensively in manuscript form before it was published in 1484. For the text, see Anon., *Ovide moralisé. Poème du commencement du quatorzième siècle publié d'après tous les manuscrits connus*, ed. C. de Boer, Martina G. de Boer, and Jeannette Th. M. van't Sant, 5 vols. (Amsterdam: Johannes Müller, 1915–1954), 3: 297 (ll. 3133–3135).

9. *Ovide moralisé*, 3: 298–303 (ll. 3196–3398).

10. Jean Seznec, *La Survivance des dieux antiques*, Studies of the Warburg Institute 11 (London: Warburg Institute, 1940) argues for continuity in the allegorical presentation of pagan gods. But even Seznec notes some changes in attitude apparent in Renaissance renderings of antiquity.

11. Ann Moss, *Ovid in Renaissance France: A Survey of the Latin Editions of Ovid and Commentaries Printed in France before 1600* (London: Warburg Institute, 1982), 27. Moss notes that commentaries omitted much of the allegory if they were reprinted after 1559, even if earlier editions had contained a great deal of it.

12. Ovid (P. Ovidius Naso), *Metamorphosis cum luculentissimis Raphaelis Regii enarrationibus* (Lyon, France: J. Mareschal, 1519), fol. 16v.

13. Ovid (P. Ovidius Naso), *Metamorphosis cum luculentissimis Raphaelis Regii enarrationibus* (Lyon, France: J. Huguetan, 1518), fol. 62v. It is possible to recognize Regius's dependence on his own gendered context. See, for instance, Valeria Finucci, *The Manly Masquerade: Masculinity, Paternity, and Castration in the Italian Renaissance* (Durham, NC: Duke University Press, 2003); Joanne M. Ferraro,

Marriage Wars in Late Renaissance Venice (Oxford: Oxford University Press, 2001); Guido Ruggiero, *The Boundaries of Eros: Sex Crimes & Sexuality in Renaissance Venice* (Oxford: Oxford University Press, 1985).

14. Ovid (P. Ovidius Naso), *P. Ovidii Nasonis Poetae Sulmonensis Opera Quae Vocantur Amatoria, cum Doctorum Virorum Commentariis partim huscusque etiam alibi editis, partim iam primum adiectis: quorum omnium Catalogum versa pagina reperies. His accesserunt Iacobi Micylli Annotationes* (Basel, Switzerland: Ioannem Hervagium, 1549), 227–29.

15. Julia Haig Gaisser, *Catullus and His Renaissance Readers* (Oxford: Clarendon Press, 1993), esp. 272–74.

16. James Hankins, *Plato in the Italian Renaissance*, 2 vols. (New York: Brill, 1990).

17. Alexander Niccholes, *A Discourse, of Marriage and Wiving: And of The greatest Mystery therein contained: How to choose a Good Wife from a bad* (London: Printed by N. O. for Leonard Becket, 1615), 6.

18. Cited in Anthony F. D'Elia, *The Renaissance of Marriage in Fifteenth-Century Italy* (Cambridge, MA: Harvard University Press, 2004), 120. D'Elia cites aversion to celibacy in a variety of writings in the fifteenth century.

19. Cited in D'Elia, *The Renaissance of Marriage*, 125.

20. Thomas Cogan, *The Haven of Health, Chiefly made for the Comfort of Students, and consequently for all those that have a care of their health, amplified upon five words of HIPPOCRATES, written Epid. 6. Labour, Meat, Drinke, Sleepe, Venus* (London: Melch Bradwood for John Norton, 1612), 242.

21. Philip Stubbs, *Anatomy of Abuses* (London: By John Kingston or Richard Iones, 1583), 59.

22. Thomas Laqueur, *Making Sex: Body and Gender from the Greeks to Freud* (Cambridge, MA: Harvard University Press, 1990).

23. For a more extended discussion of Laqueur and his critics, see Katherine Crawford, *European Sexualities, 1400–1800* (Cambridge: Cambridge University Press, 2007), 105–11.

24. P.E.H. Hair, ed. *Before the Bawdy Court: Selections from Church Court and Other Records Relating to the Correction of Moral Offences in England, Scotland, and New England, 1300–1800* (London: Elek, 1972), 147.

25. Jean Papon, *Recueil d'arrestz notables des courts souveraines de France* (Paris: Chez J. Macé, 1566), fol. 484r. The additional cases are from the 1648 edition, 1256–57.

26. Laurens Bouchel, *La Bibliothèque ou thresor du droict françois, auquel sont traictees les matieres Civiles, Criminelles, & Beneficiales, tant reglées par les Ordonnances, & Coustumes de France, que decidées par Arrests des Cours Souveraines*, 3 vols. (Paris: Chez vefve Nicolas Buon, 1639), 529. See also Michael Dalton, *The Countrey Justice, Containing the practice of Justices of the Peace out of their Sessions: Gathered for the better helpe of such Justices of Peace as have not beene conversant in the studie of the Lawes of this Realme* (London: Miles Flesher, James Haviland, and Robert Young, the assignes of Iohn More Esquire, 1630), 276. Original: 1618.

27. Katharine Park and Lorraine J. Daston, "Unnatural Conception: The Study of Monsters in Sixteenth- and Seventeenth-Century France and England," *Past and Present* 92 (August 1981): 20–54.

28. See, for instance, the cow-eared baby in Anon., *Ein wunderbarliche seltzame erschröckliche Geburt* (Augsburg, Germany: Michael Moser, 1561); Johannes Nasus,

Ecclesia Militans (Ingolstadt, Germany: Alexander Weissenhorn, 1569) reported a birth in 1536 of a child with a pig head.

29. For the monk-calf, see Philip Melanchthon, *Deuttung der zwei greulichen Figuren Papstesels zu Rom und Mönchkalbs zu Freiberg in Meissen gefunden* (Augsburg, Germany: H. Steiner, 1523).

30. Pierre Boaistuau, *Histoires prodigieuses*, ed. Gisèle Mathieu-Castellani (Geneva, Switzerland: Slatkin, 1996), 425. (Originally published in 1560.)

31. See, for instance, the analysis by Marie-Hélène Huet, "Monstrous Medicine," in *Monstrous Bodies/Political Monstrosities in Early Modern Europe*, ed. Laura Lunger Knoppers and Joan B. Landes (Ithaca, NY: Cornell University Press, 2004), 127–47.

32. Lorraine Daston and Katharine Park, *Wonders and the Orders of Nature, 1150–1750* (New York: Zone Books, 2001), esp. chapter 5.

33. Francisco Lopez de Gomara, *Histoire generalle des Indes occidentales et terres neuves, qui jusques à present ont esté descouvertes, augmentee en ceste cinquiesme edition de la description de al nouvelle Espagne, & de la grande ville de Mexique, autrement nommee Tenuctilan* (Paris: Michel Sonnius, 1584), fols. 238v–39r. Gomara was the secretary of Hernán Cortés.

34. Henri Estienne, *Apologie pour Hérodote. Satire de la société au XVIe siècle*, ed. P. Risetelhuber, vol. 2 (1879; repr. Geneva, Switzerland: Slatkine Reprints, 1969), 177–78.

35. Giovanni Marinello, *Le Medicine Partenenti allinfermità delle donne* (Venice: Gio. Bonfadino, & Compagni, 1560), fols. 53v–54r, 59v, 63v–234v. For a broad introduction to the Italian Renaissance literature of this type, see Rudolph M. Bell, *How to Do It: Guides to Good Living for Renaissance Italians* (Chicago: University of Chicago Press, 1999).

36. Michele Savonarola, *Libro della natura et virtu delle cose, che nutriscono, & delle cose non naturali, Con alcune osservationi per conservar la sanità, & alcuni quesiti bellissimi da notare* (Venice: Domenico & Gio. Battista Guerra, 1576), 261, 266, 270–71. A man who retained his sperm might be similarly damaged. See 274.

37. Guillemeau, *The Happy Deliverie*, 70.

38. Eucharius Roesslin, *The birth of mankinde, otherwise named the Womans Booke. Set forth in English by Thomas Raynalde Phisition, and by him corrected, and augmented* (London: Thomas Adams, 1604), 188. (Originally published in 1512.)

39. For documents and analysis of the case of Elena/o de Céspedes, see *Inquisitorial Inquiries: Brief Lives of Secret Jews and Other Heretics*, ed. and trans. Richard L. Kagan and Abigail Dyer (Baltimore: Johns Hopkins University Press, 2004); Israel Burshatin, "Written on the Body: Slave or Hermaphrodite in Sixteenth-Century Spain," in *Queer Iberia: Sexualities, Cultures, and Crossings from the Middle Ages to the Renaissance*, ed. Josiah Blackmore and Gregory S. Hutcheson (Durham, NC: Duke University Press, 1999), 420–56; and Lisa Vollendorf, *The Lives of Women: A New History of Inquisitorial Spain* (Nashville, TN: Vanderbilt University Press, 2005).

40. Jacques Duval, *Des hermaphrodits, accouchemens de femmes: et traitement qui est requis pour les relever en sante* (Rouen, France: David Gevffroy, 1612), a7r. For Riolan's opinion, see Jean Riolan, *Discours sur les hermaphrodits* (Paris: Ramier, 1614). For discussion of the case, see especially Katharine Park and Lorraine Daston, "The Hermaphrodite and the Orders of Nature," *Gay and Lesbian Quarterly* 1 (1995): 419–73 and Kathleen Long, "Jacques Duval on Hermaphrodites," in

High Anxiety: Masculinity in Crisis in Early Modern France, ed. Kathleen P. Long (Kirksville, MO: Truman State University Press, 2002), 107–38.

41. Thomas Dekker, *The Batchelars Banquet: or A Banquet for Batchelars: Wherein is prepared sundry daintie dishes to furnish their Table, curiously drest, and seriously served in. Pleasantly discoursing the variable humours of Women, their quicknesse of wittes, and searchable deceits* (London: Printed by T[homas] C[reede], 1603), unpaginated (chapter 9). Although anonymous, the work is attributed to Dekker, but thought by some to be the work of Robert Toste. Regardless, it is an adaptation of Antoine de la Sale (b. 1388)'s *Les quinze joies de marriage*.

42. Niccholes, *A Discourse of Marriage*, 7, 10.

43. Plato, *The Symposium*, trans. Christopher Gill (London: Penguin, 1999), 210a–12a.

44. Citations of the *De amore* are from the French critical edition. Marsilio Ficino, *Commentaire sur le banquet de Platon*, ed. and trans. Raymond Marcel (Paris: Société d'édition 'Les belles lettres,' 1956) [Hereafter *De amore*], which reproduces the Latin text from Ficino's autograph manuscript (Vatican Latin 7.705). The original date is 1469. See 146–52. On love as desire for unity with God, see also Marsilio Ficino, *The Letters of Marsilio Ficino*, trans. London School of Economic Science, 6 vols. (New York: Schocken Books, 1981, 1985), 1:37 (Ficino to Michele Mercati).

45. Ficino, *Letters*, 5: 81–83 (Ficino to Alammano Donati). Paul O. Kristeller, *The Philosophy of Marsilio Ficino*, trans. Virginia Conant (New York: Columbia University Press, 1943), 285–88. Kristeller identifies this letter as the source of the notion of Platonic love (amore Platonico): ". . . we begin meanwhile to love each other, so that apparently we have realized and perfected in ourselves that Idea of true love which Plato formulates in his work. From this Platonic love therefore a Platonic friendship arises . . ." (286).

46. For a reading more focused on the conflations, see Katherine Crawford, "Marsilio Ficino, Neoplatonism, and the Problem of Sex," *Renaissance et réforme*, 28, no. 2 (Spring 2004): 3–35.

47. Ficino, *De amore*, 184: "Amor nullo impletur aspectu corporis vel amplexu. Nullam igitur naturam corporis ardet, pulchritudinem certe sectatur. Quo fit ut ea corporeum aliquid esse non posit." On the centrality of beauty in Ficino's corpus, see especially Michael J. B. Allen, *The Platonism of Marsilio Ficino: A Study of the Phaedrus Commentary, Its Sources and Genesis* (Berkeley: University of California Press, 1984), 185–203, esp. 188.

48. For discussion, see Michael J. B. Allen, "Ficino's Theory of the Five Substances and the Neoplatonists' Parmenides," *Journal of Medieval and Renaissance Studies* 12 (1982): 19–44. See 19. Allen corrects Kristeller's reading of the hypostases in Ficino's thought.

49. Plato, *Symposium*, 210a–12a.

50. Ficino, *De amore*, 140: "Primam ipsius in deum conversionem, amoris ortum."

51. Ficino, *De amore*, 142: "Cum amorem dicimus, pulchritudinis desiderium intelligite."

52. Ficino, *De amore*, 230–39.

53. Ficino's account of the Aristophanes myth includes both the idea that the will is operative and that it is not. See *De amore*, 167–68. For more on Ficino's conception of the will, see Tamara Albertini, "Intellect and Will in Marsilio Ficino: Two Correlatives of a Renaissance Concept of Mind," in *Marsilio Ficino: His Theology, His Philosophy, His Legacy*, ed. Michael J. B. Allen and Valerie Rees, eds. (Leiden, Netherlands: Brill, 2002), 203–25, which argues for Ficino's innovative solution to the will/intellect problem. See also Kristeller, *Philosophy*, 256–88. Robert V. Mer-

rill, "The Pléiade and the Androgyne," *Comparative Literature* 1, no. 2 (Spring 1949): 97–112, see 109, argues that free will is impossible in Plato's philosophy because love is determined by God, but this is a bit reductive.

54. Pietro Bembo, *Gli Asolani di messer Pietro Bembo* (Venice: Aldo Romano, 1505); Baldesar Castiglione, *The Book of the Courtier*, trans. George Bull (London: Penguin, 1976), esp. 323–44; Leone Ebreo, *Dialoghi d'amore, composti per Leona medico, di natione hebreo, et dipoi tatto christiano* (Venice: Aldus, 1545). All three were quite popular in the Renaissance, but especially Castiglione. See Jean Festugière, *La Philosophie de l'amour de Marsile Ficin et son influence sur la literature française au XVIe siècle* (Paris: J. Vrin, 1941) and Peter Burke, *The Fortunes of the Courtier: The European Reception of Castiglione's Cortegiano* (University Park: Pennsylvania University Press, 1996).

55. Giovanni Pico della Mirandola offers a different version of the ladder of love in his *Commento a una canzone del Benivieni*, which was first published in 1519, but probably composed around 1486. After loving one beautiful body, Pico contends that the lover must make the image of love more perfect and spiritual. Then the soul distinguishes universal beauty from sensory images of love. The next step involves awareness of the power of abstraction, and particularly of ideal beauty. From there, the soul rises to approach ideal beauty intellectually, and then to merge with the Angelic Mind to achieve complete awareness. Finally, and only with God's help, the lover attains union with God. This culminating step is not available for mortal beings, according to Pico, a difference from Ficino's understanding of love as offering the ability to achieve salvation. On the context of its genesis and for an accessible modern edition, see Giovanni Pico della Mirandola, *De hominis dignitate, Heptaplus, De ente et uno*, ed. Eugenio Garin (Florence: Vallecchi Editore, 1942), 10–11, 443–581.

56. Symphorien Champier, *Nef des dames vertueuses composees par maistre Simphorien Champier Docteur en medicine contenant quatre livres* (Paris: Jehan de la Garde, 1515). The first edition was in 1503. Mario Equicola, *Libro di natura d'amore, di M. Equicola* (Venice: Pietro di Nicolini da Sabbio, 1536).

57. Marguerite de Navarre, *The Heptameron*, trans. and intro. P. A. Chilton (New York: Penguin, 1984); Antoine Héroët, *La Parfaicte amye*, ed. Christine M. Hill (1542; repr. Exeter, UK: University of Exeter, 1981); Marsilio Ficino, *Commentaire de Marsile Ficin, Florentin: sur le Banquet d'Amour de Platon: faict en François par Symon Silvius, dit J. De la Haye, Valet de Chambre de tres chrestienne Princesse Marguerite de France, Royne de Navarre* (Poitiers, France: A l'enseigne du Pelican, 1546).

58. John Ford, "'Tis Pity She's A Whore," in *'Tis Pity She's A Whore and Other Plays*, ed. Marion Lomax (Oxford: Oxford University Press, 1995), 12–26.

Chapter 4

1. Katharine Park, "Was There a Renaissance Body?" in *The Italian Renaissance in the Twentieth Century*, ed. Walter Kaiser and Michael Rocke (Florence: Olschki 2002), 21–35.

2. Maurice Merleau-Ponty, *Phénoménologie de la Perception* (Paris: Gallimard, 1945), see *Première partie: Le Corps*.

3. For discussion of Michel Foucault's theories of the function of historically specific regimes of power and their discourses of the body in the medical context, see Colin Jones and Roy Porter, eds., *Reassessing Foucault: Power, Medicine*

and the Body (London: Routledge, 1994). For an alternative reading of the Renaissance mind-set, its discourses, and its possibilities for thinking beyond its apparent limitations, see Ian Maclean, "Foucault's Renaissance Episteme Reconsidered: An Aristotelian Counterblast," *Journal of the History of Ideas* 59 (1998): 149–66.

4. Judith P. Butler, *Gender Trouble: Feminism and the Subversion of Identity* (New York: Routledge, 1990) and *Bodies That Matter: On the Discursive Limits of 'Sex'* (New York: Routledge, 1993).

5. See, for example, Gianna Pomata, *Contracting a Cure: Patients, Healers, and the Law in Early Modern Bologna* (Baltimore: Johns Hopkins University Press, 1998) and Margaret Pelling, *Medical Conflicts in Early Modern London: Patronage, Physicians, and Irregular Practitioners 1550–1640* (Oxford: Oxford University Press, 2003).

6. Marie Gaillé-Nikodimov, "Qu'est-ce que l'homme? La réponse de l'anatomiste ou la médecine comme anthropologie chez André du Laurens," *50ème Colloque International d'Etudes Humanistes: Pratique et pensée médicales à la Renaissance,* Centre d'Etudes Supérieures de la Renaissance, July 2–6, 2007. Conference presentation available online at www.cesr.uni-tours.fr.

7. See, for example, the case of rural herbalist Jeanne Lescallier, prosecuted in France in the 1570s, in Susan Broomhall, *Women's Medical Work in Early Modern France* (Manchester, UK: Manchester University Press, 2004), chapter 4.

8. Vern L. Bullough, *The Development of Medicine as a Profession: The Contribution of the Medieval University to Modern Medicine* (Basel, Switzerland: S. Karger, 1966); Alison Klairmont Lingo, "The Rise of Medical Practitioners in Sixteenth-Century France: The Case of Lyon and Montpellier" (PhD diss., University of California-Berkeley, 1980); Don Bates, ed., *Knowledge and the Scholarly Medical Traditions* (Cambridge: Cambridge University Press, 1995); Cornelius O'Boyle, *The Art of Medicine: Medical Teaching at the University of Paris, 1250–1400* (Leiden, Netherlands: Brill, 1998); Roger French, *Medicine before Science: The Business of Medicine from the Middle Ages to the Enlightenment* (Cambridge: Cambridge University Press, 2003).

9. See Charles G. Nauert, *Agrippa and the Crisis of Renaissance Thought* (Urbana: Univesity of Illinois Press, 1965) and Auguste Prost, *Corneille Agrippa: Sa vie et ses œuvres* (Nieuwkoop, Belgium: B. de Graaf, 1965), 2 vols.

10. Alison Klairmont Lingo, "Empirics and Charlatans in Early Modern France: The Genesis of the Classification of the 'Other' in Medical Practice," *Journal of Social History* 19 (1985–1986): 583–603.

11. See for example, midwifery: Hilary Marland, ed., *The Art of Midwifery: Early Modern Midwives in Europe* (London: Routledge, 1993).

12. On the experience of sickness and health in urban environments, see Margaret Pelling, *The Common Lot: Sickness, Medical Occupations and the Urban Poor in Early Modern England* (London: Longman, 1998).

13. Sandra Cavallo, *Artisans of the Body in Early Modern Italy* (Manchester, UK: Manchester University Press, 2008).

14. Martha Teach Gnudi and Jerome Pierce Webster, *The Life and Times of Gaspare Tagliocozzi, Surgeon of Bologna, 1545–1599* (New York: H. Reichner, 1950); J. F. Malgaigne, *Surgery and Ambroise Paré* (Norman: University of Oklahoma Press, 1965).

15. Pierre-André Sigal, "La Grossesse, l'accouchement et l'attitude envers l'enfant mort-né à la fin du moyen âge d'après les récits de miracles," in *Santé, Médecine et assistance au moyen âge* (Actes du 110e congrès national des sociétés savantes,

1985) (Paris: Editions du CTHS, 1987), 23–41; Robin Briggs, "Women as Victims? Witches, Judges, and the Community," *French History* 5 (1991): 438–80; David Gentilcore, *From Bishop to Witch: The System of the Sacred in Early Modern Terra d'Otranto* (Manchester, UK: Manchester University Press, 1992); Richard Kieckhefer, "The Holy and the Unholy: Sainthood, Witchcraft and Magic in Late Medieval Europe," *Journal of Medieval and Renaissance Studies* 24 (1994): 355–85; Sarah Ferber, *Demonic Possession and Exorcism in Early Modern France* (London: Routledge, 2004).

16. Carlo M. Cipolla, *Public Health and the Medical Profession in the Renaissance* (Cambridge: Cambridge University Press, 1976); Cipolla, *Fighting the Plague in Seventeenth-Century Italy* (Madison: University of Wisconsin Press, 1981).

17. Janine Bertier, "Un traité scolastique de Medicine des enfants: Le Pedenemicon de Gabriel Miron," in *Santé, Médecine et Assistance au moyen âge* (Actes du 110e congrès national des sociétés savantes, Montpellier, 1985) (Paris: Editions du CTHS, 1987), 9–22; Margaret Pelling, "Child Health as a Social Value in Early Modern England," *Social History of Medicine* 1, no. 2 (1988): 135–64; Wendy Perkins, *Midwifery and Medicine in Early Modern France: Louise Bourgeois* (Exeter, UK: University of Exeter Press, 1996); Louise Boursier, *Récit véritable de la naissance des messeigneurs et dames les enfans de France, Instruction à ma fille, et autres textes*, ed. François Rouget and Colette H. Winn (Geneva, Switzerland: Droz, 2000); Susan Broomhall, *Women's Medical Work in Early Modern France* (Manchester, UK: Manchester University Press, 2004), chapters 6 and 7; Simon de Vallambert, *Cinq Livres, de la maniere de nourrir et gouverner les enfans*, ed. Colette H. Winn (Geneva, Switzerland: Droz, 2005).

18. Colin Jones, *The Charitable Imperative: Hospitals and Nursing in Ancient Regime and Revolutionary France* (London: Routledge, 1989); Jonathan Barry and Colin Jones, eds., *Medicine and Charity before the Welfare State* (London: Routledge, 1991); Jean Imbert, *Le Droit hospitalier de l'ancien regime* (Paris: Presses universitaires de France, 1993); Daniel Hickey, *Local Hospitals in Ancient Regime France: Rationalization, Resistance, Renewal, 1530–1789* (Montreal: McGill-Queens University Press, 1997); Susan Broomhall, *Women's Medical Work*, chapter 2; John Henderson, *The Renaissance Hospital: Healing the Body and Healing the Soul* (London: Yale University Press, 2006).

19. David Harley, "Spiritual Physic, Providence and English Medicine, 1560–1640," in *Medicine and the Reformation*, ed. Ole Peter Grell and Andrew Cunningham (New York: Routledge, 1993), 101–17.

20. See Charles H. Parker, *Social Welfare and Calvinist Charity in Holland, 1572–1620* (Cambridge: Cambridge University Press, 1998) and Ole Peter Grell and Andrew Cunningham with Jon Arrizabalaga, eds., *Health Care and Poor Relief in Counter-Reformation Europe* (London: Routledge, 1999).

21. Andrew Wear, Roger French, and Iain M. Lonie, eds., *The Medical Renaissance of the Sixteenth Century* (Cambridge: Cambridge University Press, 1985), xi; Nancy G. Siraisi, *Medicine and the Italian Universities, 1250–1600* (Leiden, Netherlands: Brill, 2001).

22. Dina Bacalexi, "Trois traducteurs (N. Leoniceno, G. Cop et L. Fuchs) de Galien au XVIe siècle et leur regard sur la tradition arabe," *50ème Colloque International d'Etudes Humanistes*, July 2007.

23. Roland Antonioli, *Rabelais et la Médecine*, Etudes Rabelaisiennes, tome 12 (Geneva, Switzerland: Droz, 1976), 104–5.

24. See, for example, the following studies: Walter Pagel, *Paracelsus: An Introduction to Philosophical Medicine in the Era of the Renaissance*, 2nd rev. ed. (Basel, Switzerland: Karger, 1982) and Allen G. Debus, *English Paracelsians* (London: Oldbourne, 1968); Allen Debus, *The Chemical Philosophy: Paracelsian Science and Medicine in the Sixteenth and Seventeenth Centuries* (New York: Science History Publications, 1977); Allen Debus, *The French Paracelsians: The Chemical Challenge to Medical and Scientific Tradition in Early Modern France* (Cambridge: Cambridge University Press, 1991); Ole Peter Grell, ed. *Paracelsus: The Man and His Reputation, His Ideas and Their Transformations* (Leiden, Netherlands: Brill, 1998); Gerhild Scholz Williams and Charles D. Gunnoe Jr., eds., *Paracelsan Moments: Science, Medicine, and Astrology in Early Modern Europe* (Kirksville, OW: Truman State University Press, 2002).

25. See Françoise Wacquet, *Latin or the Empire of a Sign*, trans. John Howe (London: Verso, 2001).

26. Dominique Reulin, *Contredicts aux 'Erreurs populaires' de Laurent Joubert* (Montauban, 1580), Laurent Joubert, *La Médecine et le Régime de Santé, Des erreurs populaires et propos vulgaires* (Bordeaux, France: Simon Millanges, 1578), ed. Madeleine Tiollais, 2 vols. (Paris: L'Harmattan, 1997).

27. Natalie Zemon Davis, "Proverbial Wisdom and Popular Error," *Society and Culture in Early Modern France* (Stanford, CA: Stanford University Press, 1975), 227–667.

28. See Yasmin Haskell and Susan Broomhall, "Humanism and Medicine: A Match Made in Heaven?" *Intellectual History Review* 18, no. 1 (2008): 1–3 and volume to follow.

29. Mikhail Bakhtin, *Rabelais and His World*, trans. Helene Iswolsky (Bloomington: Indiana University Press, 1984); Gail Kern Paster, *The Body Embarrassed: Drama and the Disciplines of Shame in Early Modern Europe* (Ithaca, NY: Cornell University Press, 1993).

30. Ottavia Niccoli, *Prophecy and People in Renaissance Italy*, trans. Lydia G. Cochrane (Princeton, NJ: Princeton University Press, 1990); Jennifer Spinks, "Wondrous Monsters: Representing Conjoined Twins in Early Sixteenth-Century German Broadsheets," *Parergon* 22, no. 2 (2005): 77–112.

31. Spinks, Jennifer. "Jakob Rueff's 1554 *Trostbüchle*: A Zurich Physician Explains and Interprets Monstrous Births," *Intellectual History Review* 18, no. 1 (2008): 41–59.

32. Broomhall, *Women's Medical Work*, chapter 9.

33. Roy Porter, *Health for Sale: Quackery in England, 1660–1850* (Manchester: Manchester University Press, 1998); David Gentilcore, *Medical Charlatanism in Early Modern Italy* (Cambridge: Cambridge University Press, 2006).

34. Nancy G. Siraisi, "Oratory and Rhetoric in Renaissance Medicine," *Journal of the History of Ideas* 65, no. 2 (2004): 191–211.

35. In respect of birthing attendants, see Lianne McTavish, *Childbirth and the Display of Authority in Early Modern France* (Aldershot, UK: Ashgate, 2005).

36. Willem De Blécourt, "Witch Doctors, Soothsayers and Priests: On Cunning Folk in European Historiography and Tradition," *Social History* 19 (1994): 285–303; Marijke Gijswijt-Hofstra, Hilary Marland, and Hans de Waardt, eds., *Illness and Healing Alternatives in Western Europe* (London: Routledge, 1997); Emma Wilby, *Cunning Folk and Familiar Spirits: Shamanistic Visionary Traditions in Early Modern British Witchcraft and Magic* (Brighton, UK: Sussex Academic Press, 2006).

37. See Perkins, *Midwifery and Medicine in Early Modern France*; Boursier, *Récit véritable.*

38. Broomhall, *Women's Medical Work*, chapter 4.

39. Margaret Pelling, "Nurses and Nursekeepers: Problems of Identification in the Early Modern Period," in *The Common Lot: Sickness, Medical Occupations and the Urban Poor in Early Modern England* (London: Longman, 1998), 179–202; Naomi J. Miller and Naomi Yavneh, eds., *Maternal Measures: Figuring Caregiving in the Early Modern Period* (Aldershot, UK: Ashgate, 2000).

40. See McTavish, *Childbirth and the Display of Authority.*

41. Luis Garcia-Ballester, "The Inquisition and Minority Medical Practitioners in Counter-Reformation Spain: Judaizing and Morisco Practitioners, 1560–1610," in *Medicine and the Reformation*, ed. Ole Peter Grell and Andrew Cunningham, 156–91.

42. José Maria Lopez Piñero, "Las 'nuevas medicinas' americanas en la obra (1556–1574) de Nicolas Monardes," *Asclepio* 42, no. 1 (1990): 3–67; James Worth Estes, "The European Reception of the First Drugs from the New World," *Pharmacy in History* 37 (1995): 3–23; Donald Beecher, "The Books of Wonder of Nicholas Monardes of Seville," *Cahiers Elisabéthains* 51 (1997): 1–13; Teresa Huguet-Termes, "New World Materia Medica in Spanish Renaissance Medicine," *Medical History* 45 (2001): 359–76; Donald Beecher, "The Legacy of John Frampton: Elizabethan Trader and Translator," *Renaissance Studies* 20, no. 3 (2006): 320–39.

43. Katharine Park, "Stones, Bones and Hernias: Surgical Specialists in Fourteenth- and Fifteenth-Century Italy," in *Medicine from the Black Death to the French Disease,* ed. Roger Kenneth French, Jon Arrizabalaga, Andrew Cunningham, and Luis Garcia-Ballester (Aldershot, UK: Ashgate, 1998), 110–30.

44. See Andrew Cunningham, *The Anatomical Renaissance: The Resurrection of the Anatomical Projects of the Ancients* (Aldershot, UK: Ashgate/Scholar Press, 1997).

45. Andrea Carlino, "Le médecin et l'antiquaire. Une rencontre à Padoue vers la moitié du XVIe siècle," *50ème Colloque International d'Etudes Humanistes*, July 2007.

46. David Gentilcore, "The Church, the Devil and the Healing Activities of Living Saints in the Kingdom of Naples after the Council of Trent," in *Medicine and the Reformation*, ed. Grell and Cunningham, 134–55; Ferber, *Demonic Possession and Exorcism.*

47. See Walter Ong, *Orality and Literacy: The Technologizing of the Word* (London: Routledge, 1982); Neil Rhodes and Jonathan Sawday, *The Renaissance Computer: Knowledge Technology in the First Age of Print* (London: Routledge, 2000); Peter Burke, *A Social History of Knowledge: From Gutenberg to Diderot* (Cambridge: Polity, 2000).

48. Elizabeth Stephens, "Inventing the Bodily Interior: Ecorché Figures in Early Modern Anatomy and von Hagens' Body Worlds," *Social Semiotics* 17, no. 3 (2007): 313–26; Claudia Benthien, *Skin: On the Cultural Border between Self and the World* (New York: Columbia University Press, 2002).

49. Paul A. Russell, "Syphilis, God's Scourge or Nature's Vengeance? The German Printed Response to a Public Problem in the Early Sixteenth Century," *Archiv fur Reformationsgeschichte* 80 (1989): 286–307; Winfried Schleiner, "Moral Attitudes toward Syphilis and Its Prevention in the Renaissance," *Bulletin of the History of Medicine* 68, no. 3 (1994): 389–410; Winfried Schleiner, "Infection and Cure

through Women: Renaissance Constructions of Syphilis," *Journal of Medieval and Renaissance Studies* 24, no. 3 (1994): 499–517; Jon Arrizabalaga, John Henderson, and Roger French, *The Great Pox: The French Disease in Renaissance Europe* (New Haven, CT: Yale University Press, 1997).

50. William Clowes, *A Short and Profitable Treatise Touching the Cure of the Disease Called (Morbus Gallos)* (London, 1579).

51. This topic has been covered extensively by Thomas Laqueur, *Making Sex: Body and Gender from the Greeks to Freud* (Cambridge, MA: Harvard University Press, 1990).

52. Susan Broomhall, "Rabelais, the Pursuit of Knowledge, and Early Modern Gynaecology," *Limina: A Journal of Historical and Cultural Studies* 4 (1998): 24–34; Pollie Bromilow, "Inside Out: Female Bodies in Rabelais," *Forum for Modern Languages Studies* 44, no. 1 (2007): 1–13.

53. André du Laurens, *Toutes les Œuvres de Me. André Du Laurens*, trans. Theophile Gelée (Paris: Raphael Du Petit Val, 1621), Livre premier, Chapitre 3.

54. Livre 1, Chapitre 4.

55. Jean Céard, *La Nature et les Prodiges: L'insolite au XVIe siècle* (Geneva, Switzerland: Droz, 1977); Dudley Wilson, *Signs and Portents: Monstrous Births from the Middle Ages to the Enlightenment* (London: Routledge, 1993); Lorraine Daston and Katharine Park, *Wonders and the Order of Nature 1150–1750* (New York: Zone Books, 1998).

56. See most recently Kathleen P. Long, *Hermaphrodites in Renaissance Europe* (Aldershot, UK: Ashgate, 2006).

57. Some argue that biomedical science conducted an intentional misreading of the scientific findings in order to perpetuate control of women through knowledge of their bodies. See Stephanie E. Libbon, "Pathologizing the Female Body: Phallocentrism in Western Science," *Journal of International Women's Studies* 8, 4 (2007): 79–92.

58. See Elliott Horowitz, "The New World and the Changing Face of Europe," *Sixteenth Century Journal* 28 (1997): 1181–1201; Will Fisher, "The Renaissance Beard: Masculinity in Early Modern England," *Renaissance Quarterly* 54 (2001): 155–87.

59. Jonathan Sawday, *The Body Emblazoned: Dissection and the Human Body in Renaissance Culture* (London: Routledge, 1996) and Deanna Petherbridge and Ludmilla Jordanova, *The Quick and the Dead: Artists and Anatomy* (Berkeley: University of California Press, 1997).

60. Katharine Park, "The Criminal and the Saintly Body: Autopsy and Dissection in Renaissance Italy," *Renaissance Quarterly* 47 (1994): 1–33; Andrea Carlino, *Books of the Body: Anatomical Ritual and Renaissance Learning*, trans. John Tedeschi and Anne C. Tedeschi (Chicago: University of Chicago Press, 1999).

Chapter 5

1. Andrew Wear, *Knowledge and Practice in English Medicine, 1550–1680* (Cambridge: Cambridge University Press, 2000), 238.

2. Dorren G. Nagy, *Popular Medicine in Seventeenth-Century England* (Bowling Green, KY: Bowling Green State University Popular Press, 1988), 78.

3. Margaret Healy, *Fictions of Disease in Early Modern England: Bodies, Plagues and Politics* (New York: Palgrave, 2001), 4, 6.

4. Mary Lindeman, *Medicine and Society in Early Modern Europe* (Cambridge: Cambridge University Press, 1999), 2–3. Although he credits popular medicine with having real remedies overlooked by science in its arrogance, Roy Porter nonetheless reflects such a progress-oriented view of history as late as 1997 when he writes such comments as "Elite medicine sought to discredit health folklore" or "Once popular medicine had effectively been defeated," in *The Greatest Benefit to All Mankind: A Medical History of Humanity* (New York: W.W. Norton & Company, 1997), 37.

5. John Henry, "Doctors and Healers: Popular Culture and the Medical Profession," in *Science, Culture and Popular Belief in Renaissance Europe*, ed. S. Pumfrey, P.L. Rosse, and M. Slawinski (Manchester, UK: Manchester University Press, 1991), 198–99.

6. Notably among these are Michael C. Schoenfeldt, *Bodies and Selves in Early Modern England: Physiology and Inwardness in Spenser, Shakespeare, Herbert, and Milton* (Cambridge: Cambridge University Press, 1999); Healy, *Fictions of Disease*; Jonathan Sawday, *The Body Emblazoned: Dissection and the Human Body in Renaissance Culture* (London: Routledge, 1995); David Hillman, *Shakespeare's Entrails: Belief, Skepticism and the Interior of the Body* (New York: Palgrave, 2007); Jonathan Gil Harris, *Foreign Bodies and the Body Politic: Discourses of Social Pathology in Early Modern England* (Cambridge: Cambridge University Press, 2006); Gail Kern Paster, *The Body Embarrassed: Drama and the Disciplines of Shame in Early Modern England* (Ithaca, NY: Cornell University Press, 1993).

7. Healy, *Fictions of Disease*, 6.

8. See for instance, Vivian Nutton, "Medicine in Medieval Western Europe, 1000–1500," in *The Western Medical Tradition, 800 BC to AD 1800*, ed. L.I. Conrad, M. Neve, V. Nutton, R. Porter, and A. Wear (Cambridge: Cambridge University Press, 1995), 139–206: "From 1200 onwards the human body became more and more a part of theological discourse" part of the "unity of creation," 176.

9. In *The Anatomy of the Senses: Natural Symbols in Medieval and Early Modern Italy*, trans. Allan Cameron (Oxford: Polity Press, 1994), 26–36, Piero Camporesi recounts the role of plant life in early modern thought about the world and the body; see also 17 in the same volume for the idea of the body as a plant needing good soil and light to grow. He likewise notes the relationship between the body and vegetable creation in *The Juice of Life: The Symbolic and Magic Significance of Blood*, trans. Robert R. Barr (New York: Continuum, 1995), 76.

10. Norbert Elias, *The Civilizing Process*, vol. 1, trans. Edmund Jephcott (New York: Pantheon, 1978), 257–58. The move to a private, individualized experience of body and identity is widely observed in recent literature on the subject; in addition to the works cited in note 6, see Francis Barker, *The Tremulous Private Body: Essays on Subjection* (London: Methuen, 1984).

11. Schoenfeldt, *Bodies and Selves*, 11.

12. Schoenfeldt's discussion of Galenism is replete with references to its essentially "organic" character, for instance on 8 and 13; although Schoenfeldt does not explicitly engage with the implications of his use of this term, being more interested in the consequences of the Galenic model for the creation of agency through regulation, I see these occasional uses as significant of a certain version of the body's continuity with, and embeddedness in, a natural, communal world. Granted such a world may have never existed in perfect form, but it is fair, I think, to suggest that the Renaissance witnesses a movement away from that paradigm that involves more than just a shift in the definition of subjectivity. Language of "organicism," of the

"old Organon" and the "new Organon" (Bacon's way of celebrating a new humanist empiricism), are ideologically and etymologically connected to the notion of instrumentality, literally of how a thing (an organ) works—and at a stretch, to ideas about labor and the body's utility for it. Lurking in the transitions we register through scientific and medical discourse that describe and dictate the shift from a medieval to a Renaissance body, then, is another discourse about the body's role in an ecology of labor and subjection to exploitation.

13. Ulinka Rublack, "Fluxes: The Early Modern Body and the Emotions," trans. Pamela Selwyn, *History Workshop Journal* 53 (2002): 2.

14. Paracelsian and other approaches to the body and medicine have their advent in the Renaissance, challenging the primacy of Galenic theory; in short, Paracelsian models of the body emphasized the role of organ function over a general balance of humors and borrowed the idea of analogically applied cures (treatment of poison by poison, for example, rather than treatment by opposition as in Galenism) from folk medicine. However, any appearance of continuity with folk medicine or even religious models of macrocosm/microcosm are belied by Paracelsian chemistry and the deconstruction of the body based on functionality.

15. C. M. Woolgar, *The Senses in Late Medieval England* (New Haven, CT: Yale University Press, 2006), 2.

16. Helkiah Crooke, *Microcosmographia* (London: Printed by William Iaggard, 1616), 666.

17. John Donne, "The Extasie," *The Complete Poetry of John Donne*, ed. John T. Shawcross (New York: Anchor Books, 1967), p. 130.

18. Crooke, *Microcosmographia*, 9.

19. *John Milton: A Critical Edition of the Major Works*, ed. Stephen Orgel and Jonathan Goldberg (New York: Oxford University Press, 1991); *Paradise Lost* (3. 23, 34–6), 402–3.

20. Alan Dundes, *Interpreting Folklore* (Bloomington: Indiana University Press, 1980), 93–133.

21. Francis Bacon, *The Essayes or Counsels, Civill and Morall,* ed. Michael Kiernan (Cambridge, MA: Harvard University Press, 1985), 28.

22. Ioan P. Couliano, *Eros and Magic in the Renaissance*, trans. Margaret Cook (Chicago: University of Chicago Press, 1987), 30.

23. Bruce R. Smith, *The Acoustic World of Early Modern England: Attending to the O-Factor* (Chicago: University of Chicago Press, 1999), 104.

24. Crooke, *Microcosmographia*, 696–97.

25. Ben Jonson, *Epicoene, or The Silent Woman* (London, 1609), 2.1.

26. Cited in Bruce R. Smith, "Tuning Into London, c. 1600," in *The Auditory Culture Reader*, ed. Michael Bull and Les Black (Oxford: Berg, 2003), 129.

27. Constance Classen, David Howes, and Anthony Synnott, *Aroma: The Cultural History of Smell* (London: Routledge, 1994), 59.

28. Alain Corbin, *The Foul and Fragrant: Odor and the French Social Imagination* (Cambridge, MA: Harvard University Press, 1986), 13.

29. Classen, Howes, and Synnott, *Aroma*, 59; Terrence McLaughlin, *Coprophilia, or a Peck of Dirt* (London: Cassell, 1971), 16.

30. John Tanner, *The Temperate Man, or the Right Way of Preserving Health in Three Treastises* (London 1678), 158.

31. For more on passing wind, see Barbara C. Bowen, "The Honorable Art of Farting in Continental Renaissance Literature," in *Fecal Matters in Early Modern Litera-*

ture and Art: Studies in Scatology, ed. Jeff Parsells and Russell Garim (Burlington, VT: Ashgate Publishing, 2006), 1–13.

32. Constance Classen, "The Witch's Senses: Sensory Ideologies and Transgressive Femininities from the Renaissance to Modernity," in *The Empire of the Senses: The Sensual Culture Reader*, ed. David Howes (Oxford: Berg, 2005), 70.

33. Crooke, *Microcosmographia*, 648.

34. Carla Mazzio, "The Senses Divided: Organs, Objects, and Media in Early Modern England," in Howes, *Empire of the Senses*, 92, 96.

35. Lawrence Normand, "The Miraculous Royal Body in James VI and I, Jonson and Shakespeare, 1590–1609," in *The Body in Late Medieval and Early Modern Culture*, ed. Darryll Grantley and Nina Taunton (Aldershot, UK: Ashgate, 2000), 144.

36. William Clowes, *A Frutefull and Approved Treatise* (London, 1602), 4.

37. Georges Vigarello, *Concepts of Cleanliness: Changing Attitudes in France since the Middle Ages*, trans. Jean Birrell (Cambridge: Cambridge University Press, 1988), 16.

38. Vigarello, *Concepts of Cleanliness*, 43.

39. Brian Fagan, *The Little Ice Age: How Climate Made History 1300–1850* (New York: Basic Books, 2001), p. 106.

40. Tanner, *Temperate Man*, 160.

41. Mikhail Bakhtin, *Rabelais and His World*, trans. Helene Iswolsky (Bloomington: Indiana University Press, 1984), 281.

42. On the classical versus the grotesque body, see Peter Stallybrass, "Patriarchal Territories: The Body Enclosed," in *Rewriting the Renaissance: The Discourses of Sexual Difference in Early Modern Europe*, ed. Margaret W. Ferguson, Maureen Quilligan, and Nancy J. Vickers (Chicago: University of Chicago Press, 1986), 123–42; for a qualification of the carnivalesque body, see Schoenfeldt, *Bodies and Selves*, 11.

43. Stephen Mennell, *All Manners of Food: Eating and Taste in England and France from the Middle Ages to the Present* (Urbana: University of Illinois Press, 1995), 24–25. William Ian Miller notes that "this paradoxical method of saving by avidly consuming makes sense when any postponer of gratification was certain to see a good part of the grain he had stored ravaged by rats and birds, stolen by humans, rotted by damp, or consumed by fire." See "Gluttony," *Representations* 60 (1997): 97.

44. See Bruce Boehrer, *The Fury of Men's Gullets: Ben Jonson and the Digestive Canal* (Philadelphia: University of Pennsylvania Press, 1997), 93.

45. See Boehrer, *The Fury of Men's Gullets*, 93.

46. Michael Schoenfeldt, "Fables of the Belly in Early Modern England," in *The Body in Parts: Fantasies of Corporeality in Early Modern Europe*, ed. David Hillman and Carla Mazzio (New York: Routledge, 1997), 243.

47. Schoenfeldt, *Bodies and Selves*, 24.

48. Piero Camporesi, *The Anatomy of the Senses: Natural Symbols in Medieval and Early Modern Italy*, trans. A. Cameron (Oxford: Polity Press, 1994), 64.

49. Camporesi, *Anatomy of the Senses*, 74.

50. Tanner, *The Temperate Man*, 34–35.

51. Ibid., 14.

52. Ibid., 131.

53. See Boehrer, *The Fury of Men's Gullets*, 93.

54. William Vaughan, *Approved Directions for Health* (London, 1612), 82.

55. Tanner, *The Temperate Man*, 137.

56. Schoenfeldt, "Fables of the Belly," 254.

57. Ibid., 249.

58. Tanner, *The Temperate Man*, 143.

59. Camporesi, *Anatomy of the Senses*, 43.

60. John Tanner, *The Hidden Treasures of the Art of Physick* (London, 1659), 247.

61. See especially Healy, *Fictions of Disease*, and Harris, *Foreign Bodies*.

62. Thomas Lodge, *A Treatise of the Plague* (London, 1603), D2.

63. Laurent Joubert, *Popular Errors*, trans. Gregory David de Rocher (Tuscaloosa: University of Alabama Press, 1989), 87.

64. Wear, *Knowledge and Practice*, 5–65.

65. Linda Pollock, *With Faith and Physic: The Life of a Tudor Gentlewoman, Lady Grace Mildmay, 1552–1620* (New York: St. Martin's Press, 1993), 108.

66. See Wear, *Knowledge and Practice*, 50–51; Nagy, *Popular Medicine*, 54–78.

67. Pollock, *With Faith and Physic*, 128–29. The balm especially is unusual, although both it and the tonic, as general treatments, would tend to include more ingredients. Nevertheless, it shows a kind of *sprezzatura* applied to the arts of herbalism.

68. Robert and Michele Root-Bernstein, *Honey, Mud, Maggots, and Other Medical Marvels: The Science behind Folk Remedies and Old Wives' Tales* (New York: Houghton Mifflin, 1997), 31–43; Clifford M. Fouts, *Rhubarb, the Wondrous Drug* (Princeton, NJ: Princeton University Press, 1992), 16. Rhubarb has recently been discovered to ameliorate the toxins of cholera.

69. Gabrielle Hatfield, *Memory, Wisdom and Healing: The History of Domestic Plant Medicine* (Gloucestershire, UK: Sutton Publishing, 1999), 21.

70. Wear, *Knowledge and Practice*, 56.

71. Walter Cary, *A Breefe Treatise Called Carye's Farewell to Physick* (London, 1587), 56.

72. Keith Thomas, *Religion and the Decline of Magic* (London: Weidenfeld and Nicolson, 1971), 598.

73. Philippe Ariès, *The Hour of Our Death*, trans. Helen Weaver (New York: Alfred Knopf, 1981), 356.

74. Camporesi, *Juice of Life*, 18.

75. Ibid., 17.

76. Piero Camporesi, *The Incorruptible Flesh: Bodily Mutation and Mortification in Religion and Folklore*, trans. Tania Croft-Murray (Cambridge: Cambridge University Press, 1988), 12. Another odd application of body to body in remedies comes in the practice of drinking pus from plague wounds, according to McLaughlin, *Coprophilia*, 23.

77. Ariès, *The Hour of Our Death*, 357.

78. See Schoenfeldt, "Fables of the Belly," 245.

79. See Paul Barber, *Vampires, Burial and Death: Folklore and Reality* (New Haven, CT: Yale University Press, 1985) 13, 37, 41–42; and Wayne Bartlett and Flavia Idriceanu, *Legends of Blood: The Vampire in History and Myth* (Westport, CT: Praeger, 2006), 11.

80. Anthony Munday, *Sundry Strange and Inhumaine Murthers* (London, 1591).

81. Ralph Houlbrook, *Death, Religion and the Family in England, 1480–1750* (Oxford: Clarendon Press, 1998), 339.

82. Houlbrook, *Death, Religion and the Family*, 334–35.

Chapter 6

1. See especially Jacqueline Marie Musacchio, *The Art and Ritual of Childbirth in Renaissance Italy* (New Haven, CT: Yale University Press), 126–39.

2. See the letters of 1465 by Alessandra Macinghi negli Strozzi in *Lettere di una gentildonna fiorentina del secolo XV ai figliuoli esuli,* ed. Cesare Guasti (Florence: Sansoni, 1877), 458–59 and 463–67.

3. A useful summary of these aids to beauty is in Sandra Cavallo, "Health, Beauty and Hygiene," in *At Home in Renaissance Italy* (exhibition catalogue), ed. M. Wollheim-Adjmar and F. Dennis (London: V & A Publications, 2006), 174–87.

4. For the influence of classical rhetorical theory on Renaissance writings on art, see especially John R. Spencer, "Ut Rhetorica Pictura. A Study in Quattrocento Theory of Painting," *Journal of the Warburg and Courtauld Institutes* 20 (1957): 26–44; Michael Baxandall, *Giotto and the Orators: Humanist Observers of Painting in Italy and the Discovery of Pictorial Composition 1350–1450* (Oxford: Oxford University Press, 1986); Carl Goldstein, "Rhetoric and Art History in the Renaissance and Baroque," *Art Bulletin,* 53 (1991): 642–52.

5. Cicero, *De Oratore,* trans. E. W. Sutton, intro. H. Rackham (Cambridge, MA: Harvard University Press, 1996), 142–43.

6. Aristotle, "On the Art of Poetry," in *Classical Literary Criticism*, ed. and trans. T. S. Dorsch (Harmondsworth, UK: Penguin, 1969), 42.

7. The continuing influence of Aristotelian thinking on Renaissance aesthetics is stressed notably by David Summers, *The Judgement of Sense: Renaissance Naturalism and the Rise of Aesthetics* (Cambridge: Cambridge University Press, 1994). For a shorter overview of Renaissance aesthetic theory in relation to earlier and later thinkers, see Erwin Panofsky, *Idea: A Concept in Art Theory,* trans. Joseph J. S. Peake (New York: Harper & Row, 1968).

8. Leon Battista Alberti, *On the Art of Building in Ten Books,* trans. J. Rykwert, N. Leach, and R. Tavernor (Cambridge, MA: MIT Press, 1991), 136.

9. Vitruvius, *The Ten Books on Architecture,* trans. Morris Hicky Morgan (New York: Dover, 1960), 72–73.

10. Vitruvius, *Ten Books,* 102–4.

11. Cennino Cennini, *The Craftsman's Handbook. "Il Libro dell'arte,"* trans. D. V. Thompson (New York: Dover, 1960); Pomponius Gauricus, *De Sculptura,* ed. A. Chastel and R. Klein (Geneva, Switzerland: Droz, 1969), "De physiognomia," 128–63.

12. See Cennini, 48–49, on human proportion; for the influence on him of Byzantine canons, see Erwin Panofsky, "The History of the Theory of Human Proportions as a Reflection of the History of Styles," in *Meaning in the Visual Arts* (Garden City, NY: Doubleday, 1955), 55–107, especially 75–76.

13. Gauricus, *De sculptura,* 132–34.

14. Peter Meller, "Physiognomical Theory in Renaissance Heroic Portraits," in *The Renaissance and Mannerism: Studies in Western Art. Acts of the XXth International Congress of the History of Art* (Princeton, NJ: Princeton University Press, 1963), II, 53–69; Luba Freedman, *Titian's Portraits through Aretino's Lens* (University Park: Pennsylvania State University Press, 1995).

15. Rudolf Wittkower, *Architectural Principles in the Age of Humanism* (London: Tiranti, 1962) is the classic study of this belief as it affected Renaissance architecture.

16. Leonardo da Vinci, *Treatise on Painting*, trans. A. Philip McMahon, intro. Ludwig H. Heydenreich (Princeton, NJ: Princeton University Press, 1956), I, 278, 113.

17. For the *Adam and Eve* and Dürer's ideas on proportion, see especially Erwin Panofsky, *The Life and Art of Albrecht Dürer* (Princeton, NJ: Princeton University Press, 1971), 84–87, 260–80.

18. From Dürer's draft for an introduction to his book on human proportions (1512–1513), in *A Documentary History of Art*, vol. I, *The Middle Ages and Renaissance*, ed. Elizabeth Gilmore Holt (Princeton, NJ: Princeton University Press, 1981), 317.

19. A first book on human proportions, much indebted to Vitruvius, was abandoned around 1513; the second and third books, worked on after 1519, moved away from Vitruvius to devise thirteen basic human types: these became the bases for the *Vier Bücher von Menschlicher Proportion*, published posthumously in 1528. This was then translated into Latin in 1532–1534, and subsequently into other languages.

20. For example, "let soft lights fade imperceptibly into pleasant, delightful shadows; from this come about grace and beauty of form" (Codex Urbinas 109); McMahon, I, 413, 153.

21. Letter of Raphael to Castiglione, 1514: "*me porge una gran luce Vitruvio, ma non tanto che basti. . . . Ma essendo carestia . . . di belle donne, io mi servo di certa Idea, che mi viene nella mente*"; Vincenzo Golzio, *Raffaello nei documenti* (Farnborough, UK: Gregg International, 1971), 30–31. Raphael may be alluding to a passage toward the beginning of Cicero's *Orator* (ii, 8–10) where Phidias was said to have been guided by an inner vision of the beauty of Jupiter or Minerva, rather than by specific models; this vision is then related to Plato's Ideas; Cicero, *Orator*, trans. H. M. Hubbard (London: Heinemann, 1939), 310–12.

22. See David Summers, *Michelangelo and the Language of Art* (Princeton: Princeton University Press, 1981), especially 352–63.

23. Michael Baxandall, *Painting and Experience in Fifteenth Century Italy* (Oxford: Clarendon Press, 1972), 109–53; Martin Kemp, "Equal Excellences: Lomazzo and the Explanation of Individual Style in the Visual Arts," *Renaissance Studies* 1, no. 1 (1987), 1–26.

24. Pliny, *Historia Naturalis* I, xxxv, 53–148; *The Elder Pliny's Chapters on the History of Art*, ed. E. Sellers, trans. K. Jex-Blake (London: Macmillan), 1896, 98–171.

25. Leonardo, *Treatise on Painting*, trans. McMahon, I, 261, 108; "attend first to the movements representative of the mental attitudes of the creatures composing your narrative painting, rather than to the beauty and goodness of the parts of their bodies."

26. Patricia Emison, "Grazia," *Renaissance Studies* 5 (1991), 427–60, especially 432.

27. Giorgio Vasari, "Le vite de' più eccellenti pittori, scultori e architettori," in *Le opere di Giorgio Vasari con nuove annotazioni e commenti di Gaetano Milanesi*, vol. 4 (Florence: Le Lettere, 1998), 7–8.

28. For *maniera*, see John Shearman, "Maniera as an Aesthetic Ideal," in *The Renaissance and Mannerism: Studies in Western Art* II, 200–221.

29. Paolo Pino, "Dialogo di Pittura," in *Trattati d'arte del cinquecento*, I, ed. P. Barocchi (Bari, Italy: Laterza, 1960), "*una commensurazione e corrispondenza de' membri prodotti dalla natura senza alcuno impedimento de mali accidenti*," 98.

30. Vincenzio Danti, "Il primo libro del trattato delle perfette proporzioni" in *Trattati dell'arte del cinquecento*, I, ed. Paola Barocchi (Bari, Italy: Laterza, 1960), 209–69, strongly influenced by Aristotle's terms and concepts throughout.

31. Pino in *Trattati d'arte* . . . I, 99.

32. See Danti, "Il primo libro" in *Barocchi I*, 264–67; Frederika H. Jacobs, *Defining the Renaissance "Virtuosa"* (Cambridge: Cambridge University Press, 1997), especially 44–46, 97.

33. Ludovico Dolce in Mark W. Roskill, *Dolce's 'Aretino' and Venetian Art Theory of the Cinquecento* (Toronto, Canada: University of Toronto Press, 2000), 130–34.

34. Vasari, "Le vite," 8: "*La maniera venne poi la più bella dall'avere messo in uso il frequente ritrarre le cose più belle, e da quel più bello o mani o teste o corpi o gambe aggiugnerle insieme, e fare una figura di tutte quelle bellezze che più poteva, e metterla in uso in ogni opera per tutte le figure; che per questosi dice esser bella maniera.*"

35. Leon Battista Alberti, *On Painting and On Sculpture*, ed. and trans. C. Grayson (London: Phaidon, 1972), 55–57.

36. Quintilian, *The Institutio Oratoria of Quntilian*, trans. H. E. Butler (London: Heinemann, 1921), 292–95. For Cicero and Quintilian in relation to concepts of ornament and variety in Alberti, see especially David Summers, "Contrapposto: Style and Meaning in Renaissance Art," *Art Bulletin* 59 (1977): 336–61.

37. Baxandall, *Painting and Experience*, 122–23.

38. For decorum in antiquity and the Renaissance, see Rensselaer W. Lee, *Ut Pictura Poesis: The Humanistic Theory of Painting* (New York: Norton, 1967), 34–41, as well as items in 4.

39. Alberti, *On Painting,* 45, 87.

40. Leonardo, *Treatise on Painting* I, especially 93–97, 59–60; 382–424, 145–57.

41. Dolce in Roskill, *Dolce's "Trattato,"* 160–63, 170–79, 184–95.

42. Titian [Tiziano Vecellio], *Le lettere*, ed. C. Fabbro (Belluno: Comunità di Cadore, 1997), no. 135, 171: "*E perchè la Danae che io mandai già a Vostra Maestà, si vedeva tutte da la parte dinanzi, ho voluto in quest'altra poesia variare e farle mostrare la contraria parte, acciochè riesca il camerino . . . più grazioso a la vista. Tosto le manderò la poesia di Perseo e Andromeda, che avrà un'altra vista diversa da queste; e così Medea e Jasone . . .*"

43. Pliny, *Historia Naturalis,* I, xxxv, 66 and 98; eds. E. Sellers and K. Jex-Blake, 110–11 and 132–33; see also remarks of Cicero on the development of art away from rigidity; Cicero, *De Oratore,* 2, III, vii, 26.

44. Alberti, *On Painting*, II, 37, 74–75. For studies of *ekphrasis* in Renaissance art criticism, see Svetlana Alpers, "*Ekphrasis* and Aesthetic Attitudes in Vasari's *Lives*," *Journal of the Warburg and Courtauld Institutes* 23 (1960): 190–215; Baxandall, *Giotto and the Orators*; Norman E. Land, "Titian's Martyrdom of St. Peter Martyr and the "Limitations" of Ekphrasistic Art Criticism," *Art History* 13 (1990): 293–317 and *The Viewer as Poet: The Renaissance Response to Art* (University Park: Pennsylvania State University Press, 1994.

45. Fazio, quoted in Baxandall, *Giotto and the Orators*, 163–68.

46. Fazio in Baxandall, *Giotto and the Orators*, 167.

47. See David Summers, "*ARIA II*: The Union of Image and Artist as an Aesthetic Ideal in Renaissance Art," *Artibus et historiae* 20 (1989): 15–31; Moshe Barasch, "Character and Physiognomy: Bocchi on Donatello's St George: A Renaissance Text on Expression in Art," *Journal of the History of Ideas* 36 (1975): 413–30.

48. Vasari, "Le vite," IV, 335: "*tanta bellezza d'arie e divinità nelle figure, che grazia e vita spirano ne' fiati loro.*"

49. Letter of Dolce in Roskill, *Dolce's 'Aretino' and Venetian Art Theory of the Cinquecento*, 212–17.

50. Vasari, "Le vite," VII, 166: "*i panni straforati e finiti con bellissimo girar di lembi, e le braccia di muscoli e le mani di ossature e nervi sono a tanta bellezza e perfezzione condotte . . .*"

51. See Baxandall, *Painting and Experience*, 145–47 for *prompto* as used by Cristoforo Landino in relation to Giotto, Castagno, and Donatello; for *prestezza* as both a positive and a negative quality in the valuation of the styles of Titian and Tintoretto, see Tom Nichols, *Tintoretto: Tradition and Identity* (London: Reaktion, 1999), especially 71–73, 94–99.

52. Francesco Bocchi, "Eccellenza del San Giorgio di Donatello" in Barocchi, Trattati d'arte del cinquecento III, 125–94, at 153: "*quel vivo movimento e quello forza con l'azzione congiunta, la quale in adoperando e pronto e presta con bellezza si dimostra.*"

53. Karel van Mander, *Le Livre des Peintres*, ed. and trans. Véronique Gerard-Powell (Paris: Les Belles Lettres, 2001), 13, 15.

54. Baxandall, *Painting and Experience*, 123–24 and 145–47, remarks on *facilità* and *prompto* relating both to the action of the artist and to the characteristics of the work of art. For Vasari's identification of grace and beauty both in the artist and in his works of art, see Patricia Rubin, " 'What men saw': Vasari's Life of Leonardo da Vinci and the Image of the Renaissance Artist," *Art History* 13 (1990): 33–45; Patricia Lee Rubin, *Giorgio Vasari: Art and History* (New Haven, CT: Yale University Press, 1995), 372–79; Mary Rogers, "The Artist as Beauty," in *Concepts of Beauty in Renaissance Art*, ed. F. Ames-Lewis and M. Rogers (Aldershot, UK: Ashgate, 1998), 93–106.

55. Emison, "Grazia."

56. Vasari, "Le vite," VII, 167.

57. Anna Bryson, "The Rhetoric of Status: Gesture, Demeanour and the Image of the Gentleman in Sixteenth- and Seventeeth-Century England," in *Renaissance Bodies: The Human Figure in English Culture c. 1540–1660*, ed. L. Gent and N. Llewellyn (London: Reaktion, 1993), 145.

58. For the body in behaviour literature north of the Alps, see Bryson, "The Rhetoric of Status."

59. Sharon Fermor, "Movement and Gender in Sixteenth-Century Italian Painting," in *The Body Imaged: The Human Form and Visual Culture since the Renaissance*, ed. K. Adler and M. Pointon (Cambridge: Cambridge University Press, 1993), 129–45.

60. Baldassar Castiglione, *Il Libro del Cortegiano*, ed. A. Quondam (Milan: Grazanti, 1981), especially Book 1, XIX-XLI, 49–89.

61. Castiglione, *Il Libro del Cortegiano*, Book 1, XXVII, 63, based on Pliny, *Naturalis historiae* XXXV, LXXX.

62. See note 54.

63. Castiglione, *Il Libro del Cortegiano*, Book 3, especially IV-IX, 264–72.

64. Giovanni della Casa, *Galateo*, ed. G. Manganelli and C. Milanini (Milan: Rizzoli, 1984).

65. della Casa, *Galateo*, V, XXVIII-XXIX, 66, 129–30, 133.

66. della Casa, *Galateo*, XXVIII, 126.

67. For studies of other portraits expressing ideals of male beauty, see Mary Vaccaro, "Beauty and Identity in Parmigianino's Portraits," in *Fashioning Identities in Renaissance Art,* ed. M. Rogers (Aldershot, UK: Ashgate, 2000), 107–18; Rogers, "The Artist as Beauty"; David Alan Brown and Jane Van Nimmen, *Raphael and the Beautiful Banker. The Story of the Bindo Altoviti Portrait* (New Haven, CT: Yale University Press, 2005), especially 17–29.

68. Very different interpretations have been made of the dancing peasants in Brueghel and other sixteenth-century northern artists, but there is general agreement that their movements would have been seen as clumsy and boorish, as mentioned by van Mander, 188.

69. Agnolo Firenzuola, *On the Beauty of Women,* trans. and ed. K. Eisenbichler and J. Murray (Philadelphia: Philadelphia University Press, 1992). See also Elizabeth Cropper, "On Beautiful Women: Parmigianino, Petrarchismo, and the Vernacular Style," *Art Bulletin* 58 (1976): 374–94 and Mary Rogers, "The Decorum of Women's Beauty: Trissino, Firenzuola, Luigini and the Representation of Women in Sixteenth-Century Painting," *Renaissance Studies* 2 (1988): 47–88.

70. Firenzuola, *On the Beauty of Women,* 35: *"una occulta proporzione, e da una misura che non è ne' nostri libri . . . un non so che."*

71. Basic studies of the impact of Florentine Neoplatonism on art are André Chastel, *Art et humanisme à Florence au temps de Laurent le Magnifique: études sur la Renaissance* (Paris: Presses universitaires de France, 1959); Erwin Panofsky, "The Neoplatonic Movement in Florence and North Italy (Bandinelli and Titian)" and "The Neoplatonic Movement and Michelangelo," both in *Studies in Iconology* (New York: Harper & Row, 1962), 129–230.

72. Some of the few attempts to explore what effect Ficino's Neoplatonist ideas might have had on the style, as opposed to the philosophy, of artists are Chastel, *Art et humanisme à Florence au temps de Laurent le Magnifique,* especially 187, 308, and David Hemsoll, "Beauty as an Aesthetic and Artistic Ideal in Late Fifteenth-Century Florence," in *Concepts of Beauty in Renaissance Art,* edited by Frances Ames-Lewis and Mary Rogers (London: Ashgate Publishing, 1998), 66–79.

73. Panofsky, "The Neoplatonic Movement and Michelangelo," 180: Michelangelo adopted Neoplatonism "in its entirety . . . as a metaphysical justification of his own self."

74. Codex Urbinas 36; *Treatise on Painting,* trans. McMahon I, 280, 113.

75. The major proponents of the idea that Neoplatonic concepts were all-pervasive in Michelangelo's art and thought were Charles de Tolnay, *Michelangelo* (Princeton, NJ: Princeton University Press, 1943–1960), and Panofsky in "The Neoplatonic Movement and Michelangelo."

76. De Holanda's well-known account of Michelangelo's contempt for Netherlandish painting is in Francisco de Holanda, *Dialogues with Michelangelo,* trans. C.B. Holroyd, ed. David Hemsoll (London: Pallas Athene, 2006), 46–47; there is a current critical tendency to take this account more seriously than used to be the case.

77. Especially the "Idea del Tempio della Pittura" of 1590 by Lomazzo, in Gian Paolo Lomazzo, *Scritti sulle arti,* ed. Roberto Paolo Ciardi (Florence: Marchi and Bertolli, 1973), I, 310–15.

78. For this diffusion of Neoplatonism, see Panofsky, "The Neoplatonic Movement in Florence and North Italy."

79. Castiglione, *Il Libro del Cortegiano,* Book 4, LI-LXVII, 426–47; Pietro Bembo, *Gli Asolani,* in Pietro Bembo, *Opere in volgare,* ed. M. Marti (Florence: Sansoni,

1961), 9–163. For the much-discussed relationship between Bernardo Bembo, Ginevra de' Benci, and her portrait by Leonardo, see the bibliography in David Alan Brown, ed., *Virtue and Beauty: Leonardo's "Ginevra de' Benci" and Renaissance Portraits of Women*, exhibition catalogue (Princeton, NJ: Princeton University Press, 2001).

80. Michelangelo, *The Poems*, ed. and trans. C. Ryan (London: Dent, 1996), such as No. 105, 99: "It was not something mortal my eyes saw when in your beautiful eyes I found complete peace . . . if my soul had not been created by God equal to myself, then indeed it would wish for nothing more than external beauty, which pleases the eyes; but because this is so deceptive, the soul rises above to beauty's universal form"; see also especially Nos. 88–108, 83–101.

81. See Leonard Forster, *The Icy Fire: Five Studies in European Petrarchism* (Cambridge: Cambridge University Press, 1969).

82. Symptomatic is the repeated quoting of passages on beauty in the poetry of Petrarch and others in the book on "beauty aids" (in the modern sense) by the medical doctor Giovanni Marinelli, *Gli ornamenti delle donne* (Venice, 1562).

83. Elizabeth Cropper, "The Beauty of Women: Problems in the Rhetoric of Renaissance Portraiture," in *Rewriting the Renaissance: The Discourses of Sexual Difference in Renaissance Europe*, ed. M. W. Ferguson, M. Quilligan, and N. J. Vickers (Chicago: University of Chicago Press, 1986), 176.

84. Among the many discussions of this aspect of the *paragone* are Cropper, "The Beauty of Women" and Vaccaro, "Beauty and Identity."

85. The connection between the poetic and philosophical culture of Medicean Florence has been made ever since the 1890s; the most recent detailed study is Charles Dempsey, *The Portrayal of Love: Botticelli's "Primavera" and Humanist Culture at the Time of Lorenzo the Magnificent* (Princeton, NJ: Princeton University Press, 1992).

86. The description of the idealized Simonetta Vespucci in the *Stanze per la giostra* begins: "*Candida è ella, e candid la vesta, ma pur di rose e fior dipinta e d'erba; lo inanellato crin dall'aurea testa scende in la fronte umilmente superba*"; Angelo Poliziano, *Poesie italiane*, ed. M. Luzi and S. Orlando (Milan: Rizzoli, 1976), 53.

87. See Andrew Morrall, "Defining the Beautiful in Early Renaissance Germany," in *Concepts of Beauty in Renaissance Art*, ed. F. Ames-Lewis and M. Rogers (Aldershot, UK: Ashgate, 1998), 80–92, in relation to German art.

88. François Villon, *Selected Poems*, ed. and trans. P. Dale (Harmondsworth, UK: Penguin, 1978), 80–81.

89. For example, Botticelli's *Madonna of the Magnificent*, in the Uffizi, Florence, or the *Madonna del Libro* in the Poldi-Pezzoli Museum, Milan, both with subtle effects of gold and light symbolism; see Ronald Lightbown, *Botticelli, Life and Work* (London: Elek, 1978), I, 53–55.

90. "*Vergine bella, che di sol vestita, coronata di stelle, al sommo Sole piacesti sì che 'n te sua luce ascosa*"; poem to the Virgin in Francesco Petrarca, *Rime. Trionfi e poesie latine*, ed. F. Neri, G. Martellotti, E. Bianchi, and N. Sapegno (Milan: Ricciardi, 1951), CCCLXVI, 472.

91. Castiglione, *Il Libro del Cortegiano*, Book 4, LII, 428. For the fusion of secular and Christian ideals of beauty in relation to female saints, see Mary Rogers, "Reading the Female Body in Venetian Renaissance Art," in *New Interpretations of Venetian Renaissance Painting*, ed. Francis Ames-Lewis (London: Birkbeck College, 1994), especially 82–89.

92. Mary Vaccaro, "Resplendent Vessels: Parmigianino at Work in the Steccata," in *Concepts of Beauty in Renaissance Art*, 134–46. Such a fusion of secular and sacred sources in relation to images of Christ or male saints deserves further exploration.

93. See Forster, frontispiece and ix–x.

94. For Elizabethan Petrarchism, see Forster, 122–47 and illustration 4; for Van Dyck, Elise Goodman, "Woman's Supremacy over Nature: Van Dyck's 'Portrait of Elena Grimaldi,'" *Artibus et historiae* 30 (1994): 129–43; for Terborch, Alison McNeil Kettering, "Terborch's Ladies in Satin," *Art History* 16:1 (1993): 95–124.

95. The key passage is in Vasari ed. Milanesi, IV, 10–11. From the antique works, artists gained qualities of *gagliardezza, leggiadria, grazia,* and *prontezza,* gaining a *"bellezza nuova e più viva."*

96. Francisco de Holanda, *Dialogues with Michelangelo*, 30.

97. van Mander, *Le Livre des Peintres*, 192.

98. Paula Nuttall, *From Flanders to Florence: The Impact of Netherlandish Painting 1400–1500* (New Haven, CT: Yale University Press, 2007); Bernard Aikema and Beverly Louise Brown, eds., *Il Rinascimento a Venezia e la pittura del Nord ai tempi di Bellini, Dürer, Tiziano,* exhibition catalogue (Venice: Bompiani, 1999).

99. Pliny, *Historia Naturalis*, I, xxxv, 112–17, 144–49.

100. For Dürer's comments on these, see Holt, *A Documentary History of Art*, 331, 339, 340.

Chapter 7

1. Traditional views of Renaissance are explained in the classic work by Jacob Burckhardt, *The Civilization of the Renaissance in Italy*, trans. S.C.G. Middlemore (New York: Harper, 1958) and later endorsed by most critics.

2. In *Rabelais and His World*, Mikhail Bakhtin (trans. H. Iswolsky, Bloomington: Indiana University Press, 1984) opposes the features of the grotesque body to those of the classical body. The classical body is closed, static, and contained, whereas the grotesque one is connected with "those parts of the body that are open to the outside . . . the open mouth, the genital organs, the breasts, the phallus, the pot-belly, the nose" (26).

3. In *The Portable Machiavelli*, ed. Peter Bondanella, Mark Musa (New York, Viking Penguin, 1979), 59–60. "*Omè! Fu' per cadere in terra morto, tanta era bructa quella femina. E' se la vedeva prima un ciuffo di capelli fra bianchi et neri, cioé canuticci, et benché l'avessi el cocuzolo del capo calvo, per la cui calvitie ad lo scoperto si vedeva passeggiare qualche pidochio, nondimeno e pochi capelli et rari le aggiugnevono con le barbe loro infino su le ciglia; . . . in ogni puncta delle ciglia di verso li ochi haveva un mazeto di peli di lendini; . . . piene le lagrimatoie di cispa et e nipitelli dispillicciati; . . . et l'una delle nari tagliata, piene di mocci; . . . nel labbro di sopra haveva la barba lunghetta, ma rara; . . . et come prima aperse la bocca, n'uscí un fiato sí puzolente, che trovandosi offesi da questa peste due porte di due sdegnosissimi sensi*" (Nicolò Machiavelli, *Opere*, ed. Franco Gaeta, vol. 3, Turin: Unione Tipografico-Editrice Torinese, 1984, 321–22).

4. Lorenzo Venier (1510–1550) was a notable of an illustrious Venetian family who held some administrative posts in the Venetian Republic and composed two obscene poems against prostitutes: *La puttana errante* and *La Zaffetta* (1531). Lorenzo was a close friend of Pietro Aretino who sent a copy of *Puttana errante*

to Federico Gonzaga, marquis of Mantua, along with his own burlesque poem penned to accompany Venier's poem. Although the name of the wondering whore is never mentioned in the poem, she is believed to be Venetian courtesan Elena Ballerina, who was the subject of invective in *La Zaffetta* as well.

5. Unless otherwise stated all translations are my own. "*Una vecchia parlò dopo costei,/Ch'in tutte le masselle ha quattro denti;/L'unghie ha d'un palmo de le mani e piei,/Pute 'l suo fiato più ch'otto conventi,/La barb'ha d'uomo, e gli occhi de'giudei,/come valigie le poppe pendenti,/e i capei radi d'un biancaccio giallo/ Qual di carrett'ha la cod'un cavallo*" (*Puttana errante* Canto III, XXXI, 71).

6. "*Fronte verde, occhi zalli,/Naso rovan, masselle crespe e guanze,/Recchie d'ogni hora carghe de buganze,/Bocca piena de zanze./Fia spuzzolente, denti bianchi, e bei,/A par delle cegie e dei cavei,/ . . . /Quella magra desfatta,/Anzi, secca, incandìa, arsa destrutta,/Quella, che nome in ossi sta redutta/Che cazze dalla brutta,/Quella, che spesso, i putti per la via/Tio in fallo per la morte stravestia.*" For the complete version of the original poem in Venetian dialect and its English translation see Patrizia Bettella, "Antonfrancesco Grazzini and Maffio Venier. Poetry on Prostitutes" *In Dialogue with the Other Voice*, eds. Julie D. Campbell, Maria Galli Stampino (Toronto: Centre for Renaissance and Reformation Studies), forthcoming.

7. For more on witches, see my book, *The Ugly Woman: Transgressive Aesthetic Models in Italian Poetry from the Middle Ages to the Baroque* (Toronto, Canada: University of Toronto Press, 2005), chapter 3.

8. "*O dolce e bela serore, / o uoci de sole inrazè,/o massele inverzelè/pì che no fo mè basta o persuto salò, / o lavri rosè, o biè dinti da ravolò,/o boca inmelata, / . . . / o tete, a' fassè pur contento / per grandeza ogni vacaro, / o piè biè grande da vetolaro, / o gambe grosse ben norì, o bel lacheto, / tondo, grosso, bianco e neto, / che ogni botazo è picolo a piè de ti! /. . . . / o pieto bianco e scolorìo, / com fo mè ravo in campo, / o corpo giusioso e santo, / o braze ben da sapa o da baìle / o man da lavorar / ben mile bughè int'un dì.*" The *Betìa* is included in the collected edition of Beolco, A. (Ruzante), *Teatro*, ed. Alvise Zorzi (Turin, Italy: Einaudi, 1967). The quote is from Act III, 312.

9. Among Leonardo's grotesque, mostly sketches in red pencil and many housed in the Royal Collection of Her Majesty Queen Elizabeth II, we find for example *A Bald Fat Man with a Broken Nose in Right Profile* (ca. 1485–1490), *A Grotesque Old Man Leaning on a Stick and a Man's Back* (ca. 1510–1515). The grotesque was later cultivated by Leonardo's pupils, such as Francesco Melzi, who was believed to have copied his *Grotesque Old Woman* from Leonardo (Leonardo 1491–1493, Melzi 1510). The Academy of Val di Blenio, founded in 1560 by Giovan Paolo Lomazzo, became the place of dissemination of both theoretical and pictorial figurations of grotesque realism. Such realism reflected an interest for the comic/grotesque shared by Leonardo and Lomazzo, a theme that was well represented in Italian literature in rustic poetry in many regions such as Tuscany and Veneto.

10. "*Nelle istorie debbe essere homini di varie complessioni, età, incarnationi, attitudini, grassezze, magrezze; grossi, sottili, grandi, piccoli, grassi, magni, fieri, civili, vechi, giovani, forti et muscolosi, debboli . . .*" (Quoted in Michael W. Kwakkelstein, "Leonardo da Vinci's Grotesque Heads and the Breaking of the Physiognomic Mould," *Journal of the Warburg and Courtauld Institutes* 54 (1991): 131.

11. Martin Clayton, *Leonardo da Vinci: The Divine and the Grotesque* (London: Royal Collection Enterprise, 2002). For a depiction of grotesques monstrous and evil in art see Umberto Eco, *On Ugliness* (New York: Rizzoli, 2007), chapters 3 and 4.

12. Bernardo Bellincioni (1452–1492) was a Florentine poet who spent time at the Medici court in Florence and later in Milan, where he was court poet of Lodovico Sforza, the patron of Leonardo da Vinci. Bellincioni composed some sonnets that allow us to get a glimpse of Leonardo's art. In one sonnet (from *Rime*, 1493) he praised Leonardo's *Portrait of a Lady with Ermine*. Leonardo also designed stage machinery for the production of Bellincioni's *Il Paradiso* (1490), a play to entertain guests at the wedding of Gian Galeazzo Sforza and Isabel of Aragon. Besides more traditional poetry, Bellincioni also composed comic realistic poetry, which may have inspired Leonardo for his grotesques.

13. Martin Clayton, *Leonardo da Vinci: The Divine and the Grotesque*, 82. Another type of other that appears in Leonardo's grotesque drawings is the gypsy in *A Man Tricked by Gypsies* (1493). Here one gypsy reads the palm of the man in the center while her accomplice steals his purse. Gypsies were a familiar subject in paintings but only at least a century later. They arrived in Western Europe from the Balkans in the early fifteenth century and soon acquired the reputation as fortune-tellers and thieves. Clayton also gives evidence to the fact that, in the late fifteenth century, gypsies were banned from the city of Milan because of their thefts and crime (see Clayton, 96).

14. "*Qualora voleva dipingere qualche figura, considerava prima la sua qualità et la sua natura, cioé se deveva ella essere nobile o plebea, gioiosa o severa, turbata o lieta, vecchia o giovane, irata o di animo tranquillo, buona o malvagia; et poi, conosciuto l'esser suo, se n'andava ove egli sapea che si ragunassero persone di tal qualità et osservava diligentemente e i lor visi, le lor maniere, gli abiti et i movimenti del corpo, e trovata cosa che gli paresse atta a quel che far voleva, la riponeva collo stile al suo libriccinio che sempre egli teneva a cintola*" (C. G. Giraldi, *Discorso dei romanzi*, ed. L. Benedetti, G. Monorchio, and E. Musacchio (Bologna: Italy, Millenium, 1999), 231.

15. Ibid., 214.

16. This *Head of a Cretin with Goiter*, also known as *Ser Caputagn Nasotra* or *Grotesque head* ("Captain Big Nose," second half of sixteenth century), inspired by or perhaps copied from one of Leonardo's grotesques, was attributed to one member of the Academy of Val di Blenio. For more on this drawing, housed at the Pinacoteca Ambrosiana in Milan, and on the academy see *Rabisch, Il grottesco nell'arte del Cinquecento. L'accademia della Val di Blenio, Lomazzo e l'ambiente milanese.* Milan, Italy: Skira, 1998, 121–83.

17. This style of depiction of the disproportionate, disfigured, and ugly individual, called "comico figurativo" was adopted by many of Leonardo's followers in Lombardy, particularly by the Academy of Val di Blenio, which became the place of dissemination of theoretical and pictorial figurations of grotesque realism.

18. Georges Vigarello, "The Upward Training of the Body from the Age of Chivalry to Courtly Civility." In *Fragments for a History of the Human Body, Part Two*, 149–99 (New York: Zone, 1989), 151.

19. "Ma come la Natura tutte quante/di pura terra fe', così vanno / di quella ornate dal capo alle piante; . . . 'e i capei folti, bosco da pidocchi, / e gli denti smaltati di ricotta, / e le pope, che van fin a' ginocchi" (Bettella, *The Ugly Woman*, 112).

20. Vigarello, 1989, 151.

21. Baldesar Castiglione, *The Book of the Courtier*, trans. George Bull (London: Penguin, 1976), 147.

22. Giovanni della Casa, *Galateo*, trans. Konrad Eisenbichler and Kenneth R. Bartlett (Toronto: Centre for Reformation and Renaissance Studies, 1986), 9.

23. della Casa, *Galateo*, chapter 29.

24. Ibid., 12.

25. Ibid., 60.

26. P. Magli, "The Face and the Soul," in *Fragments for a History of the Human Body, Part Two*, 87–127 (New York: Zone, 1989), 87.

27. "*luxuriosa animalia sunt, hircus, porcus, . . . asinum, simia*"; see Giambattista della Porta, *De Humana Phisiognomonia Ioannis Baptistae Portae Neapolitani* (Urselli, Italy: Cornelio Sutori, 1601), 496.

28. "*Niger color meticulosae dolosaeque mentis indicium*" (della Porta, 197).

29. J. Schiesari, "The Face of Domestication, Physiognomy, Gender Politics, and Humanism's Others," in *Women, 'Race' and Writing in the Early Modern Period*, ed. M. Hendricks and P. Parker, 55–70 (London: Routledge, 1994), 60.

30. Brucioli, A. *A Commentary upon the Canticle of Canticles* (London: R. Field for Tho. Man, 1598). Brucioli contrasts the blackness and fairness in relation to the Church "made blacke and duskish, that is to say deformed & unhandsome . . . but it is not without beautiful and welfavoure by reason of her vertues" (p. B3), and later associates blackness with oppression and sins and talks about St. Paul's blackness as deformity. All the negative aspects of blackness, however, in a Christian view, are overcome by the fairness of heavenly things.

31. Kate Lowe, "The Stereotyping of Black Africans in Renaissance Europe," in *Black Africans in Renaissance Europe*, ed. T. F. Earle and K.J.P. Lowe, 17–47 (Cambridge: Cambridge University Press, 2005).

32. This fact is reported in the article by Kate Lowe.

33. Alvise Cadamosto, *The Voyages of Cadamosto and Other Documents on Western Africa Second Half of the Fifteenth Century* (London: Hakluyt Society, 1937), 13.

34. Leo Africanus lived in Morocco and accompanied his uncle on diplomatic missions in Maghreb and Timbuktu. He was captured by pirates in the Mediterranean and sold as a slave. In 1520 he was freed and baptized by Pope Leo X. The Pope asked him to write an account of his knowledge of the African continent. His *Cosmographia dell'Africa* of 1526 was later published with the title *Descrittione dell'Africa* in Giovanni Battista Ramusio's collection *Delle Naviagationi e viaggi* published in Venice in 1555. The book was greatly successful and published numerous times. It was a precious source of information about Africa and was translated into English by John Pory as *A Geographical Historie of Africa* in 1600.

35. K. F. Hall, *Things of Darkness: Economies of Race and Gender in Early Modern England* (Ithaca, NY: Cornell University Press, 1995).

36. Hall, 36.

37. Anu Korhonen, "Washing the Ethiopian White: Conceptualising Black Skin in Renaissance England," in *Black Africans in Renaissance Europe*, ed. T. F. Earle and K.J.P. Lowe, 94–112 (Cambridge: Cambridge University Press, 2005), 98.

38. Paul H.D. Kaplan, "Isabella d'Este and Black African Women," in *Black Africans in Renaissance Europe*, ed. T. F. Earle and K.J.P. Lowe, 125–54 (Cambridge: Cambridge University Press, 2005).

39. When Isabella of Aragon married Giangaleazzo Sforza in 1488, among her servants were some dark slaves. We know from a letter to Isabella d'Este from Tedora Angelini, damigella of the Gonzaga, that Anna Sforza at the court of Ferrara also had a black slave in her retinue. Slaves of all ethnic origin, including Ethiopians, and Numidians (the Berber kingdom of Algeria and Tunisia), were kept at the court of Cardinal Ippolito de' Medici.

40. Pisanello's illustration of the Arthurian epic (1447–1448) in Mantua's Palazzo Ducale depicts a youthful black figure. At the Palazzo Schifanoia in Ferrara Francesco del Cossa's painting shows Duke Borso d'Este approached by an older black man in position of subordination with bent knees. Andrea Mantegna's frescoes for the *Camera Picta* (1474) has a dark face looking down; it is not clear if this is a male or female figure, but it's mainly believed to be female. Mantegna depicted black Africans numerous times before and during his stay at Isabella's d'Este court. Famous is the *Adoration of the Magi*, which includes a group of African retainers and the black Magus.

41. Alessandro Luzio and Rodolfo Renier, "Buffoni, Nani e Schiavi dei Gonzaga," *Nuova Antologia* 36 (1891): 112–46. Luzio and Renier document Isabella's passion for *moretti* who were kept at her court, not to carry out domestic work but rather as a form of amusement and a status symbol.

42. Kaplan, "Isabella d'Este and Black African Women."

43. "*Una moretta che non habia più di quatro anni*" (Quoted in Luzio and Renier, "Buffoni," 140).

44. "*De negreza et de fateze ne satisfà più che non havessimo saputo desiderare*" (Letter of July 1491, in Luzio and Renier, "Buffoni," 141).

45. "*Se così è che 'l [moretto] sia de summa beleza, apostatelo per nui, . . . ma non essendo ben negro et bene proporzionato, non ne fati mercato né conventione alcuna*" (Quoted in Luzio and Renier, "Buffoni," 141).

46. "*Una mora che pochissimo tempo è che fu presa in Barbaria . . . poi havere da 16 a 17 anni ed è bellissima persona . . . ben fatta quanto è possibile . . . bello volto, salvo che ha el labro de sotto della bocca grosso*" (Quoted in Luzio and Renier, "Buffoni," 145).

47. Dark slaves were favorites of noble Renaissance ladies and courtesans. Particular interest in dark slaves is shown in the desire to portray them in official paintings. For more on the connection between slaves and buffoons at the Court of Isabella d'Este see Luzio and Renier, "Buffoni."

48. Elise Goodman, "Woman's Supremacy over Nature: Van Dyck's Portrait of Elena Grimaldi," *Artibus et historiae* 30 (1994): 129–43. Goodman examines this theme in Anthony Van Dyck's *Portrait of Marchesa Elena Grimaldi* (1623) where the lady is accompanied by a young black retinue, holding a parasol for her and placed behind her to enhance her status. He serves as a dark foil for her whiteness. Goodman examines other portraits of Van Dyck and of other contemporaries, where the aristocratic *dama* is depicted with a young Moor servant (Rubens, Titian, Pierre Mignard). Because Van Dyck was working for the Genoese family of Grimaldi, it is possible that the painter had seen this black boy in the Genoese African community, as Genoese were active in slave trade and had black household servants.

49. Quoted in Goodman, "Woman's Supremacy over Nature," 141.

50. "*Bruna sei tu, ma bella / qual vergine viola;/e del tuo vago sembiante io sì m'appago / che non disdegno signoria d'ancella,*" Torquato Tasso, Rime (Rome: Salerno, 1994), 328.

51. "*Ecco la notte al mio bel sole intorno, / ch'abbellisce con l'ombre i suoi splendori*"; quoted in *Marino e i marinisti*, ed. Giuseppe Guido Ferrero (Milan: Ricciardi, 1954), 684.

52. For more on unconventional beauty and the dark lady in Italian baroque poetry see chapter 4 of Patrizia Bettella, *The Ugly Woman*. Blackness is also depicted as irrationality and violence. The black African par excellence in Renaissance literature is the Moor of Venice, who appears in Shakespeare's *Othello*. The story of

the Moor of Venice derives from a novella in Giambattista Giraldi Cintio's collection *Hecatommiti* (III, 7 published in 1565). This collection was very popular
and frequently reprinted in Italy and later in French and Spanish translation. The
Moor of Venice provides an example of the taboo represented by the possibility of
miscegenation: because an honorable black man is married to a white lady, their
union could produce a progeny of mixed blood, possibly black in color. This event
would be inconceivable for Renaissance mentality.

53. The idealized image of the good savages with beautiful and graceful bodies is found
in many travel narratives about the New World: for example, Antonio Pigafetta, one
of the members of Magellan's expedition of circumnavigation, in *Notizie del nuovo
mondo* (*News from the new world*) reported that the indigenous people of Brazil
lived in a state of nature and were easy to convert. Amerigo Vespucci (*Letter of the
Newly found Islands*, sent in 1504 to Pier Soderini) also notes the state of nature
and lack of religion in the native people of Brazil, whose land is compared to earthly
paradise. Also Giovanni da Verrazzano describes North American natives with the
highest praise for their physical beauty. See R. Romeo, *Le scoperte americane nella
coscienza italiana del Cinquecento* (Milan, Italy: Ricciardi, 1971), 13–15.

54. This passage is from L. Firpo, *Prime relazioni di navigatori italiani sulla scoperta
dell'America. Colombo, Vespucci, Verrazzano* (Turin, Italy: Unione Tipografico-
Editrice Torinese, 1966), 40.

55. Michael Householder uncovers a wider range of portrayals of indigenous American women that goes beyond the Eve-like innocence and monstrous Amazonian
sexual appetite. In Pietro Martire and Richard Eden, Householder finds indigenous women's courageous resistance to Spanish cruelty. See Michael Householder,
"Eden's Translations: Women and Temptation in Early America," *Huntington Library Quarterly* 70, no. 1 (2007): 11–36.

56. This account, found in Michele da Cuneo's account of Columbus's second voyage
in 1495 (see Firpo, *Prime relazioni*, 52), speaks volumes about the brutal treatment
inflicted by European conquistadors on native Americans, particularly women.

57. T. J. Cachey, Jr. "Between Humanism and New Historicism Rewriting the New
World Encounter," *Annali d'italianistica* 10 (1992): 33–35.

58. As Cachey "Between Humanism," notes, Martire, influenced by his humanistic
background, ennobles and sentimentalizes the female gender in his narrative. On
the contrary, as noted earlier, Vespucci describes savage women as cannibals.

59. Jerry Brotton, *The Renaissance Bazaar: From the Silk Road to Michelangelo* (Oxford: Oxford University Press, 2003), 37.

60. Bronwen Wilson, *The World in Venice: Print, the City, and Early Modern Identity*
(Toronto, Canada: Toronto University Press, 2005), 254.

61. Wilson, *The World in Venice*, examines the impact of printed images in Renaissance Venice (and in European society in general) for the formation of modern
subjectivity. She claims that costume is used more than race to classify people and
that costume books contributed to profound change in identity formation in a way
that initiates the development of racial and ethnic stereotypes based on physical
appearance. Wilson's claim is well suited to the representation of the Turk circulating in Venice and other parts of Europe, but it does not hold true for the portrayal
of Africans and Native Americans, where the naked body and the dark color of the
skin is the defining element of otherness.

62. For more on Jews in Renaissance Italy see Barbara Wisch, "Vested Interest: Redressing Jews on Michelangelo's Sistine Ceiling," *Artibus et Historiae* 24 (2003):

143–72 and Robert Bonfil, *Gli Ebrei in Italia all'epoca del Rinascimento* (Florence, Italy: Sansoni, 1991).

63. See for example Gloria Allaire, "Noble Saracens or Muslim Enemy? The Changing Image of the Saracen in Late Medieval Italian Literature," in *Western Views of Islam in Medieval and Early Modern Europe*, ed. D.R. Blanks and M. Frassetto, 173–84 (New York: St. Martin's Press, 1999).

64. D.J. Vitkus, "Early Modern Orientalism: Representation of Islam in Sixteenth- and Seventeenth-Century Europe," in *Western Views of Islam in Medieval and Early Modern Europe*, ed. D.R. Blanks and M. Frassetto, 207–30 (New York: St. Martin's Press, 1999). As Vitkus notes, there is a demonization and distortion of Islam in romances and chivalric legends of conflict between Christian and Saracens. Tasso's *Liberata,* also set in the distant past of the first crusade, conveys European unity and Christian superiority over the infidels at a time when the Catholic Church was under threat by Protestantism and by the close proximity of the Ottoman Empire. The poem depicts rebellious characters who challenge religious orthodoxy, such as the Ethiopian warrior-woman Clorinda, daughter of the king of Ethiopia. Despite being born from black parents Clorinda does not have black skin because, when she was conceived, her mother was under the influence of a painting of Saint George freeing a white princess from the dragon. Although Clorinda's parents were Christian, because of the color of her skin, she was raised by an Egyptian servant; she was never baptized and grew up in Egypt. Clorinda, who fought with the infidels, converts to Christianity just before dying in battle.

65. Piero Camporesi, *The Land of Hunger* (Cambridge, UK: Polity Press, 1996), 36.

66. "*Quant'è bella giovinezza / Che si fugge tuttavia!*"

67. "*Ipsa senectus morbus est*" This quote from *Adagia,* 1507, chil. II, cent. VI. Prov. XXXVII is in Daniel Schäfer, "Medical Representations of Old Age in the Renaissance: The Influence of Non-Medical Texts," in *Growing Old in Early Modern Europe: Cultural Representations*, ed. Erin Campbell, 12–19 (Aldershot, UK: Ashgate, 2006), 13.

68. Erin Campbell, " 'Unenduring' Beauty: Gender and Old Age in Early Modern Art and Aesthetics," in *Growing Old in Early Modern Europe: Cultural Representations*, ed. Erin Campbell, 153–67 (Aldershot, UK: Ashgate, 2006), 159.

69. "*Una vecchia mi vagheggia/vizza e secca insino all'osso/ . . . / ell'ha logra la gingiva,/ tanto biascia fichi secchi*" (see Bettella, *The Ugly Woman,* 76).

70. "*Vecchia ritrosa, perfida e maligna/inimica d'ogni ben, invidiosa,/e strega incantatrice e maliosa*" (see Bettella, *The Ugly Woman,* 68).

71. Cited in R.J.C. Young, *Postcolonialism: An Historical Introduction* (Malden, MA: Blackwell, 2001), 354.

Chapter 8

1. Jack Goody, *Representations and Contradictions: Ambivalence Towards Images, Theatre, Fiction, Relics and Sexuality* (Oxford: Blackwell, 1997).

2. Goody, *Representations and Contradictions,* 91; Labourt, J., 1955, *Saint Jerome, Lettres,* Vol. 4, Paris, 201 (cited in Goody, 91).

3. John Calvin, *Treatise on Relics,* http://www.godrules.net/library/calvin/176calvin4.htm.

4. Thomas More, *A Dialogue of Comfort against Tribulation*, introduction and notes by Frank Manley (New Haven, CT: Yale University Press, 1977), 98.

5. Walter Howard Frere and William McClure Kennedy, eds., *Visitation Articles and Injunctions of the Period of the Reformation* (London: Longmans, Green, 1910), 38–39, 218, 224.

6. Desiderius Erasmus, *The Colloquies of Erasmus*, trans. Craig R. Thompson (Chicago: University of Chicago Press, 1965), 288–89, 293–97, 301.

7. *Statutes of the Realm*, ed. A. Luders III, 448, 450. See also L. E. Whatmore, "The Sermon against the Holy Maid of Kent and Her Adherents, Delivered at Paul's Cross, November the 23rd, 1533, and at Canterbury, December the 7th," *The English Historical Review* 58, no. 232 (1943): 463–75, 469–70; on Barton, see Diane Watt, *Secretaries of God: Women Prophets in Late Medieval and Early Modern England* (Cambridge: D. S. Brewer, 1997); and Ethan Shagan, "The Anatomy of Opposition in Early Reformation England: The Case of Elizabeth Barton, the Holy Maid of Kent," in *Popular Politics in the English Reformation* (Cambridge: Cambridge University Press, 2003), 61–88.

8. Desiderius Erasmus, *Collected Works of Erasmus* (Toronto, Canada: University of Toronto Press, 2003).

9. Michael Kunze, *Highroad to the Stake: A Tale of Witchcraft*, trans. William E. Yuill (Chicago: University of Chicago Press, 1987), 276.

10. Robert Muchembled, *A History of the Devil: From the Middle Ages to the Present* (Oxford: Polity, 2003), 221.

11. There have been a plethora of academic works on the topic of the rise of disgust. See William Miller, *The Anatomy of Disgust* (Cambridge, MA: Harvard University Press, 1997); Martha C. Nussbaum, *Hiding from Humanity: Disgust, Shame, and the Law* (Princeton, NJ: Princeton University Press, 2004); William A. Cohen and Ryan Johnson, eds., *Filth: Dirt, Disgust and Modern Life* (Minneapolis: University of Minnesota Press, 2005); Emily Cockayne, *Hubbub: Filth, Noise, and Stench in England, 1600–1770* (New Haven, CT: Yale University Press, 2007). On smell, see especially Ruth Brown, "Middens and Miasma: A Portrait of Seventeenth Century Village Life in Banburyshire," *Cake and Cockhorse* 16, no. 1 (2003): 2–8; Mark Jenner, "Civilization and Deodorization? Smell in Early Modern English Culture" in *Civil Histories: Essays Presented to Sir Keith Thomas*, ed. P. Burke, B. H. Harrison, and P. Slack, 127–44 (Oxford: Oxford University Press, 2000); and Dominique Laporte, *The History of Shit*, trans. Nadia Benabid and Rodolphe El-Khoury (Cambridge, MA: MIT Press, 2000).

12. James Calfhill, *An Answer to John Martiall's Treatise on the Cross*, ed. for the Parker Society by Richard Gibbings (Cambridge: University Press, 1846), 17.

13. Keith Thomas, *Religion and the Decline of Magic* (London: Penguin, 1971), 60.

14. Ibid., 583.

15. Ibid., 83.

16. Eamon Duffy, *The Stripping of the Altars* (New Haven, CT: Yale University Press, 1992), 384–85.

17. Ronald Hutton, *The Stations of the Sun: A History of the Ritual Year in Britain* (Oxford: Oxford University Press, 1996), 100. Ritual and its disposal of the bodies of the faithful attracted criticism too, and by the end of the Reformation era was often regarded as synonymous with fraud: see Edward Muir, *Ritual in Early Modern Europe* (Cambridge: Cambridge University Press, 2005), 175.

18. Duffy, *The Stripping of the Altars*, 186.

19. Thomas Malory, *Le Morte D'Arthur*, ed. S.H.A. Shepherd (New York: W.W. Norton, 2003), 3–20.

20. Elisabeth Bronfen, *Over Her Dead Body: Death, Femininity and the Aesthetic* (Manchester, UK: Manchester University Press, 1992), 12.

21. Janet Knepper, "A Bad Girl Will Love You to Death: Excessive Love in the Stanzaic Morte Arthur and Malory," in *On Arthurian Women: Essays in Memory of Maureen Fries*, ed. B. Wheeler and F. Tolhurst (Dallas, TX: Scriptorium Press, 2001), 229–44. I am grateful to Carolyne Larrington for supplying this reference. See also Geraldine Heng, "Enchanted Ground: The Feminine Subtext in Malory," in *Arthurian Women*, ed. T.S. Fenster (New York: Routledge, 2000), 97–114.

22. *Le Haut Livre Du Graal: Perlesvaus*, ed. W.A. Nitze and T.A. Jenkins (Chicago: University of Chicago Press, 1932); available in translation as *The High Book of the Grail: A Translation of the Thirteenth Century Romance of Perlesvaus* by Nigel Bryant (Cambridge, UK: D.S. Brewer, 1978). This quotation comes from the Norton edition of *Morte d'Arthur*: taken from Nitze and Atkinson-Jenkins, 343–45, lines 8312–78. This translation is by S.H.A. Shepherd.

23. St. Agnes, in particular, cannot be burned, and her hair grows miraculously when she is taken to a brothel.

24. *Second Royal Injunctions of Henry VIII* (1538); John Colet, in Erasmus, *Colloquies*, 305, 308, 310. See also Peter Marshall, "Forgery and Miracles in the Reign of Henry VIII," *Past and Present* 178 (2003): 39–73, and "The Rood of Boxley, the Blood of Hailes and the Defence of the Henrician Church," *Journal of Ecclesiastical History* 46 (1995): 689–96. On the Protestant horror of the dead, see Andrew Spicer, "Defyle not Christ's kirk with your carrion: Burial and the Development of Burial Aisles in Post-Reformation Scotland," in *The Place of the Dead: Death and Rememebrance in Late Medieval and Early Modern Europe,* ed. Bruce Gordon and Peter Marshall (Cambridge: Cambridge University Press, 2000), 149–69.

25. Richard L. Greene, ed. *The Early English Carols* (Oxford: Clarendon Press, 1935).

26. Caroline Walker Bynum, *Jesus as Mother: Studies in the Spirituality of the High Middle Ages* (Berkeley: University of California Press, 1982).

27. Richard Rambuss, *Closet Devotions* (Durham, NC: Duke University Press, 1998).

28. Alan Stewart, *Close Readers: Humanism and Sodomy in Early Modern England* (Princeton, NJ: Princeton University Press, 1997).

29. John Donne, "The Relic," in *The Elegies and the Songs and Sonnets*, ed. Helen Gardner (Oxford: Clarendon, 1965).

30. R. Po-chia Hsia, *The Myth of Ritual Murder: Jews and Magic in Reformation Germany* (New Haven, CT: Yale University Press, 1988), 10.

31. Bynum, *Holy Feast*, 142.

32. Jacques de Vitry, on Mary of Oignies, cited by Bynum, *Holy Feast*, 59.

33. Bynum, *Holy Feast*, 63. The Song of Songs is itself a text about longing and desire.

34. Carlo Ginzburg, *Ecstasies: Deciphering the Witches' Sabbath*, trans. Raymond Rosenthal, ed. Gregory Elliot (London: Hutchinson Radius, 1990), 75.

35. H.C. Erik Midelfort, *Witch Hunting in Southwestern Germany, 1562–1684: The Social and Intellectual Foundations* (Stanford, CA: Stanford University Press, 1972), 25–26.

36. Johannes Nider, *Formicarius*, cited in Walter Stephens, *Demon Lovers: Witchcraft, Sex, and the Crisis of Belief* (Chicago: Chicago University Press, 2003), 241.

37. Hsia, 228. See Johann Weyer, *De praestigiis daemonum* (Frankfurt: Nicolaus Basseum, 1586).

38. Ljubica Stefan, Excerpts from *From Fairy Tale to Holocaust*, 1993, http://www.hic.hr/books/from-fairytale/part-01.htm. On Hansel and Gretel, see Maria Tatar, *Off with Their Heads! Fairy Tales and the Culture of Childhood* (Princeton, NJ: Princeton University Press, 1992), 208–10.

39. Miri Rubin, *Gentile Tales: The Narrative Assault on Late Medieval Jews* (New Haven, CT: Yale University Press, 1999).

40. Ibid., 26.

41. Ibid., 35.

42. Ibid., 36.

43. See my discussion of Bessie Dunlop in "Losing Babies, Losing Women: Attending to Women's Stories in Scottish Witchcraft Trials," in *Culture and Change: Attending to Early Modern Women*, ed. Margaret Mikesell and Adele Seeff (Newark: University of Delaware Press, 2003).

44. Jacques Guillemeau, *Child-birth, or, the Happy Deliuerie of Women.* (Amsterdam: Da Capo Press, 1972).

45. Erasmus, *Colloquies*, 288–89, 293–97, 301.

46. Rubin, *Gentile Tales*, 163.

47. Of all body parts, right hand reliquaries were the most common because the hand mimicked a bishop's blessing gesture. See Thomas P. F. Hoving, "A Newly Discovered Reliquary of St. Thomas Becket," *Gesta* 4 (Spring 1965): 28–30.

48. B. Taylor, "The Hand of St James," *Berkshire Archaeological Journal* 75 (1994–1997): 97–102.

Chapter 9

1. Erasmus of Rotterdam, *A lytell boke of good maners for children [De civilitate morum puerilium] . . . with interpretacion of the same in to the vulgare Englysshe tonge by Robert Whytynton laureate poete* (London, 1532) sigs. A2v-A3r. All citations are to this edition.

2. Norbert Elias, *The Civilizing Process: The History of Manners*, trans. E. Jephcott (Oxford: Basil Blackwell, 1978), 73–80.

3. Elias, *The Civilizing Process*, 73–79.

4. Desiderius Erasmus, *Praise of Folly*, trans. B. Radice (Harmondsworth, UK: Penguin, 1971), 104.

5. Desiderius Erasmus, *The Education of a Christian Prince*, ed. L. Jardine (Cambridge: Cambridge University Press, 1997), 17.

6. Juan Luis Vives, *A Very Frutefull and Pleasant Boke Called the Instruction[n] of a Christen Woma[n]*, trans. R. Hyrd (London, 1529), title page.

7. Stephen Greenblatt, *Renaissance Self-Fashioning from More to Shakespeare* (Chicago: University of Chicago, 1980), 162.

8. Jacob Burckhardt, "The Civilization of the Renaissance in Italy," 1860, in *The Haven of Health*, ed. T. Cogan (London, 1584); Greenblatt, *Renaissance Self-Fashioning*, 162.

9. For an extended discussion of this see, Margaret Healy, *Fictions of Disease in Early Modern England: Bodies, Plagues and Politics* (New York: Palgrave, 2001), 18–49.

10. Michel Foucault, *The Use of Pleasure: The History of Sexuality*, vol. 2, trans. R. Hurley (London: Penguin, 1992), 101.

11. Thomas Paynell, *Regimen sanitatis Salerni* (London, 1528); Thomas Elyot, *The Castel of Health* (London, 1534). All citations are to these editions.

12. Rudolph Wittkower, *Architectural Principles in the Age of Humanism* (London: Academy Editions, 1977), 14.

13. William Bullein, *Bullein's Bulwarke of Defence* (London, 1562), f.lxvij.r. All citations are to this edition.

14. Thomas Newton, *The Touchstone of Complexions* (London, 1576), f.4r. All citations are to this edition.

15. Thomas Cogan, *The Haven of Health* (London, 1584), f.2v. All citations are to this edition.

16. Plato, *The Republic of Plato*, trans. F. MacDonald Cornford (Oxford: Oxford University Press, 1945), 298–99.

17. Justus Lipsius, *Six bookes of politickes or civil doctrine*, trans. W. Jones (London, 1594), 150; *Two Bookes Of Constancie*, trans. J. Stradling (London: Richard Johnes, 1595), 7, 80.

18. Edmund Spenser, "A Letter to the Right Noble, and Valorous, Sir Walter Raleigh Knight," in *The Faerie Queene*, ed. T.P. Roche (New Haven, CT: Yale University Press), 15. All citations are to this edition.

19. Helkiah Crooke, *Microcosmographia* (London, 1618), 270–72.

20. On body heat and hierarchies in the classical context, see Richard Sennet, *Flesh and Stone: The Body and the City in Western Civilization* (London: Penguin, 2002), especially 34.

21. Anon., "A Homily on the State of Matrimony" (1562), in *Certain Sermons or Homilies Appointed to Be Read in Churches* (London, 1563), 265r.

22. T.E., *The Law's Resolution of Women's Rights* (London, 1632), 68–70.

23. Louis Montrose, "The Work of Gender in the Discourse of Discovery" in *New World Encounters*, ed. S. Greenblatt, 177–217 (California: University of California, 1993). See also, Mary C. Fuller, "Raleigh's Fugitive Gold," in Greenblatt, *New World Encounters*, 218–40.

24. Montrose, "Work of Gender," 180.

25. Ibid., 181.

26. Michel de Montaigne, "Of the Caniballes" trans. John Florio, 1603, in *Montaigne's Essays*, 3 vols., ed. Ernest Rhys (London: Everyman, J. M. Dent & Sons Ltd, 1910), vol. 1, Ch. 30, 215–29; 215–16. All citations are to this edition.

27. Ibid., 223

28. Ibid.

29. Ibid., 229

30. Sir Walter Raleigh, *The Discoverie of the Large, Rich and Bewtiful Empyre of Guiana*, trans. N.L. Whitehead (Manchester, UK: Manchester University Press, 1997), 196–98. All citations are to this edition.

31. John Knox, *The First Blast of the Trumpet against the Monstrous Regiment of Women* (London, 1558), f.2r.

32. See, for example, Andrew Belsey and Catherine Belsey, "Icons of Divinity: Portraits of Elizabeth I," in *Renaissance Bodies: The Human Figure in English Culture c. 1540–1660* (Wiltshire, UK: Reaktion Books, 1990), 11–35, especially 11–13.

33. Mary Douglas, *Purity and Danger: An Analysis of the Concepts of Pollution and Taboo* (London: Routledge, 1996), 116.

34. Mark Johnson, "Preface," *The Body in the Mind: The Bodily Basis of Meaning, Imagination, and Reason* (Chicago: University of Chicago Press, 1987), xiv–xv.

35. Charles Webster, "Alchemical and Paracelsian Medicine," in *Health, Medicine and Mortality in the Sixteenth Century*, ed. C. Webster (Cambridge: Cambridge University Press, 1979), 314.

36. R. Bostocke, *Auncient and Later Phisike* (London, 1585), 127.

37. On the coexistence of Galenism and Paracelsianism see Healy, *Fictions of Disease*, 26–59.

38. On representations of early modern syphilis, see Healy, *Fictions of Disease*, 123–87.

39. On gender stereotypes of syphilis and Shakespeare's Venus and Adonis, see Margaret Healy, "Anxious and Fatal Contacts: Taming the Contagious Touch," in *Sensible Flesh: On Touch in Early Modern Culture*, ed. E. Harvey (Philadelphia: University of Pennsylvania Press, 2003), 22–38.

40. On Fracastoro's poem see Healy, "Anxious and Fatal Contacts," 35–37.

41. Desiderius Erasmus, "The Young Man and the Harlot," in *Tudor Translations of the Colloquies of Erasmus*, ed. D. Spurgeon (New York: Scholars Facsimiles and Reprints, 1972).

42. In F. J. Furnivall, *Four Supplications, 1529-1553*, Early English Text Society (London: N. Trubner, 1871), 6.

43. Giambattista Vico, *The New Science of Giambattista Vico*, ed. T. Goddard Bergin and M. Harold Fisch (Ithaca, NY: Cornell University Press, 1970), 405.

44. Maurice Merleau-Ponty, *Phenomenology of Perception* (London: Routledge Classics, 2002), 146.

45. George Lakoff and Mark Johnson, *Metaphors We Live By* (Chicago: University of Chicago Press, 1980).

46. Michael Lambek, "Body and Mind in Mind, Body and Mind in Body: Some Anthropological Interventions in a Long Conversation," in *Body and Persons: Comparative Perspectives from Africa and Melanesia*, ed. M. Lambek and A. Strathern (Cambridge: Cambridge University Press, 1998), 112.

Chapter 10

1. Leo Steinberg, *The Sexuality of Christ in Renaissance Art and in Modern Oblivion*, 2nd ed. (Chicago: University of Chicago Press, 1996).

2. Michele Savonarola, *Il trattato ginecologico-pediatrico in volgare; ad mulieres ferrarienses de regimine pregnantium et noviter natorum usque ad septennium*, ed. L. Belloni (Milan, Italy: Società italiana ostetricia e ginecologia, 1952).

3. Eucharius Rösslin, *When Midwifery Became the Male Physician's Province: The Sixteenth Century Handbook: The Rose Garden for Pregnant Women and Midwives*, trans. and ed. W. Arons (Jefferson, NC: McFarland, 1994).

4. Rösslin, *Midwifery*, 34.

5. Aphra Behn, "Epitaph on the Tombstone of a Child," in *Miscellany, Being a Collection of Poems by Several Hands* (London: J. Hindmarsh, 1685). http://rpo.library.utoronto.ca/poem/143.html.

6. Philippe Ariès, *Centuries of Childhood: A Social History of Family Life,* trans. R. Baldick (New York: Knopf, 1962).

7. Lawrence Stone, *The Family, Sex and Marriage in England, 1500-1800* (New York: Harper & Row, 1977); Steven E. Ozment, *Flesh and Spirit: Private Life in Early Modern Germany* (New York: Viking, 1999). A recent review of this debate is Margaret L. King, "Concepts of Childhood: What We Know and Where We Might Go," *Renaissance Quarterly* 60, no. 2 (2007): 371–407.

8. For admissions to the Innocenti, see Richard C. Trexler, *Power and Dependence in Renaissance Florence*, 1: *The Children of Florence* (Binghamton, NY: Medieval & Renaissance Texts & Studies, 1993), 13, Table 1. See also Philip Gavitt, *Charity and Children in Renaissance Florence: The Ospedale degli Innocenti, 1410–1536* (Ann Arbor: University of Michigan Press, 1990).

9. Francesco Bianchi, *La Ca' Di Dio di Padova nel Quattrocento: Riforma e Governo di un ospedale per l'infanzia abbandonata* (Venice: Istituto veneto di scienze, lettere ed arti, 2005), 215, Table A; Nicholas Terpstra, *Abandoned Children of the Italian Renaissance: Orphan Care in Florence and Bologna* (Baltimore: Johns Hopkins University Press, 2005), 92–93, 98, Graphs 2.1, 2.2, 2.4.

10. Jacob A. Riis, *How the Other Half Lives: Studies among the Tenements of New York* (New York: Charles Scribner's Sons, 1890), 16.2. http://www.yale.edu/amstud/inforev/riis/title.html.

11. See especially Trexler, "Infanticide in Florence: New Sources and First Results," in *Power and Dependence*, 1: *The Children of Florence*; Peter C. Hoffer and N.E.H. Hull, *Murdering Mothers: Infanticide in England and New England, 1558–1803* (New York: New York University Press, 1981); Deborah A. Symonds, *Weep Not for Me: Women, Ballads, and Infanticide in Early Modern Scotland* (University Park: Pennsylvania State University Press, 1997).

12. Charles Dickens, *The Adventures of Oliver Twist* (New York: Oxford University Press, 1978), chapter 2.

13. As displayed in tabular format by David Herlihy, *Medieval Households* (Cambridge, MA: Harvard University Press, 1985), 150–55, Table 6.5.

14. Christiane Klapisch-Zuber, "Female Celibacy and Service in Florence in the Fifteenth Century," in *Women, Family, and Ritual in Renaissance Italy* (Chicago: University of Chicago Press, 1985), 165–77.

15. Anonymous, *Lazarillo de Tormes: His Fortunes and Adversities*, bilingual Spanish/English edition, trans. and ed. G. Stephen Staley, 2003. http://www.4olin.com/ (accessed January 19, 2008).

16. For apprentices, see especially Ilana Krausman Ben-Amos, *Adolescence and Youth in Early Modern England* (New Haven, CT: Yale University Press, 1994); Barbara A. Hanawalt, *Growing Up in Medieval London: The Experience of Childhood in History* (New York: Oxford University Press, 1993).

17. Leon Battista Alberti, *De Iciarchia*, in Alberti, *Opere volgari*, ed. C. Grayson (Bari: Giuseppe Laterza e Figli, 1960–1973), 185–286; Alberti, *The Family in Renaissance Florence (I libri della famiglia)*, trans. and ed. R. Neu Watkins (Columbia: University of South Carolina Press, 1969).

18. For the marriages of Henry VIII, see especially Antonia Fraser, *The Wives of Henry VIII* (New York: Knopf, 1992); D. M. Loades, *Henry VIII and His Queens* (Stroud, UK: A. Sutton, 1996); Retha Warnicke, *The Rise and Fall of Anne Boleyn: Family Politics at the Court of Henry VIII* (New York: Cambridge University Press, 1989); Alison Weir, *The Children of Henry VIII* (New York: Ballantine Books, 1997).

19. For European schooling in this period, see especially, Robert Black, *Humanism and Education in Medieval and Renaissance Italy: Tradition and Innovation in*

Latin Schools from the Twelfth to the Fifteenth Century (Cambridge: Cambridge University Press, 2001); Paul F. Grendler, *Schooling in Renaissance Italy: Literacy and Learning, 1300–1600* (Baltimore: Johns Hopkins University Press, 1989); Nicholas Orme, *From Childhood to Chivalry: The Education of the English Kings and Aristocracy, 1066–1530* (London: Methuen, 1984); Orme, *English Schools in the Middle Ages* (London: Methuen, 1973).

20. Black, *Humanism and Education;* Grendler, *Schooling in Renaissance Italy;* Craig Kallendorf, trans. and ed., *Humanist Educational Treatises* (Cambridge, MA: Harvard University Press, 2002).

21. Desiderius Erasmus, *On Education for Children,* trans. and ed. B.C. Verstraete, 26: 291–346 in *Collected Works of Erasmus,* ed. J.K. Sowards (Toronto: University of Toronto Press, 1985), 305.

22. Desiderius Erasmus, *The Praise of Folly,* trans. B. Radice (Baltimore: Penguin, 1971), 49. http://www.stupidity.com/erasmus/declam13.htm.

23. Cameo portraits of the elderly conqueror Dandolo in Jonathan Phillips, *The Fourth Crusade and the Sack of Constantinople* (New York: Viking, 2004), 58–59, and Donald Queller, *The Fourth Crusade: The Conquest of Constantinople, 1201–1204* (Philadelphia: University of Pennsylvania Press, 1977), 9–10. For age as a requirement for office in Venice, see Robert Finlay, "The Venetian Republic as a Gerontocracy: Age and Politics in the Renaissance," *Journal of Medieval and Renaissance Studies* 8 (1978): 157–78.

24. For Fedele, see Diana Robin's introduction to Cassandra Fedele, *Letters and Orations,* trans. and ed. D. Robin (Chicago: University of Chicago Press, 2000).

25. Petrarch (Francesco Petrarca), *To Posterity.* http://petrarch.petersadlon.com/read_letters.html?s=pet01.html (accessed 7 January 2008). For the experience of aging in the Renaissance, see also Creighton Gilbert, "When Did a Man in the Renaissance Grow Old?" *Studies in the Renaissance* 14 (1967): 7–32.

26. Desiderius Erasmus, *The Praise of Folly,* trans. B. Radice (Baltimore: Penguin, 1971), 13. http://www.stupidity.com/erasmus/declam13.htm.

27. Michel de Montaigne, *The Complete Essays of Montaigne,* trans. D.M. Frame (Stanford: Stanford University Press, 1958), III.13, 835.

28. Montaigne, *Complete Essays,* III.13, 837.

29. Poggio Bracciolini, *On Avarice,* trans. B.G. Kohl and E.B. Welles, 241–89 in Kohl and R.G. Witt, eds., *The Earthly Republic: Italian Humanists on Government and Society* (Philadelphia: University of Pennsylvania Press, 1978), 251.

30. James A. Connor, *Kepler's Witch: An Astronomer's Discovery of Cosmic Order Amid Religious War, Political Intrigue, and the Heresy Trial of His Mother* (San Francisco: HarperSanFrancisco, 2004).

31. David Herlihy, *Medieval Households* (Cambridge, MA: Harvard University Press, 1985), 154–55.

32. Desiderius Erasmus, *The Christian Widow,* trans. J. Tolbert Roberts, in *Erasmus on Women,* ed. E. Rummel (Toronto: University of Toronto Press, 1996), 193.

33. David Chambers and Brian Pullan, eds., *Venice: A Documentary History, 1450–1630* (Oxford: Blackwell, 1992), 120–23.

34. Chambers and Pullan, *Venice: A Documentary History,* 127.

35. Leah Lydia Otis, *Prostitution in Medieval Society: The History of an Urban Institution in Languedoc* (Chicago: University of Chicago Press, 1985), 42.

36. Jacques Rossiaud, *Medieval Prostitution,* trans. L.G. Cochrane (Chicago: University of Chicago Press, 1988), 131–32.

37. Cathy Santore, "Julia Lombardo, 'Somtuosa Meretrize': A Portrait by Property," *Renaissance Quarterly* 41 (1988): 44–83, especially 46–47.

38. For Franco, see Margaret Rosenthal, *The Honest Courtesan: Veronica Franco* (Chicago: University of Chicago Press, 1992). For the *Catalogue,* see Margaret L. King, *Women of the Renaissance* (Chicago: University of Chicago Press, 1991).

39. Rosenthal, *Honest Courtesan,* 128.

40. Ibid., 133.

41. Jacob Rader Marcus, *The Jew in the Medieval World: A Source Book, 315-1791,* rev. ed. Marc Saperstein (Detroit: Wayne State University Press, 1999), 170–72. http://books.google.com/books?id=PCalmtflYtEC.

42. Rosemary Horrox, ed., *The Black Death* (Manchester: Manchester University Press, 1994), 221–22. http://books.google.com/books?id=1O_PX2wVD0sC.

43. For the Jews of Italy, see especially Robert C. Davis and Benjamin C. I. Ravid, *The Jews of Early Modern Venice* (Baltimore: Johns Hopkins University Press, 2001); Brian S. Pullan, *The Jews of Europe and the Inquisition of Venice, 1550–1670* (Totowa, NJ: Barnes & Noble, 1983); Cecil Roth, *The Jews in the Renaissance* (Philadelphia: Jewish Publication Society of America, 1959); David B. Ruderman, *The World of a Renaissance Jew: The Life and Thought of Abraham Ben Mordecai Farissol* (Cincinnati: Hebrew Union College Press, 1981); Moses A. Shulvass, *The Jews in the World of the Renaissance* (Leiden, Netherlands: Brill, 1973); Kenneth R. Stow, *Theater of Acculturation: The Roman Ghetto in the Sixteenth Century* (Seattle: University of Washington Press, 2001).

44. R. Po-chia Hsia, *Trent 1475: Stories of a Ritual Murder Trial* (New Haven, CT: Yale University Press, 1992).

45. Gene Brucker, ed., *The Society of Renaissance Florence: A Documentary Study* (New York: Harper & Row, 1971), 242.

46. Robert Finlay, "The Foundation of the Ghetto: Venice, the Jews, and the War of the League of Cambrai," *Proceedings of the American Philosophical Society* 126, no. 2 (1982): 140–54; David Ravid, "The Religious, Economic and Social Background and Context of the Establishment of the Ghetto of Venice," in *Gli Ebrei a Venezia,* ed. Gaetano Cozzi, 211–59 (Milan: Edizioni di communità, 1987).

47. Chambers and Pullan, *Venice: A Documentary History,* 338.

48. Roberta Curiel, Bernard Dov Cooperman, and Graziano Arici, *The Venetian Ghetto* (New York: Rizzoli, 1990), 99.

49. For this argument, see Penny Schine Gold, *The Lady & the Virgin: Image, Attitude, and Experience in Twelfth-Century France* (Chicago: University of Chicago Press, 1985); Andrew Martindale, "The Child in the Picture: A Medieval Perspective," in ed. Diana Wood, *The Church and Childhood: Papers Read at the 1993 Summer Meeting and the 1994 Winter Meeting of the Ecclesiastical History Society* (Oxford: Published for the Ecclesiastical History Society by Blackwell Publishers, 1994), 197–233; Geraldine A. Johnson, "Sculpture and the Family in Fifteenth-Century Florence," in ed. Giovanni Ciappelli and Patricia Rubin, *Art, Memory, and Family in Renaissance Florence* (Cambridge: Cambridge University Press, 2000), 215–33; Geraldine A. Johnson, "Beautiful Brides and Model Mothers: The Devotional and Talismanic Functions of Early Modern Marian Reliefs," in ed. A. L. McClanan and K. Rosoff Encarnación, *The Material Culture of Sex, Procreation, and Marriage in Premodern Europe* (New York: Palgrave, 2002), 135–62.

50. Martindale, "The Child in the Picture," 226.

51. J.B. Ross, "The Middle-Class Child in Urban Italy, Fourteenth to Early Sixteenth Century," in *The History of Childhood*, ed. L. DeMause (New York: Psychohistory Press, 1974), 199; Ross's searching questions deserve full quotation: "Could these religious pictures represent a secular fantasy of maternal intimacy which the artists themselves probably never knew or could not have observed even in the infancy of their own children? Do they suggest an attempt by adult males to blot out the deprivation suffered in the years with the *balia* and to compensate for this loss by picturing in loving detail the various forms of intimacy between mother and child? And did these pictures afford some emotional compensation for 'the fellowship in feeling' they could not or did not experience to the women whose suckling babies were out of reach?"

52. Geraldine Johnson, "Sculpture and the Family," 220, 221.

53. Klapisch-Zuber comments on the obsession of the learned with maternal lactation in her "Blood Parents and Milk Parents: Wet Nursing in Florence, 1300–1530," in *Women, Family and Ritual,* 132–64 at 161–62. Valerie A. Fildes, *Wet Nursing: A History from Antiquity to the Present* (Oxford: Basil Blackwell, 1988), presents an overview of the literature at 32–111.

54. Fildes, *Wet Nursing,* 9.

55. Suzanne Dixon, *The Roman Mother* (Norman: University of Oklahoma Press, 1988), 115; Valerie A. Fildes, *Breasts, Bottles, and Babies: A History of Infant Feeding* (Edinburgh: Edinburgh University Press, 1986), 30.

56. Francesco Barbaro, "On Wifely Duties," in *The Earthly Republic: Italian Humanists on Government and Society*, ed. B.G. Kohl and R.G. Witt, 223–24 (Philadelphia: University of Pennsylvania Press, 1978).

57. Juan Luis Vives, *The Education of a Christian Woman: A Sixteenth-Century Manual,* trans. and ed. C. Fantazzi (Chicago: University of Chicago Press, 2000), 53.

58. Desiderius Erasmus, *Colloquies*, ed. and trans. C. R. Thompson, in *Collected Works of Erasmus* (Toronto: University of Toronto Press, 1997), 595–96.

59. For Gouge's advice to mothers, see Patricia Nardi, *Mothers at Home: Their Role in Childrearing and Instruction in Early Modern England* (dissertation, City University of New York, 2007), 47–50.

60. Quoted in Janet Lynne Golden, *A Social History of Wet Nursing in America: From Breast to Bottle* (Cambridge: Cambridge University Press, 1996), 11.

61. Fildes, *Wet Nursing,* passim; Klapisch-Zuber, "Blood Parents and Milk Parents."

62. Patricia Crawford, "The Construction and Experience of Maternity in Seventeenth-Century England," in *Women as Mothers in Preindustrial England,* ed. Valerie Fildes (London: Routledge, 1990), 11–12, 23–24; Patricia Crawford and Sara Mendelson, *Women in Early Modern England, 1550–1720* (Oxford: Clarendon Press, 1998), 154–56; Lisa Klein, "Lady Anne Clifford as Mother and Matriarch: Domestic and Dynastic Issues in Her Life and Writings," *Journal of Family History* 21 (2001): 29; Alan Macfarlane, *The Family Life of Ralph Josselin, a Seventeenth-Century Clergyman; an Essay in Historical Anthropology* (Cambridge: University Press, 1970), 86–87. Elizabeth Clinton wrote a whole pamphlet (1622) urging her daughters to nurse their own children: *The Countesse of Lincolnes Nurserie* (Oxford: Iohn Lichfield, and Iames Short printers to the famous Vniversitie, 1622; reprint Norwich, NJ: Theatrum Orbis Terrarum, 1975).

63. Felicity Nussbaum, *Torrid Zones: Maternity, Sexuality, and Empire in Eighteenth-Century English Narratives* (Baltimore: Johns Hopkins University Press, 1995), 25.

64. Rudolf Bell, *Holy Anorexia* (Chicago: University of Chicago Press, 1985), 154.

65. Bell, *Holy Anorexia,* 98; Marvello Craveri, *Sante e streghe: biografie e documenti dal XIV Al XVII secolo,* 2nd ed. (Milan: Feltrinelli economica, 1981), 114.

66. Bell, *Holy Anorexia,* 91; Richard Kieckhefer, *Unquiet Souls: Fourteenth-Century Saints and Their Religious Milieu* (Chicago: University of Chicago Press, 1984), 188; Elizabeth Petroff, *Medieval Women's Visionary Literature* (Oxford: Oxford University Press, 1986), 42.

67. Kieckhefer, *Unquiet Souls,* 26, 117.

68. Bell, *Holy Anorexia,* 22–53.

69. Craveri, *Sante e streghe,* 70.

70. Teresa of Avila, *The Life of Saint Teresa of Avila by Herself,* trans. J.M. Cohen (New York: Viking Penguin, 1957), 210, chapter 29.

71. Dante Alighieri, *Divine Comedy,* trans. A. Mandelbaum (New York: Knopf, 1995), *Paradiso* 17:87–89. http://dante.ilt.columbia.edu/comedy/.

72. Niccolò Machiavelli, *History of Florence and of the Affairs of Italy: From the Earliest Times to the Death of Lorenzo the Magnificent,* trans. M.W. Dunne (Colonial Press, 1901), 49. http://books.google.com/books?id=q-gXH2bioqcC&dq=machiavelli+history+florence.

73. For Gattamelata, see especially Giovanni Eroli, *Erasmo Gattamelata Da Narni; suoi monumenti e sua famiglia* (Rome: Coi tipi del Salviucci, 1877); also Margaret L. King, *Death of the Child Valerio Marcello* (Chicago: University of Chicago Press, 1994), 82 and 87 for his equestrian monument.

74. For Colleoni's monument, see King, *Death of the Child,* 82 and 87.

75. For Sforza, see William Pollard Urquhart, *Life and Times of Francesco Sforza, Duke of Milan, with a Preliminary Sketch of the History of Italy* (Edinburgh: London, W. Blackwood and Sons, 1852); for his military enterprise, Maria Nadia Covini, *L'esercito del duca: organizzazione militare e istituzioni al tempo degli Sforza (1450-1480)* (Rome: Istituto storico italiano per il Medioevo, 1998); for his patronage of letters, Gary Ianziti, *Humanistic Historiography under the Sforzas: Politics and Propaganda in Fifteenth-Century Milan* (Oxford: Clarendon Press, 1988).

76. King, *Death of the Child,* 31.

77. Ibid., 198–99.

78. Baldassare Castiglione, *The Book of the Courtier: An Authoritative Text Criticism,* trans. Charles Singleton, ed. D. Javitch (New York: W.W. Norton, 2002), 13, no. 1.5.

79. Castiglione, *Courtier,* 22, no. 1:14.

80. Ibid., 25, no. 1.17.

81. Ibid., 32, no. 1.26.

82. Ibid., 150, no. 3:4.

BIBLIOGRAPHY

Aikema, Bernard, and Beverly Louis Brown, eds. *Il Rinascimento a Venezia e la Pittura del Nord ai Tempi di Bellini, Dürer, Tiziano*, exhibition catalogue. Venice: Bompiani, 1999.

Alberti, Leon Battista. "De Iciarchia." In *Opere volgari*, edited by C. Grayson, 185–286. Vol. 3. Bari, Italy: Giuseppe Laterza e Figli, 1960–1973.

Alberti, Leon Battista. *The Family in Renaissance Florence* [I libri della famiglia]. Translated and edited by R.N. Watkins. Columbia: University of South Carolina Press, 1969.

Alberti, Leon Battista. *On the Art of Building in Ten Books*. Translated by J. Rykwert, N. Leach, and R. Tavernor. Cambridge, MA: MIT Press, 1991.

Alberti, Leon Battista. *On Painting and on Sculpture*. Translated and edited by C. Grayson. London: Phaidon, 1972.

Albertini, Tamara. "Intellect and Will in Marsilio Ficino: Two Correlatives of a Renaissance Concept of Mind." In *Marsilio Ficino: His Theology, His Philosophy, His Legacy*, edited by M.J.B. Allen and V. Rees, 203–25. Leiden, Netherlands: Brill, 2002.

Allaire, G. "Noble Saracens or Muslim Enemy? The Changing Image of the Saracen in Late Medieval Italian Literature." In *Western Views of Islam in Medieval and Early Modern Europe*, edited by D.R. Blanks and M. Frassetto, 173–84. New York: St. Martin's Press, 1999.

Allen, Michael J.B. "Ficino's Theory of the Five Substances and the Neoplatonists' Parmenides." *Journal of Medieval and Renaissance Studies* 12 (1982): 19–44.

Allen, Michael J.B. *The Platonism of Marsilio Ficino: A Study of the Phaedrus Commentary, Its Sources and Genesis*. Berkeley: University of California Press, 1984.

Alpers, Svetlana L. "Ekphrasis and Aesthetic Attitudes in Vasari's Lives." *Journal of the Warburg and Courtauld Institutes* 23 (1960): 190–215.

Anon. *Ein wunderbarliche seltzame erschröckliche Geburt*. Augsburg, Germany: Michael Moser, 1561.

Anon. "A Homily on the State of Matrimony." In *Certain Sermons or Homilies Appointed to Be Read in Churches*. London, 1563.

Anon. *Lazarillo de Tormes: His Fortunes and Adversities*. Translated and edited by G. S. Staley, 2003. http://www.4olin.com/.

Anon. *Ovide moralisé. Poème du Commencement du Quatorzième siècle publié d'après tous les Manuscrits Connus*. Edited by C. de Boer, M. G. de Boer, and J.T.M. van't Sant. 5 vols. Amsterdam: Johannes Müller, 1915–1954.

Anon. "A Warning to Be Ware" in *The Minor Poems of the Vernon MS*. Part II, edited by F. J. Furnivall. London: Kegan Paul, 1901.

Antonioli, Roland. *Rabelais et la Médecine*. Tome 12. *Etudes Rabelaisiennes*. Geneva, Switzerland: Droz, 1976.

Ariès, Philippe. *Centuries of Childhood: A Social History of Family Life*. Translated by Robert Baldick. New York: Knopf, 1962.

Ariès, Philippe. *The Hour of Our Death*. Translated by Helen Weaver. New York: Alfred Knopf, 1981.

Ariès, Philippe. *Images of Man and Death*. Translated by Janet Lloyd. Cambridge, MA: Harvard University Press, 1985.

Ariès, Philippe. *Western Attitudes toward Death: From the Middle Ages to the Present*. Translated by Patricia M. Ranum. Baltimore: Johns Hopkins University Press, 1974.

Arnulf of Orléans. "Allegoriae." In *Arnolfo d'Orléans, un cultore di Ovidio nel secolo XII*, edited by Fausto Ghisalberti, *Memorie del Reale Istituto lombardo di scienze e lettere* 24. Milan: Libraio del R. Istituto Lombardo di scienze e lettere, 1932.

Aristotle. "On the Art of Poetry." In *Classical Literary Criticism*, edited and translated by T. S. Dorsch. Harmondsworth, UK: Penguin, 1969.

Arons, Wendy, trans. *Eucharius Rösslin: When Midwifery Became the Male Physician's Province*. Jefferson, NC: McFarland, 1994.

Arrizabalaga, Jon, John Henderson, and Roger French. *The Great Pox: The French Disease in Renaissance Europe*. New Haven, CT: Yale University Press, 1997.

Bacalexi, Dina. "Trois traducteurs (N. Leoniceno, G. Cop et L. Fuchs) de Galien au XVIe siècle et leur regard sur la tradition arabe," *50ème Colloque International d'Etudes Humanistes*, July 2007.

Bacon, Francis. *The Essayes or Counsels Civill & Morall*. London: J. M. Dent, 1900.

Bacon, Francis. *The Essayes or Counsels, Civill and Morall*, edited by M. Kiernan. Cambridge, MA: Harvard University Press, 1985.

Bakhtin, Mikhail. *Rabelais and His World*. Translated by H. Iswolsky. Bloomington: Indiana University Press, 1984.

Barasch, Moshe. "Character and Physiognomy: Bocchi on Donatello's St George: A Renaissance Text on Expression in Art." *Journal of the History of Ideas* 36 (1975): 413–30.

Barbaro, Francesco. "On Wifely Duties." In *The Earthly Republic: Italian Humanists on Government and Society*, edited by B. G. Kohl and R. G. Witt, 223–24. Philadelphia: University of Pennsylvania Press, 1978.

Barber, Paul. *Vampires, Burial and Death: Folklore and Reality*. New Haven, CT: Yale University Press, 1985.

Barker, Francis. *The Tremulous Private Body: Essays on Subjection*. London: Methuen, 1984.

Barry, Jonathan, and C. Jones, eds. *Medicine and Charity before the Welfare State*. London: Routledge, 1991.

Bartlett, Robert. *The Hanged Man: A Story of Miracle, Memory, and Colonialism in the Middle Ages*. Princeton, NJ: Princeton University Press, 2004.

Bartlett, Wayne, and Flavia Idriceanu. *Legends of Blood: The Vampire in History and Myth*. Westport, CT: Praeger, 2006.

Bates, Don, ed. *Knowledge and the Scholarly Medical Traditions*. Cambridge: Cambridge University Press, 1995.

Baxandall, Michael. *Giotto and the Orators: Humanist Observers of Painting in Italy and the Discovery of Pictorial Composition 1350–1450*. Oxford: Oxford University Press, 1986.

Baxandall, Michael. *Painting and Experience in Fifteenth Century Italy*. Oxford: Clarendon Press, 1972.

Becker, Lucinda. "The Absent Body: Representations of Dying Early Modern Women in a Selection of Seventeenth-Century Diaries." *Women's Writing* 8, no. 2 (2001): 251–62.

Beecher, Donald. "The Books of Wonder of Nicholas Monardes of Seville." *Cahiers Elisabéthains* 51 (1997): 1–13.

Beecher, Donald. "The Legacy of John Frampton: Elizabethan Trader and Translator." *Renaissance Studies* 20, no. 3 (2006): 320–39.

Behn, Aphra. "Epitaph on the Tombstone of a Child." In *Miscellany, Being a Collection of Poems by Several Hands*. London: J. Hindmarsh, 1685. http://rpo.library.utoronto.ca/poem/143.html.

Bell, Rudolph M. *Holy Anorexia*. Chicago: University of Chicago Press, 1985.

Bell, Rudolph M. *How to Do It: Guides to Good Living for Renaissance Italians*. Chicago: University of Chicago Press, 1999.

Belsey, Andrew, and Catherine. "Icons of Divinity: Portraits of Elizabeth I." In *Renaissance Bodies: The Human Figure in English Culture c. 1540–1660*. Wiltshire, UK: Reaktion Books, 1990.

Bembo, Pietro. *Gli Asolani di messer Pietro Bembo*. Venice: Aldo Romano, 1505.

Bembo, Pietro, *Gli Asolani*, in *Pietro Bembo, Opere in volgare*, ed. Mario Marti (Florence: Sansoni, 1961), 9–163.

Ben-Amos, Ilana Krausman. *Adolescence and Youth in Early Modern England*. New Haven, CT: Yale University Press, 1994.

Benthien, Claudia. *Skin: On the Cultural Border between Self and the World*. New York: Columbia University Press, 2002.

Beolco, A. (Ruzante). *Teatro*. Edited by Alvise Zorzi. Turin, Italy: Einaudi, 1967.

Bertelli, Sergio. *The King's Body: Sacred Rituals of Power in Medieval and Early Modern Europe*. Translated by R. Burr Litchfield. University Park: Pennsylvania State University Press, 2001.

Bertier, Janine. "Un traité scolastique de Medicine des enfants: Le Pedenemicon de Gabriel Miron." In *Santé, Médecine et Assistance au moyen âge*, 9–22. Paris: Editions du CTHS, 1987.

Bettella, Patrizia. "Antonfrancesco Grazzini and Maffio Venier. Poetry on Prostitutes." In *Dialogue with the Other Voice*, edited by Julie D. Campbell, Maria Galli Stampino. Toronto: Centre for Renaissance and Reformation Studies, forthcoming.

Bettella, Patrizia. *The Ugly Woman: Transgressive Aesthetic Models in Italian Poetry from the Middle Ages to the Baroque*. Toronto: University of Toronto Press, 2005.

Beverwijck, J. van. *Treatise on the Stone*. Amsterdam, 1652.

Bianchi, Francesco. *La Ca' Di Dio di Padova nel Quattrocento: riforma e governo di un ospedale per l'infanzia abbandonata*. Venice: Istituto veneto di scienze, lettere ed arti, 2005.

Bisha, N. " 'New Barbarians' or Worthy Adversary? Humanist Construct of the Ottoman Turks in Fifteenth-Century Italy." In *Western Views of Islam in Medieval and*

Early Modern Europe, edited by D.R. Blanks and M. Frassetto, 185–205. New York: St. Martin's Press, 1999.

Black, Robert. *Humanism and Education in Medieval and Renaissance Italy: Tradition and Innovation in Latin Schools from the Twelfth to the Fifteenth Century*. Cambridge: Cambridge University Press, 2001.

Blanks, David R., and Michael Frassetto, eds. *Western Views of Islam in Medieval and Early Modern Europe*. New York: St. Martin's Press, 1999.

Blécourt, Willem de. "Witch Doctors, Soothsayers and Priests: On Cunning Folk in European Historiography and Tradition." *Social History* 19 (1994): 285–303.

Blumenfeld-Kosinski, Renate. *Not of Woman Born: Representations of Caesarean Birth in Medieval and Renaissance Culture*. Ithaca, NY: Cornell University Press, 1990.

Boaistuau, Pierre. *Histoires prodigieuses*. Edited by G. Mathieu-Castellani. Geneva, Switzerland: Slatkin, 1996.

Bocchi, Francesco. "Eccellenza del San Giorgio di Donatello." In *Trattati d'arte del Cinquecento,* edited by Paola Barocchi, vol. 3, 125–94. Bari, Italy: Laterza, 1962.

Boehrer, Bruce. *The Fury of Men's Gullets: Ben Jonson and the Digestive Canal*. Philadelphia: University of Pennsylvania Press, 1997.

Boethius, Anicius Manlius Severinus. *The Consolation of Philosophy*. Edited and translated by W. Anderson. Carbondale: Southern Illinois University Press, 1963.

Bonfil, Robert. *Gli Ebrei in Italia all'epoca del Rinascimento*. Florence: Sansoni, 1991.

Boon, Sonja. "Last Rites, Last Rights: Corporeal Abjection as Autobiographical Performance in Suzanne Curchod Nicker's Des inhumations précipitées (1790)," *Eighteenth-Century Fiction* 21 (2008): 89–107.

Bostocke, R. *Auncient and Later Phisike*. London, 1585.

Bouchel, Laurens. *La Bibliothèque ou thresor du droict françois, auquel sont traictees les matieres Civiles, Criminelles, & Beneficiales, tant reglées par les Ordonnances, & Coustumes de France, que decidées par Arrests des Cours Souveraines*, 3 vols. Paris: Chez vefve Nicolas Buon, 1639.

Boursier, Louise. *Récit véritable de la naissance des messeigneurs et dames les enfans de France, Instruction à ma fille, et autres textes*. Edited by François Rouget and Colette H. Winn. Geneva, Switzerland: Droz, 2000.

Bowen, Barbara C. "The Honorable Art of Farting in Continental Renaissance Literature." In *Fecal Matters in Early Modern Literature and Art: Studies in Scatology,* edited by J. Parsells and R. Garim, 1–13. Burlington, VT: Ashgate, 2006.

Bracciolini, Poggio. *On Avarice*. Translated by B. G. Kohl and E. B. Welles. In *The Earthly Republic: Italian Humanists on Government and Society*, edited by B.G. Kohl and R. G. Witt, 241–89. Philadelphia: University of Pennsylvania Press, 1978.

Briggs, Robin. "Women as Victims? Witches, Judges, and the Community." *French History* 5 (1991): 438–80.

Bromilow, Pollie. "Inside Out: Female Bodies in Rabelais." *Forum for Modern Languages Studies* 44, no. 1 (2007): 1–13.

Bronfen, Elisabeth. *Over Her Dead Body: Death, Femininity and the Aesthetic*. Manchester, UK: Manchester University Press, 1992.

Broomhall, Susan. "Rabelais, the Pursuit of Knowledge, and Early Modern Gynaecology." *Limina: A Journal of Historical and Cultural Studies* 4 (1998): 24–34.

Broomhall, Susan. *Women's Medical Work in Early Modern France*. Manchester, UK: Manchester University Press, 2004.

Brotton, Jerry. *The Renaissance Bazaar: From the Silk Road to Michelangelo*. Oxford: Oxford University Press, 2003.

Brown, D. A. and J. Van Nimmen. *Raphael and the Beautiful Banker: The Story of the Bindo Altoviti Portrait*. New Haven, CT: Yale University Press, 2005.

Brown, David Allen, ed. *Virtue and Beauty: Leonardo's "Ginevra de' Benci" and Renaissance Portraits of Women*, exhibition catalogue. Princeton, NJ: Princeton University Press, 2001.

Brown, Ruth. "Middens and Miasma: A Portrait of Seventeenth Century Village Life in Banburyshire." *Cake and Cockhorse* 16, no. 1 (2003): 2–8.

Brucioli, A. *A Commentary upon the Canticle of Canticles*. London: R. Field for Tho. Man, 1598.

Brucker, Gene A., ed. *The Society of Renaissance Florence: A Documentary Study*. New York: Harper & Row, 1971.

Bryant, Nigel. *The High Book of the Grail: A Translation of the Thirteenth Century Romance of Perlesvaus*. Cambridge, UK: D. S. Brewer, 1978.

Bryson, Anna. "The Rhetoric of Status: Gesture, Demeanour and the Image of the Gentleman in Sixteenth- and Seventeeth-Century England." In *Renaissance Bodies: The Human Figure in English Culture c. 1540–1660*, edited by L. Gent and N. Llewellyn, 136–53. London: Reaktion, 1993.

Bullein, William. *Bullein's Bulwarke of Defence*. London, 1562.

Bullein, William. *A Newe Booke Entituled the Governement of Healthe*. London, 1558.

Bullough, Vern L. *The Development of Medicine as a Profession: The Contribution of the Medieval University to Modern Medicine*. Basel, Switzerland: S. Karger, 1966.

Burckhardt, Jacob. *The Civilization of the Renaissance in Italy*. Translated by S.C.G. Middlemore. New York: Harper, 1958.

Burckhardt, Jacob. "The Civilization of the Renaissance in Italy, 1860." In *The Haven of Health*, edited by T. Cogan. London, 1584.

Burke, Peter. *The Fortunes of the Courtier: The European Reception of Castiglione's Cortegiano*. University Park: Pennsylvania University Press, 1996.

Burke, Peter. *A Social History of Knowledge: From Gutenberg to Diderot*. Cambridge, UK: Polity, 2000.

Burshatin, Israel. "Written on the Body: Slave or Hermaphrodite in Sixteenth-Century Spain." In *Queer Iberia: Sexualities, Cultures, and Crossings from the Middle Ages to the Renaissance*, edited by J. Blackmore and G. S. Hutcheson. Durham, NC: Duke University Press, 1999.

Butler, Judith P. *Bodies That Matter: On the Discursive Limits of "Sex."* New York: Routledge, 1993.

Butler, Judith P. *Gender Trouble: Feminism and the Subversion of Identity*. New York: Routledge, 1990.

Bynum, Caroline Walker. *Jesus as Mother: Studies in the Spirituality of the High Middle Ages*. Berkeley: University of California Press, 1982.

Bynum, Caroline Walker. *The Resurrection of the Body in Western Christianity, 200–1336*. New York: Columbia University Press, 1995.

Cachey, T. J., Jr. "Between Humanism and New Historicism Rewriting the New World Encounter." *Annali d'italianistica* 10 (1992): 28–46.

Caciola, Nancy. "Spirits Seeking Bodies: Death, Possession and Communal Memory in the Middle Ages." In *The Place of the Dead: Death and Remembrance in Late Medieval and Early Modern Europe*, edited by B. Gordon and P. Marshall, 66–86. Cambridge: Cambridge University Press, 2000.

Cadamosto, Alvise. *The Voyages of Cadamosto and Other Documents on Western Africa Second Half of the Fifteenth Century.* London: Hakluyt Society, 1937.

Calfhill, James. *An Answer to John Martiall's Treatise on the Cross.* Edited for the Parker Society by Richard Gibbings. Cambridge: University Press, 1846.

Calvin, John, *Treatise on Relics,* http://www.godrules.net/library/calvin/176calvin4.htm.

Campbell, Erin. "'Unenduring' Beauty: Gender and Old Age in Early Modern Art and Aesthetics." In *Growing Old in Early Modern Europe: Cultural Representations,* edited by Erin Campbell, 153–67. Aldershot, UK: Ashgate, 2006.

Camporesi, Piero. *The Anatomy of the Senses: Natural Symbols in Medieval and Early Modern Italy.* Translated by A. Cameron. Oxford: Polity Press, 1994.

Camporesi, Piero. *The Incorruptible Flesh: Bodily Mutation and Mortification in Religion and Folklore.* Translated by Tania Croft-Murray. Cambridge: Cambridge University Press, 1988.

Camporesi, Piero. *The Juice of Life: The Symbolic and Magic Significance of Blood.* Translated by R. R. Barr. New York: Continuum, 1995.

Camporesi, Piero. *The Land of Hunger.* Cambridge, UK: Polity Press, 1996.

Cantor, David, ed. *Reinventing Hippocrates.* Aldershot, UK: Ashgate, 2002.

Carboni, Stefano, ed. *Venice and the Islamic World: 828–1797.* New York: Metropolitan Museum of Art, 2007.

Carlino, Andrea. *Books of the Body: Anatomical Ritual and Renaissance Learning.* Translated by John Tedeschi and Anne C. Tedeschi. Chicago: University of Chicago Press, 1999.

Carlino, Andrea. "Le médecin et l'antiquaire. Une rencontre à Padoue vers la moitié du XVIe siècle," *50ème Colloque International d'Etudes Humanistes,* July 2007.

Cary, Walter. *A Breefe Treatise Called Carye's Farewell to Physick.* London, 1587.

Casa, Giovanni della. *Galateo.* Edited by G. Manganelli and C. Milanini. Milan: Rizzoli, 1984.

Casa, Giovanni della. *Galateo.* Translated by Konrad Eisenbichler and Kenneth R. Bartlett. Toronto: Centre for Reformation and Renaissance Studies, 1986.

Castiglione, Baldesar. *The Book of the Courtier.* Translated by George Bull. London: Penguin, 1976.

Castiglione, Baldassar. *The Book of the Courtier: An Authoritative Text Criticism.* Translated by C. Singleton and edited by D. Javitch. New York: W. W. Norton, 2002.

Castiglione, Baldassar. *Il Libro del Cortegiano.* Edited by A. Quondam. Milan: Garzanti, 1981.

Castiglione, Baldassar. *Il libro del Cortegiano.* Translated by George Bull. New York: Penguin, 2006.

Cavallo, Sandra. *Artisans of the Body in Early Modern Italy.* Manchester, UK: Manchester University Press, 2008.

Cavallo, Sandra. "Health, Beauty and Hygiene." In *At Home in Renaissance Italy,* exhibition catalogue, edited by M. Wollmar-Ajmar and F. Dennis, 174–87. London: V & A Publications, 2006.

Céard, Jean. *La Nature et les Prodiges: L'insolite au XVIe siècle.* Geneva: Droz, 1977.

Cennino Cennini. *The Craftsman's Handbook. "Il Libro dell'arte."* Translated by D. V. Thompson. New York: Dover, 1960.

Chambers, D., and B. S. Pullan, eds. *Venice: A Documentary History, 1450–1630.* Oxford: Basil Blackwell, 1992.

Champier, Symphorien. *Nef des dames vertueuses composees par maistre Simphorien Champier Docteur en medicine contenant quatre livres*. Paris: Jehan de la Garde, 1515.

Chance, Jane. *Medieval Mythography from Roman North Africa to the School of Chartres, A.D. 433–1177*. Gainesville: University of Florida Press, 1994.

Charlton, C.A.C. *The Urological System*. Harmondsworth, UK: Penguin, 1973.

Chartier, Roger, ed. *A History of Private Life: Passions of the Renaissance*. Cambridge, MA: Belknap Press, 1989.

Chastel, André. *Art et humanisme à Florence au temps de Laurent le Magnifique: études sur la Renaissance*. Paris: Presses Universitaires de France, 1959.

Cicero. *Brutus: Orator*. Translated by G. L. Hendrickson and H. M. Hubbard. London: Heinemann, 1939.

Cicero. *De Oratore*. Translated by E. W. Sutton. Cambridge, MA: Harvard University Press, 1996.

Cipolla, Carlo M. *Fighting the Plague in Seventeenth-Century Italy*. Madison: University of Wisconsin Press, 1981.

Cipolla, Carlo M. *Public Health and the Medical Profession in the Renaissance*. Cambridge: Cambridge University Press, 1976.

Classen, C., D. Howes and A. Synnott. *Aroma: The Cultural History of Smell*. London: Routledge, 1994.

Classen, C., D. Howes, and A. Synnott. "The Witch's Senses: Sensory Ideologies and Transgressive Femininities from the Renaissance to Modernity." In *The Empire of the Senses: The Sensual Culture Reader*, edited by D. Howes, 70–84. Oxford: Berg, 2005.

Clayton, Martin. *Leonardo da Vinci: The Divine and the Grotesque*. London: Royal Collection Enterprise, 2002.

Clement of Alexandria. "Stromata." In *The Ante-Nicene Fathers. Translations of the Writings of the Fathers down to A.D. 325*. Alexander Roberts and James Donaldson, 9 vols. New York: Charles Scribner's Sons, 1896–1926.

Clinton, Elizabeth. *Countess of Lincoln: The Countesse of Lincolnes Nurserie*. Oxford: Iohn Lichfield, and Iames Short printers to the famous Vniversitie, 1622.

Clowes, William. *A Frutefull and Approved Treatise*. London, 1602.

Clowes, William. *A Short and Profitable Treatise Touching the Cure of the Disease Called (Morbus Gallos)*. London, 1579.

Cockayne, Emily. *Hubbub: Filth, Noise, and Stench in England, 1600–1770*. New Haven, CT: Yale University Press, 2007.

Cody, Lisa Forman. *Birthing the Nation: Sex, Science, and the Conception of Eighteenth-Century Britons*. Oxford: Oxford University Press, 2005.

Cody, Lisa Forman. "Living and Dying in Georgian London's Lying-In Hospitals." *Bulletin of the History of Medicine* 78, no. 2 (2004): 309–48.

Cogan, Thomas. *The Haven of Health, Chiefly made for the Comfort of Students, and consequently for all those that have a care of their health, amplified upon five words of HIPPOCRATES, written Epid. 6. Labour, Meat, Drinke, Sleepe, Venus*. London: Melch Bradwood for John Norton, 1612.

Cohen, William A., and Ryan Johnson, eds. *Filth: Dirt, Disgust and Modern Life*. Minneapolis: University of Minnesota Press, 2005.

Comfort, William Wistar. "The Saracens in Italian Epic Poetry." *Periodical of the Modern Language Association* 59 (1944): 882–910.

Connor, James A. *Kepler's Witch: An Astronomer's Discovery of Cosmic Order Amid Religious War, Political Intrigue, and the Heresy Trial of His Mother*. San Francisco: HarperSanFrancisco, 2004.

Cook, Harold J. *Trials of an Ordinary Doctor: Joannes Groenevelt in Seventeenth-Century London*. Baltimore: John Hopkins University Press, 1985.

Corbin, Alain. *The Foul and Fragrant: Odor and the French Social Imagination*. Cambridge, MA: Harvard University Press, 1986.

Cotta, John. *Short Discoverie of the Unobserved Dangers of severall sorts of ignorant and unconsiderate practisers of physicke in England*. London, 1612.

Couliano, Ioan P. *Eros and Magic in the Renaissance*. Translated by M. Cook. Chicago: University of Chicago Press, 1987.

Covini, Maria Nadia. *L'esercito del Duca: Organizzazione Militare ei Istituzioni al Tempo degli Sforza (1450–1480)*. Rome: Istituto storico italiano per il Medioevo, 1998.

Craveri, Marcello. *Sante e Streghe: Biografie e Documenti dal XIV Al XVII Secolo*. 2nd ed. Milan: Feltrinelli economica, 1981.

Crawford, Katherine. *European Sexualities, 1400–1800*. Cambridge: Cambridge University Press, 2007.

Crawford, Katherine. "Marsilio Ficino, Neoplatonism, and the Problem of Sex." *Renaissance et Réforme*, 28 (Spring), 2004.

Crawford, Patricia. "The Construction and Experience of Maternity in Seventeenth-Century England." In *Women as Mothers in Preindustrial England*, edited by V. Fildes, 3–38. London: Routledge, 1990.

Cressy, David. *Birth, Marriage, and Death: Ritual, Religion, and the Life-Cycle in Tudor and Stuart England*. Oxford: Oxford University Press, 1997.

Crooke, Helkiah. *Microcosmographia*. London: Printed by William Iaggard, 1616.

Cropper, Elizabeth. "The Beauty of Women: Problems in the Rhetoric of Renaissance Portraiture." In *Rewriting the Renaissance: The Discourses of Sexual Difference in Renaissance Europe*, edited by M.W. Ferguson, M. Quilligan and N. J. Vickers, 175–90. Chicago: University of Chicago Press, 1986.

Cropper, Elizabeth. "On Beautiful Women: Parmigianino, Petrarchismo, and the Vernacular Style." *Art Bulletin* 58 (1976): 374–94.

Culpeper, Nicholas. *The English Physician*. London: Printed for the benefit of the Commonwealth of England, 1652.

Cunningham, Andrew. *The Anatomical Renaissance: The Resurrection of the Anatomical Projects of the Ancients*. Aldershot, UK: Ashgate/Scolar Press, 1997.

Curiel, R., B.D. Cooperman, and G. Arici. *The Venetian Ghetto*. New York: Rizzoli, 1990.

Dalton, Michael. *The Countrey Justice, Containing the practice of Justices of the Peace out of their Sessions: Gathered for the better helpe of such Justices of Peace as have not beene conversant in the studie of the Lawes of this Realme*. London: Miles Flesher, James Haviland, and Robert Young, the assignes of Iohn More Esquire, 1630.

Dante Alighieri. *The Divine Comedy of Dante Alighieri*. Translated by John Aitken Carlyle and Philip H. Wicksteed. New York: Modern Library, 1932.

Dante Alighieri. *The Divine Comedy*. Translated by A. Mandelbaum. New York: Knopf; 1995. http://dante.ilt.columbia.edu/comedy/.

Danti, Vincenzio. "Il primo libro del trattato delle perfette proporzioni." In *Trattati d'arte del Cinquecento fra Manierismo e Controriforma*, edited by Paola Barocchi, 209–69. Bari, Italy: Laterza, 1960.

Daston, Lorraine, and Katharine Park. *Wonders and the Order of Nature 1150–1750*. New York: Zone Books, 1998.

Davis, Natalie Zemon. "Proverbial Wisdom and Popular Error." In *Society and Culture in Early Modern France*, 227–667. Stanford, CA: Stanford University Press, 1975.

Davis, R. C., and B. C. I. Ravid. *The Jews of Early Modern Venice*. Baltimore: Johns Hopkins University Press, 2001.

De Beer, E. S., ed. *The Diary of John Evelyn*. Oxford: Clarendon Press, 1955.

Debus, Allen G. *The Chemical Philosophy: Paracelsian Science and Medicine in the Sixteenth and Seventeenth Centuries*. New York: Science History Publications, 1977.

Debus, Allen G. *English Paracelsians*. London: Oldbourne, 1968.

Debus, Allen G. *The French Paracelsians: The Chemical Challenge to Medical and Scientific Tradition in Early Modern France*. Cambridge: Cambridge University Press, 1991.

Defoe, Daniel. *A Journal of the Plague Year*. Edited by L. Landa. Oxford: Oxford University Press, 1990.

Dekker, Thomas. *The Batchelars Banquet: or a Banquet for Batchelars: Wherein is prepared sundry daintie dishes to furnish their Table, curiously drest, and seriously served in. Pleasantly discoursing the variable humours of Women, their quicknesse of wittes, and searchable deceits*. London: Printed by T[homas] C[reede], 1603.

Dekker, Thomas. *London Looke Backe*. London, 1630.

Dekker, Thomas. *Newes from Graves-end*. London, 1604.

Dekker, Thomas. *The Wonderfull Yeare*. London, 1603.

D'Elia, Anthony F. *The Renaissance of Marriage in Fifteenth-Century Italy*. Cambridge, MA: Harvard University Press, 2004.

Dempsey, Charles. *The Portrayal of Love. Botticelli's "Primavera" and Humanist Culture at the Time of Lorenzo the Magnificent*. Princeton, NJ: Princeton University Press, 1992.

De Tolnay, Charles. *Michelangelo*. Princeton, NJ: Princeton University Press, 1943–1960.

Dickens, Charles. *The Adventures of Oliver Twist*. New York: Oxford University Press, 1978.

Dionis, Pierre. *Traité général des accouchemens*. Paris: Charles-Maurice d'Houry, 1714.

Dixon, Suzanne. *The Roman Mother*. Norman: University of Oklahoma Press, 1988.

Dobbie, B. M. Willmott. "An Attempt to Estimate the True Rate of Maternal Mortality, Sixteenth to Eighteenth Centuries." *Medical History* 26 (1982): 79–90.

Dolce, Ludovico. "Dialogo della Pittura." In *Dolce's "Aretino" and Venetian Art Theory of the Cinquecento*, edited by M. W. Roskill, 81–195. Toronto: University of Toronto Press, 2000.

Donne, John. "The Extasie," *The Complete Poetry of John Donne*, edited by John T. Shawcross. New York: Anchor Books, 1967.

Donne, John. "The Relic." In *The Elegies and The Songs and Sonnets*. Edited by Helen Gardner. Oxford: Clarendon, 1965.

Dormandy, Thomas. *The Worst of Evils: The Fight against Pain*. New Haven, CT: Yale University Press, 2006.

Douglas, Mary. *Purity and Danger: An Analysis of the Concepts of Pollution and Taboo*. London: Routledge, 1996.

DuBruck, E. E. and B. I. Gusick, eds. *Death and Dying in the Middle Ages*. New York: Peter Lang, 1999.

Duclow, Donald F. "Dying Well: The *Ars moriendi* and the Dormition of the Virgin." In *Death and Dying in the Middle Ages*, edited by E. E. DuBruck and B. I. Gusick, 379–429. New York: Peter Lang, 1999.

Duffy, Eamon. *The Stripping of the Altars*. New Haven, CT: Yale University Press, 1992.

Du Laurens, André. *Toutes les Œuvres de Me. André Du Laurens*. Translated by Theophile Gelée. Paris: Raphael Du Petit Val, 1621.

Dundes, Alan. *Interpreting Folklore*. Bloomington: Indiana University Press, 1980.

Duval, Jacques. *Des hermaphrodits, accouchemens de femmes: et traitement qui est requis pour les relever en sante*. Rouen, France: David Gevffroy, 1612.

E., T. *The Law's Resolution of Women's Rights*. London, 1632.

Ebreo, Leone. *Dialoghi d'amore, composti per Leona medico, di natione hebreo, et dipoi tatto christiano*. Venice: Aldus, 1545.

Echinger-Maurach, C. "Michelangelo's Monument for Julius II in 1534." *The Burlington Magazine* 145 (May 2003): 336–44.

Eco, Umberto. *On Ugliness*. New York: Rizzoli, 2007.

Eisenstein, Elizabeth. *The Printing Press as an Agent of Change*. 2 vols. Cambridge: Cambridge University Press, 1979.

Elias, Norbert. *The Civilizing Process*, vol. I. Translated by E. Jephcott. New York: Pantheon, 1978.

Elyot, Thomas. *The Castel of Helth*. London, 1534.

Emison, Patricia. "Grazia." *Renaissance Studies* 5 (1991): 427–60.

Epstein, Steven A. *Speaking of Slavery: Color, Ethnicity, and Human Bondage in Italy*. Ithaca, NY: Cornell University Press, 2005.

Equicola, Mario. *Libro di natura d'amore, di M. Equicola*. Venice: Pietro di Nicolini da Sabbio, 1536.

Erasmus, Desiderius. *Collected Works of Erasmus*. Toronto: University of Toronto Press, 2003.

Erasmus, Desiderius. "Colloquies." In *Collected Works of Erasmus*. Translated and edited by C. R. Thompson, 59–618. Toronto: University of Toronto Press, 1997.

Erasmus, Desiderius. *The Colloquies of Erasmus*. Translated by C. R. Thompson. Chicago: University of Chicago Press, 1965.

Erasmus, Desiderius. *The Education of a Christian Prince*. Edited by L. Jardine. Cambridge: Cambridge University Press, 1997.

Erasmus, Desiderius. *Encomium matrimonii*. Frobenius: 1518.

Erasmus, Desiderius. *Epistle 1759 to John Francis*. English translation in *Opuscula selecta Neerlandicorum de Arte Medica* XVII.

Erasmus, Desiderius. *Erasmus on Women*. Edited by E. Rummel. Toronto: University of Toronto Press, 1996.

Erasmus, Desiderius. "Festina lente." In *Erasmus on His Times: A Shortened Version of The Adages of Erasmus*. Translated by Margaret Mann Phillips. Cambridge: Cambridge University Press, 1967.

Erasmus, Desiderius. *A lytell boke of good maners for children [De civilitate morum puerilium] . . . with interpretacion of the same in to the vulgare Englysshe tonge by Robert Whytynton laureate poete*. London: 1532.

Erasmus, Desiderius. "On Education for Children." In *Collected Works of Erasmus*. Edited by J. K. Sowards, 291–346. Toronto: University of Toronto Press, 1985.

Erasmus, Desiderius. *The Praise of Folly*. Translated by B. Radice. Harmondsworth, UK: Penguin 1971.

Erasmus, Desiderius. "The Young Man and the Harlot." In *Tudor Translations of the Colloquies of Erasmus*. Edited by D. Spurgeon. New York: Scholars Facsimiles and Reprints, 1972.

Eroli, Giovanni. *Erasmo Gattamelata da Narni: Suoi Monumenti e Sua Famiglia*. Rome: Coi tipi del Salviucci, 1877.

Estes, James Worth. "The European Reception of the First Drugs from the New World." *Pharmacy in History* 37 (1995): 3–23.

Estienne, Charles, and Jean Liébault. *Maison rustique*. London: Arnold Hatfield for Norton and Bill, 1606.

Estienne, Henri. *Apologie pour Hérodote. Satire de la société au XVIe siècle*. Edited by P. Risetelhuber. 2 vols. (1879; repr. Geneva, Switzerland: Slatkine Reprints, 1969).

Eusebius of Caesarea. *Eusebium Pamphili De evangelica praep[ar]atione Latiunu[m] ex Graeco beatissime pater iussu tuo effeci*. Venice: Leonhardus, 1473.

Evenden, Doreen. *The Midwives of Seventeenth-Century London*. Cambridge: Cambridge University Press, 2000.

Fagan, Brian. *The Little Ice Age: How Climate Made History 1300–1850*. New York: Basic Books, 2001.

Fedele, Cassandra. *Letters and Orations*. Translated and edited by D. Robin. Chicago: University of Chicago Press, 2000.

Ferber, Sarah. *Demonic Possession and Exorcism in Early Modern France*. London: Routledge, 2004.

Fermor, Sharon. "Movement and Gender in Sixteenth-Century Italian Painting." In *The Body Imaged: The Human Form and Visual Culture since the Renaissance*, edited by K. Adler and M. Pointon, 129–45. Cambridge: Cambridge University Press, 1993.

Ferraro, Joanna M. *Marriage Wars in Late Renaissance Venice*. Oxford: Oxford University Press, 2001.

Ferrero, Giuseppe Guido, ed. *Marino e i marinisti*. Milan, Italy: Ricciardi, 1954.

Festugière, Jean. *La Philosophie de l'amour de Marsile Ficin et son influence sur la literature française au XVIe siècle*. Paris: J. Vrin, 1941.

Ficino, Marsilio. *Commentaire de Marsile Ficin, Florentin: sur le Banquet d'Amour de Platon: faict en François par Symon Silvius, dit J. De la Haye, Valet de Chambre de tres chrestienne Princesse Marguerite de France, Royne de Navarre*. Poitiers, France: A l'enseigne du Pelican, 1546.

Ficino, Marsilio. *Commentaire sur le banquet de Platon*. Edited and translated by R. Marcel. Paris: Société d'édition 'Les belles lettres,' 1956.

Ficino, Marsilio. *The Letters of Marsilio Ficino*. Translated by the London School of Economic Science. 6 vols. New York: Schocken Books, 1981, 1985.

Fildes, Valerie A. *Breasts, Bottles, and Babies: A History of Infant Feeding*. Edinburgh: Edinburgh University Press, 1986.

Fildes, Valerie A. *Wet Nursing: A History from Antiquity to the Present*. Oxford: Basil Blackwell, 1988.

Filippini, Nadia Maria. "The Church, the State and Childbirth: The Midwife in Italy during the Eighteenth Century." In *The Art of Midwifery: Early Modern Midwives in Europe*, edited by H. Marland, 152–75. London: Routledge, 1993.

Finlay, Robert. "The Foundation of the Ghetto: Venice, the Jews, and the War of the League of Cambrai." *Proceedings of the American Philosophical Society* 126, no. 2 (1982): 140–54.

Finlay, Robert. "The Venetian Republic as a Gerontocracy: Age and Politics in the Renaissance." *Journal of Medieval and Renaissance Studies* 8 (1978): 157–78.

Finucci, Valeria. *The Manly Masquerade: Masculinity, Paternity, and Castration in the Italian Renaissance*. Durham, NC: Duke University Press, 2003.

Firenzuola, Agnolo. *On the Beauty of Women*. Translated and edited by K. Eisenbichler and J. Murray. Philadelphia: Philadelphia University Press, 1992.

Firmianus, Lucius Caelius. *Lucii Coelii Lactantii Firmiani, Opera quae extant, Ad fidem MSS. Recognita et Commentariis illustrate*. Oxford: Theatro Sheldoniano, 1684.

Firpo, L. *Prime relazioni di navigatori italiani sulla scoperta dell'America. Colombo, Vespucci, Verazzano*. Turin, Italy: Unione Tipografico-Editrice Torinese, 1966.

Fisher, Will. "The Renaissance Beard: Masculinity in Early Modern England." *Renaissance Quarterly* 54 (2001): 155–87.

Flandrin, Jean-Louis. *Families in Former Times: Kinship, Household and Sexuality*. Translated by Richard Southern. Cambridge: Cambridge University Press, 1979.

Ford, John, "'Tis Pity She's A Whore." In *'Tis Pity She's A Whore and Other Plays*. Edited by Marion Lomax. Oxford: Oxford University Press, 1995.

Forster, Leonard. *The Icy Fire: Five Studies in European Petrarchism*. Cambridge: Cambridge University Press, 1969.

Fortes, Meyer. "Foreword." In *Social Anthropology and Medicine*, edited by J. B. Loudon. London: Academic Press, 1979.

Foucault, Michel. *History of Sexuality*. Vol. 1, *An Introduction*. Translated by Robert Hurley. New York: Vintage, 1978.

Foucault, Michel. *The History of Sexuality*. Vol. 2, *The Use of Pleasure*. Translated by R. Hurley. London: Penguin, 1992.

Fouts, Clifford M. *Rhubarb, the Wondrous Drug*. Princeton, NJ: Princeton University Press, 1992.

Franco, Pierre. *Chirurgie*. Edited by E. Nicaise. Geneva, Switzerland: Slatkine Reprints, 1975.

Franco, Pierre. *Traité des hernies contenant une ample déclaration de toutes leurs espèces, et autres excellentes parties de la chirurgie*. Lyon, France: Thibaud Payan, 1561.

Fraser, Antonia. *The Wives of Henry VIII*. New York: Knopf, 1992.

Freedman, Luba. *Titian's Portraits through Aretino's Lens*. University Park: Pennsylania State University Press, 1995.

French, Roger. *Dissection and Vivisection in the European Renaissance*. Aldershot, UK: Ashgate, 1999.

French, Roger. *Medicine before Science: The Business of Medicine from the Middle Ages to the Enlightenment*. Cambridge: Cambridge University Press, 2003.

Frere, Walter Howard, and William McClure Kennedy, eds., *Visitation Articles and Injunctions of the Period of the Reformation*. London: Longmans, Green, 1910.

Fulgentius Placiades, C. *Iulii Hygini Augusti Liberti, Fabularum Liber, Ad Omnium poëtarum lectionem mire necessarius & ante hac nunquam excusus*. Basel, Switzerland: Joan Hervagium, 1535.

Fuller, Mary C. "Raleigh's Fugitive Gold." In *New World Encounters*, edited by S. Greenblatt, 218–40. California: University of California, 1993.

Furnivall, F. J. *Four Supplications, 1529–1553*. Early English Text Society. London: N. Trubner, 1871.

Gaillé-Nikodimov, Marie. "Qu'est-ce que l'homme? La réponse de l'anatomiste ou la médecine comme anthropologie chez André du Laurens." *50ème Colloque International d'Etudes Humanistes: Pratique et pensée médicales à la Renaissance*, Centre d'Etudes Supérieures de la Renaissance, July 2–6, 2007.

Gaisser, Julia Haig. *Catullus and His Renaissance Readers*. Oxford: Clarendon Press, 1993.

Garcia-Ballester, Luis. "The Inquisition and Minority Medical Practitioners in Counter-Reformation Spain: Judaizing and Morisco Practitioners, 1560–1610." In *Medicine and the Reformation*, edited by Ole Peter Grell and Andrew Cunningham, 156–91. New York: Routledge, 1993.

Garzoni, Tomaso. *La Piazza Universale di tutte le Professioni del Mondo*. Venice: Michiel Miloc, 1665.

Gaukroger, Stephen. *Descartes: An Intellectual Biography*. Oxford: Oxford University Press, 1995.

Gaukroger, Stephen. *The Emergence of a Scientific Culture: Science and the Shaping of Modernity*. Oxford: Clarendon Press, 2006.

Gauricus, Pomponius. *De Sculptura*. Edited by A. Chastel and R. Klein. Geneva, Switzerland: Droz, 1969.

Gavitt, Philip. *Charity and Children in Renaissance Florence: The Ospedale Degli Innocenti, 1410–1536*. Ann Arbor: University of Michigan Press, 1990.

Gelbart, Nina Rattner. *The King's Midwife: A History and Mystery of Madame du Coudray*. Berkeley: University of California Press, 1998.

Gélis, Jacques. *La sage-femme ou le médecin: une nouvelle conception de la vie*. Paris: Fayard, 1988.

Gélis, Jacques. *Les enfants des limbes: mort-nés et parents dans l'Europe chrétienne*. Paris: Louis Audibert, 2006.

Gentilcore, David. "The Church, the Devil and the Healing Activities of Living Saints in the Kingdom of Naples after the Council of Trent." In *Medicine and the Reformation*, edited by Ole Peter Grell and Andrew Cunningham, 134–55. New York: Routledge, 1993.

Gentilcore, David. *From Bishop to Witch: The System of the Sacred in Early Modern Terra d'Otranto*. Manchester, UK: Manchester University Press, 1992.

Gentilcore, David. *Medical Charlatanism in Early Modern Italy*. Cambridge: Cambridge University Press, 2006.

Gijswijt-Hofstra, Marijke, Hilary Marland, and Hans de Waardt, eds. *Illness and Healing Alternatives in Western Europe*. London: Routledge, 1997.

Gilbert, Creighton. "When Did a Man in the Renaissance Grow Old?" *Studies in the Renaissance* 14 (1967): 7–32.

Gilman, Sander. *Disease and Representation: Images of Illness from Madness to AIDS*. Ithaca, NY: Cornell University Press, 1988.

Ginzburg, Carlo. *Ecstasies: Deciphering the Witches' Sabbath*. Translated by R. Rosenthal. Edited by G. Elliot. London: Hutchinson Radius, 1990.

Giraldi, C. G. *Discorso dei romanzi*. Edited by L. Benedetti, G. Monorchio, and E. Musacchio. Bologna, Italy: Millenium, 1999.

Giraldi, Giambattista Cinzio. *De gli hecatommithi*. Monte Regale, Italy: L. Torrentino, 1565.

Gnudi, Martha Teach, and Jerome Pierce Webster. *The Life and Times of Gaspare Tagliocozzi, Surgeon of Bologna, 1545–1599*. New York: H. Reichner, 1950.

Gold, Penny Schine. *The Lady & the Virgin: Image, Attitude, and Experience in Twelfth-Century France*. Chicago: University of Chicago Press, 1985.

Golden, Janet Lynne. *A Social History of Wet Nursing in America: From Breast to Bottle*. Cambridge: Cambridge University Press, 1996.

Goldstein, Carl. "Rhetoric and Art History in the Renaissance and Baroque." *Art Bulletin* 53 (1991): 642–52.

Golzio, Vincenzo. *Raffaello nei documenti*. Farnborough, UK: Gregg International, 1971.

Gomara, Francisco Lopez de. *Histoire generalle des Indes occidentales et terres neuves, qui jusques à present ont esté descouvertes, augmentee en ceste cinquiesme edition de la description de al nouvelle Espagne, & de la grande ville de Mexique, autrement nommee Tenuctilan*. Paris: Michel Sonnius, 1584.

Goodman, Elise. "Woman's Supremacy over Nature: Van Dyck's Portrait of Elena Grimaldi." *Artibus et historiae* 30 (1994): 129–43.

Goody, Jack. *Representations and Contradictions: Ambivalence towards Images, Theatre, Fiction, Relics and Sexuality*. Oxford: Blackwell, 1997.

Gordon, Bruce, and Peter Marshall, eds. *The Place of the Dead: Death and Remembrance in Late Medieval and Early Modern Europe*. Cambridge: Cambridge University Press, 2000.

Gowing, Laura. *Common Bodies: Women, Touch and Power in Seventeenth-Century England*. New Haven, CT: Yale University Press, 2003.

Graunt, John. *Natural and Political Observations Made upon the Bills of Mortality*. Tho. Roycroft, 1662.

Grayling, A. C. *Descartes: The Life of René Descartes and Its Place in His Times*. London: Free Press, 2005.

Greenblatt, Stephen. *Renaissance Self-Fashioning from More to Shakespeare*. Chicago: University of Chicago Press, 1980.

Greene, Richard L., ed. *The Early English Carols*. Oxford: Clarendon Press, 1935.

Gregg, Charles T. *Plague: An Ancient Disease in the Twentieth Century*. Albuquerque: University of New Mexico Press, 1985.

Grell, Ole Peter, ed. *Paracelsus: The Man and His Reputation, His Ideas and Their Transformations*. Leiden, Netherlands: Brill, 1998.

Grell, Ole Peter. "Plague, Prayer and Physic: Helmontian Medicine in Restoration England." In *Religio Medici: Medicine and Religion in Seventeenth-Century England*, edited by Ole Peter Grell and Andrew Cunningham. Aldershot, UK: Scholar Press, 1996.

Grell, Ole Peter, and Andrew Cunningham, eds. *Medicine and the Reformation*. New York: Routledge, 1993.

Grell, Ole Peter, Andrew Cunningham with Jon Arrizabalaga, eds. *Health Care and Poor Relief in Counter-Reformation Europe*. London: Routledge, 1999.

Grendler, Paul F. *Schooling in Renaissance Italy: Literacy and Learning, 1300–1600*. Baltimore: Johns Hopkins University Press, 1989.

Guillemeau, Jacques. *Child-birth, or the Happy Deliverie of Women*. Amsterdam: Da Capo Press, 1972.

Guillemeau, Jacques. *De l'heureux accouchement des femmes*. Paris: Nicolas Buon, 1609.

Hair, P.E.H., ed. *Before the Bawdy Court: Selections from Church Court and Other Records Relating to the Correction of Moral Offences in England, Scotland, and New England, 1300–1800*. London: Elek, 1972.

Halkin, Leon E. *Erasmus: A Critical Biography*. Translated by John Tonkin. Oxford: Blackwell, 1993.

Hall, K. F. *Things of Darkness: Economies of Race and Gender in Early Modern England*. Ithaca, NY: Cornell University Press, 1995.

Halperin, David M. "Forgetting Foucault: Acts, Identities, and the History of Sexuality." *Representations* 63 (1998): 93–120.

Hanawalt, Barbara A. *Growing Up in Medieval London: The Experience of Childhood in History*. New York: Oxford University Press, 1993.

Hankins, James. *Plato in the Italian Renaissance*. 2 vols. Leiden, Netherlands: Brill, 1990.

Harding, Vanessa. "Whose Body?: A Study of Attitudes towards the Dead Body in Early Modern Paris." In *The Place of the Dead: Death and Remembrance in Late Medieval and Early Modern Europe*, edited by B. Gordon and P. Marshall, 170–87. Cambridge: Cambridge University Press, 2000.

Harley, David. "Spiritual Physic, Providence and English Medicine, 1560–1640." In *Medicine and the Reformation*, edited by Ole Peter Grell and Andrew Cunningham, 101–17. New York: Routledge, 1993.

Harris, Jonathan Gil. *Foreign Bodies and the Body Politic: Discourses of Social Pathology in Early Modern England*. Cambridge: Cambridge University Press, 2006.

Haskell, Yasmin, and Susan Broomhall. "Humanism and Medicine: A Match Made in Heaven?" *Intellectual History Review* 18, no. 1 (2008): 1–3.

Hatfield, Gabrielle. *Memory, Wisdom and Healing: The History of Domestic Plant Medicine*. Gloucestershire, UK: Sutton Publishing, 1999.

Healy, Margaret. "Anxious and Fatal Contacts: Taming the Contagious Touch." In *Sensible Flesh: On Touch in Early Modern Culture*, edited by E. Harvey, 22–38. Philadelphia: University of Pennsylvania Press, 2003.

Healy, Margaret. "Defoe's Journal and the English Plague Writing Tradition." *Literature and Medicine* 22 (2003): 25–44.

Healy, Margaret. "Fashioning Civil Bodies and 'Others': Cultural Representations." In *A Cultural History of the Human Body in the Renaissance*, edited by Linda Kalof and William Bynum. Oxford: Berg, 2010.

Healy, Margaret. *Fictions of Disease in Early Modern England: Bodies, Plagues and Politics*. New York: Palgrave, 2001.

Healy, Margaret. "Journeying with the 'Stone': Montaigne's Healing Travel Journal." *Literature and Medicine* 24 (2005): 231–49.

Helmont, Johannes Baptista van. *Oriatrike or Physick Refined*. Translated by J.C. of Oxon. London, 1662.

Hemsoll, David. "Beauty as an Aesthetic and Artistic Ideal in Late Fifteenth-Century Florence." In *Concepts of Beauty in Renaissance Art*, edited by Francis Ames-Lewis and Mary Rogers, 66–79. Aldershot: Ashgate, 1998.

Henderson, John. *The Renaissance Hospital: Healing the Body and Healing the Soul*. London: Yale University Press, 2006.

Heng, Geraldine. "Enchanted Ground: The Feminine Subtext in Malory." In *Arthurian Women*, edited by T. S. Fenster. New York: Routledge, 2000.

Henry, John. "Doctors and Healers: Popular Culture and the Medical Profession." In *Science, Culture and Popular Belief in Renaissance Europe*, edited by S. Pumfrey, P. L. Rosse, and M. Slawinski, 191–221. Manchester, UK: Manchester University Press, 1991.

Herbert, Mary Sidney, Countess of Pembroke. "From *A Discourse of Life and Death.*" In *Women's Writing of the Early Modern Period 1588–1688: An Anthology,* edited by Stephanie Hodgson-Wright, 7–14. New York: Columbia University Press, 2002.

Herlihy, David. *Medieval Households.* Cambridge, MA: Harvard University Press, 1985.

Herlihy, David, and Christiane Klapisch-Zuber. *Tuscans and Their Families: A Study of the Florentine Catasto of 1427.* New Haven, CT: Yale University Press, 1985.

Héroët, Antoine. *La Parfaicte amye,* edited by Christine M. Hill. Exeter, UK: University of Exeter, 1981.

Hickey, Daniel. *Local Hospitals in Ancient Regime France: Rationalization, Resistance, Renewal, 1530–1789.* Montreal: McGill-Queens University Press, 1997.

Hillman, David. *Shakespeare's Entrails: Belief, Skepticism and the Interior of the Body.* New York: Palgrave, 2007.

Hodgen, Margaret T. *Early Anthropology in the Sixteenth and Seventeenth Centuries.* Philadelphia: University of Pennsylvania Press, 1964.

Hodgson-Wright, Stephanie, ed. *Women's Writing of the Early Modern Period 1588–1688: An Anthology.* New York: Columbia University Press, 2002.

Hoffer, Peter C., and N.E.H. Hull. *Murdering Mothers: Infanticide in England and New England, 1558–1803.* New York: New York University Press, 1981.

Holanda, Francisco de. *Dialogues with Michelangelo.* Translated by C. B. Holroyd and edited by D. Hemsoll. London: Pallas Athene, 2006.

Holt, Elizabeth Gilmore, ed. *A Documentary History of Art.* Vol. I, *The Middle Ages and Renaissance.* Princeton, NJ: Princeton University Press, 1981.

Horowitz, Elliott. "The New World and the Changing Face of Europe." *Sixteenth Century Journal* 28 (1997): 1181–1201.

Horrox, Rosemary, ed. *The Black Death.* Manchester, UK: Manchester University Press, 1994. http://books.google.com/books?id=1O_PX2wVD0sC.

Houlbrook, Ralph. *Death, Religion and the Family in England, 1480–1750.* Oxford: Clarendon Press, 1998.

Householder, Michael. "Eden's Translations: Women and Temptation in Early America." *Huntington Library Quarterly* 70, no. 1 (2007): 11–36.

Hoving, Thomas P. F. "A Newly Discovered Reliquary of St. Thomas Becket." *Gesta* 4 (Spring 1965): 28–30.

Hsia, R. Po-chia. *The Myth of Ritual Murder: Jews and Magic in Reformation Germany.* New Haven, CT: Yale University Press, 1988.

Hsia, R. Po-chia. *Trent 1475: Stories of a Ritual Murder Trial.* New Haven, CT: Yale University Press, 1992.

Huet, Marie-Hélène. "Monstrous Medicine." In *Monstrous Bodies/Political Monstrosities in Early Modern Europe,* edited by Laura Lunger Knoppers and Joan B. Landes. Ithaca, NY: Cornell University Press, 2004.

Huguet-Termes, Teresa. "New World Materia Medica in Spanish Renaissance Medicine." *Medical History* 45 (2001): 359–76.

Huizinga, Johan, trans. *Erasmus and the Age of Reformation.* London: Phoenix Press, 2002.

Hutton, Patrick H. *Philippe Ariès and the Politics of French Cultural History.* Amherst: University of Massachusetts Press, 2004.

Hutton, Ronald. *The Stations of the Sun: A History of the Ritual Year in Britain.* Oxford: Oxford University Press, 1996.

Ianziti, Gary. *Humanistic Historiography under the Sforzas: Politics and Propaganda in Fifteenth-Century Milan*. Oxford: Clarendon Press, 1988.

Imbert, Jean. *Le Droit hospitalier de l'ancien regime*. Paris: Presses universitaires de France, 1993.

Imhof, Arthur E. *Lost Worlds: How Our European Ancestors Coped with Everyday Life and Why Life Is So Hard Today*. Translated by Thomas Robisheaux. Charlottesville: University Press of Virginia, 1996.

Immel, A., and M. Witmore, eds. *Childhood and Children's Books in Early Modern Europe, 1550–1800*. New York: Routledge, 2006.

Jacobs, Frederika H. *Defining the Renaissance "Virtuosa."* Cambridge: Cambridge University Press, 1997.

Jardine, Lisa. *Ingenious Pursuits: Building on the Scientific Revolution*. London: Abacus, 2000.

Jenner, Mark. "Civilization and Deodorization? Smell in Early Modern English Culture." In *Civil Histories: Essays Presented to Sir Keith Thomas*, edited by P. Burke, B. H. Harrison, and P. Slack, 127–44. Oxford: Oxford University Press, 2000.

Jocelin, Elizabeth. *The Mothers Legacie to her Unborn Childe*. London: John Haviland, 1624.

Johns, Adrian. *The Nature of the Book: Print and Knowledge in the Making*. Chicago: University of Chicago Press, 1998.

Johnson, Geraldine A. "Beautiful Brides and Model Mothers: The Devotional and Talismanic Functions of Early Modern Marian Reliefs." In *The Material Culture of Sex, Procreation, and Marriage in Premodern Europe,* edited by A. L. McClanan and K. R. Encarnación, 135–62. New York: Palgrave, 2002.

Johnson, Geraldine A. "Sculpture and the Family in Fifteenth-Century Florence." In *Art, Memory, and Family in Renaissance Florence*, edited by G. Ciappelli and P. Rubin, 215–33. Cambridge: Cambridge University Press, 2000.

Johnson, Mark. *The Body in the Mind: The Bodily Basis of Meaning, Imagination, and Reason*. Chicago: University of Chicago Press, 1987.

Jones, Colin. *The Charitable Imperative: Hospitals and Nursing in Ancient Regime and Revolutionary France*. London: Routledge, 1989.

Jones, Colin, and Roy Porter, eds. *Reassessing Foucault: Power, Medicine and the Body*. London: Routledge, 1994.

Jonson, Ben. *Epicoene, or the Silent Woman*. London, 1609.

Joubert, Laurent. *La Médecine et le Régime de Santé, Des erreurs populaires et propos vulgaires* (Bordeaux: Simon Millanges, 1578). Edited by Madeleine Tiollais. 2 vols. Paris: L'Harmattan, 1997.

Joubert, Laurent. *Popular Errors*. Translated by G. D. de Rocher. Tuscaloosa: University of Alabama Press, 1989.

Kagan, Richard L., and Abigail Dyer, eds. and trans. *Inquisitorial Inquiries: Brief Lives of Secret Jews and Other Heretics*. Baltimore: Johns Hopkins University Press, 2004.

Kallendorf, Craig, ed. and trans. *Humanist Educational Treatises*. Cambridge, MA: Harvard University Press, 2002.

Kaplan, Paul H. D. "Isabella d'Este and Black African Women." In *Black Africans in Renaissance Europe*, edited by T. F. Earle and K.J.P. Lowe, 125–54. Cambridge: Cambridge University Press, 2005.

Kelly, John. *The Great Mortality: An Intimate History of the Black Death*. London: Fourth Estate, 2005.

Kemp, Martin. "Equal Excellences: Lomazzo and the Explanation of Individual Style in the Visual Arts." *Renaissance Studies* 1 (1987): 1–26.

Kern Paster, Gail. *The Body Embarrassed: Drama and the Disciplines of Shame in Early Modern Europe*. Ithaca, NY: Cornell University Press, 1993.

Kieckhefer, Richard. "The Holy and the Unholy: Sainthood, Witchcraft and Magic in Late Medieval Europe." *Journal of Medieval and Renaissance Studies* 24 (1994): 355–85.

Kieckhefer, Richard. *Unquiet Souls: Fourteenth-Century Saints and Their Religious Milieu*. Chicago: University of Chicago Press, 1984.

King, Helen. "As If None Understood the Art That Cannot Understand Greek: The Education of Midwives in Seventeenth-Century England." In *The History of Medical Education in Britain*, edited by V. Nutton and R. Porter, 184–98. Amsterdam: Rodopi, 1995.

King, Margaret L. "Concepts of Childhood: What We Know and Where We Might Go." *Renaissance Quarterly* 60, no. 2 (2007): 371–407.

King, Margaret L. *The Death of the Child Valerio Marcello*. Chicago: University of Chicago Press, 1994.

King, Margaret L. *Women of the Renaissance*. Chicago: University of Chicago Press, 1991.

Kite, Charles. *An Essay for the Recovery of the Apparently Dead*. London: C. Dilly, 1788.

Klapisch-Zuber, Christiane. "Blood Parents and Milk Parents: Wet Nursing in Florence, 1300–1530." In *Women, Family, and Ritual in Renaissance Italy*, 132–64. Chicago: University of Chicago Press, 1985.

Klapisch-Zuber, Christiane. "Female Celibacy and Service in Florence in the Fifteenth Century." In *Women, Family, and Ritual in Renaissance Italy*, 165–77. Chicago: University of Chicago Press, 1985.

Klein, Lisa. "Lady Anne Clifford as Mother and Matriarch: Domestic and Dynastic Issues in Her Life and Writings." *Journal of Family History* 21 (2001): 18–39.

Knepper, Janet. "A Bad Girl Will Love You to Death: Excessive Love in the Stanzaic *Morte Arthur* and Malory." In *On Arthurian Women: Essays in Memory of Maureen Fries*, edited by B. Wheeler and F. Tolhurst, 229–44. Dallas, TX: Scriptorium Press, 2001.

Knox, John. *The First Blast of the Trumpet against the Monstrous Regiment of Women*. London, 1558.

Kohl, B.G., and R.G. Witt, eds. *The Earthly Republic: Italian Humanists on Government and Society*. Philadelphia: University of Pennsylvania Press, 1978.

Korhonen, Anu. "Washing the Ethiopian White: Conceptualising Black Skin in Renaissance England." In *Black Africans in Renaissance Europe*, edited by T.F. Earle and K.J.P. Lowe, 94–112. Cambridge: Cambridge University Press, 2005.

Kravitsky, Peter. "Erasmus' Medical Milieu." *Bulletin of the History of Medicine* 47 (1973): 113–54.

Kristeller, Paul O. *The Philosophy of Marsilio Ficino*. Translated by Virginia Conant. New York: Columbia University Press, 1943.

Kumar, Parveen, and Michael Clark, eds. *Clinical Medicine*. London: Saunders, 2002.

Kunze, Michael. *Highroad to the Stake: A Tale of Witchcraft*. Translated by W.E. Yuill. Chicago: University of Chicago Press, 1987.

Kwakkelstein, Michael W. "Leonardo da Vinci's Grotesque Heads and the Breaking of the Physiognomic Mould." *Journal of the Warburg and Courtauld Institutes* 54 (1991): 127–36.

Laget, Mireille. "La césarienne ou la tentation de l'impossible: XVIIe et XVIIIe siècle." *Annales de bretagne et des pays de l'ouest* 86 (1979): 177–89.

Lakoff, G., and M. Johnson. *Metaphors We Live By*. Chicago: University of Chicago Press, 1980.

Lambek, Michael. "Body and Mind in Mind, Body and Mind in Body: Some Anthropological Interventions in a Long Conversation." In *Body and Persons: Comparative Perspectives from Africa and Melanesia,* edited by M. Lambek and A. Strathern. Cambridge: Cambridge University Press, 1998.

Land, Norman E. "Titian's Martyrdom of St. Peter Martyr and the Limitations of Ekphrasistic Art Criticism." *Art History* 13 (1990): 293–317.

Land, Norman E. *The Viewer as Poet. The Renaissance Response to Art*. University Park: Pennsylvania State University Press, 1994.

Langham, William. *The Garden of Health*. London: Deputies of Christopher Barker, 1597[1598].

Laporte, Dominique. *The History of Shit*. Translated by N. Benabid and R. El-Khoury. Cambridge: MIT Press, 2000.

Laqueur, Thomas. *Making Sex: Body and Gender from the Greeks to Freud*. Cambridge, MA: Harvard University Press, 1990.

Latham, Robert, and William Matthews. *The Diary of Samuel Pepys: A New and Complete Transcription*. London: G. Bell and Sons Ltd., 1970–1972.

Lee, Rensselaer W. *Ut pictura poesis: The Humanistic Theory of Painting*. New York: W. W. Norton, 1967.

Leeuwenhoek, Antoni van. *A Part of the 69th Letter of January 4th 1692 To the Royal Society in London*. Trans. into modern English in *Opuscula Selecta Neerlandicorum de Arte Medica* XVII, 177–83.

Le Fournier, André. *La decoration dhumaine nature & aornement des dames*. Lyon, France: Thibault Payen, 1533.

Le Goff, Jacques. *The Birth of Purgatory*. Translated by Arthur Goldhammer. Chicago: University of Chicago Press, 1984.

Leonardo da Vinci. *Treatise on Painting* [Codex Urbinas Latinus 1270]. Translated by A. P. McMahon. Princeton, NJ: Princeton University Press, 1956.

Levret, André. *L'art des accouchemens*. Paris: Alexandre Le Prieur, 1761.

Libbon, Stephanie E. "Pathologizing the Female Body: Phallocentrism in Western Science." *Journal of International Women's Studies* 8, no. 4 (2007): 79–92.

Lightbown, Ronald. *Botticelli, Life and Work*. London: Elek, 1978.

Lindeman, Mary. *Medicine and Society in Early Modern Europe*. Cambridge: Cambridge University Press, 1999.

Lingo, Alison Klairmont. "Empirics and Charlatans in Early Modern France: The Genesis of the Classification of the 'Other' in Medical Practice." *Journal of Social History* 19 (1985–1986): 583–603.

Lingo, Alison Klairmont. "The Rise of Medical Practitioners in Sixteenth-Century France: The Case of Lyon and Montpellier." PhD diss., University of California-Berkeley, 1980.

Lipsius, Justus. *Sixe bookes of politickes or civil doctrine*. Translated by W. Jones. London, 1594.

Lipsius, Justus. *Two Bookes of Constancie*. Translated by J. Stradling. London: Richard Johnes, 1595.

Llewellyn, Nigel. "Honour in Life, Death and in the Memory: Funeral Monuments in Early Modern England." *Transactions of the Royal Historical Society*, 6th series, 6 (1996): 179–200.

Loades, D. M. *Henry VIII and His Queens*. Stroud, UK: A. Sutton, 1996.

Lodge, Thomas. *A Treatise of the Plague*. London, 1603.

Lomazzo, Gian Paolo. *Scritti sulle arti*. Edited by Roberto Paolo Ciardi. Florence: Marchi and Bertolli, 1973.

Long, Kathleen. *Hermaphrodites in Renaissance Europe*. Aldershot, UK: Ashgate, 2006.

Long, Kathleen. "Jacques Duval on Hermaphrodites." In *High Anxiety: Masculinity in Crisis in Early Modern France*, edited by Kathleen P. Long, 107–38. Kirksville, MO: Truman State University Press, 2002.

Lopez Piñero, José Maria. "Las 'nuevas medicinas' americanas en la obra (1556–1574) de Nicolas Monardes. *Asclepio* 42, no. 1 (1990): 3–67.

Loudon, Irvine. *Death in Childbirth: An International Study of Maternal Care and Maternal Mortality 1800–1950*. Oxford: Clarendon, 1992.

Loudon, Irvine. *The Tragedy of Childbed Fever*. New York: Oxford University Press, 2000.

Lowe, Kate. "The Stereotyping of Black Africans in Renaissance Europe." In *Black Africans in Renaissance Europe*, edited by T. F. Earle and K. J. P. Lowe, 17–47. Cambridge: Cambridge University Press, 2005.

Luders, A., ed. *The Statutes of the Realm: Printed by Command of His Majesty King George the Third, in Pursuance of an Address of the House of Commons of Great Britain, From Original Records and Authentic Manuscripts,* 11 vols. (London: Record Commission, 1810–1828).

Luzio, Alessandro, and Rodolfo Renier. "Buffoni, Nani e Schiavi dei Gonzaga." *Nuova Antologia* 36 (1891): 112–46.

MacDonald, Katherine. "Humanistic Self-Representation in Giovan Battista Della Porta's *Della Fisionomia dell'uomo*: Antecedents and Innovation." *Sixteenth Century Journal* 36 (2005): 397–414.

Macfarlane, Alan. *The Family Life of Ralph Josselin, a Seventeenth-Century Clergyman; An Essay in Historical Anthropology*. Cambridge: Cambridge University Press, 1970.

Machiavelli, Niccolò. *History of Florence and of the Affairs of Italy: From the Earliest Times to the Death of Lorenzo the Magnificent*. New York: M. W. Dunne, 1901. http://books.google.com/books?id=q-gXH2bioqcC&dq=machiavelli+history+florence.

Machiavelli, Niccolò. *Opere*. Edited by F. Gaeta, Vol. III. Turin, Italy: Unione Tipografico-Editrice Torinese, 1984.

Machiavelli, Nicolò. *The Portable Machiavelli*. Edited by P. Bondanella and M. Musa. New York: Viking Penguin, 1979.

Macinghi negli Strozzi, Alessandra. *Lettere di una gentildonna fiorentina del secolo XV ai figliuoli esuli*. Edited by C. Guasti. Florence: Sansoni, 1877.

Mack, Arien, ed. *In Time of Plague: The History and Social Consequences of Lethal Epidemic Disease*. New York: New York University Press, 1991.

Mack, Charles R. "Montaigne in Italy: Of Kidney Stones and Thermal Spas." *Renaissance Papers* (1991): 105–24.

Maclean, Ian. "Foucault's Renaissance Episteme Reconsidered: An Aristotelian Counterblast." *Journal of the History of Ideas* 59, no. 1 (1998): 149–66.

Magli, P. "The Face and the Soul." In *Fragments for a History of the Human Body, Part Two*, 87–127. New York: Zone, 1989.

Malgaigne, J. F. *Surgery and Ambroise Paré*. Norman: University of Oklahoma Press, 1965.

Malory, Thomas. *Le Morte D'Arthur*. Edited by S.H.A. Shepherd. New York: W.W. Norton, 2003.

Mander, Karel van. *Le Livre des Peintres*. Edited and translated by Véronique Gerard-Powell. Paris: Les Belles Lettres, 2001.

Marais, Marin. *Le Labyrinthe et autres histoires*. San Lorenzo del Escorial, Spain: Glossa, 2000.

Marcus, J. R. and M. Saperstein. *The Jew in the Medieval World: A Source Book, 315–1791*. Cincinnati: Hebrew Union College Press, 1999. http://books.google.com/books?id=PCalmtflYtEC.

Marinello, Giovanni. *Delle medicine partenenti all infermità delle donne*. Venice: Gio. Bonfadino, & Compagni, 1560.

Marinelli, Giovanni. *Gli ornamenti delle donne*. Venice: de' Franceschi, 1562.

Marland, Hilary, ed. *The Art of Midwifery: Early Modern Midwives in Europe*. London: Routledge, 1993.

Marland, Hilary, trans. and ed. *"Mother and Child Were Saved" The Memoirs (1693–1740) of the Frisian Midwife Catharina Schrader*. Amsterdam: Rodopi, 1987.

Marland, Hilary. "Stately and Dignified, Kindly and God-fearing: Midwives, Age and Status in the Netherlands in the Eighteenth Century." In *The Task of Healing: Medicine, Religion and Gender in England and the Netherlands 1450–1800*, edited by H. Marland and M. Pelling, 271–305. Rotterdam, Netherlands: Erasmus Publishing, 1996.

Marshall, Peter. *Beliefs and the Dead in Reformation England*. Oxford: Oxford University Press, 2002.

Marshall, Peter. "Forgery and Miracles in the Reign of Henry VIII." *Past and Present* 178 (2003): 39–73.

Marshall, Peter. "The Rood of Boxley, the Blood of Hailes and the Defence of the Henrician Church." *Journal of Ecclesiastical History* 46 (1995): 689–96.

Martindale, Andrew. "The Child in the Picture: A Medieval Perspective." In *The Church and Childhood: Papers Read at the 1993 Summer Meeting and the 1994 Winter Meeting of the Ecclesiastical History Society*, edited by D. Wood, 197–233. Oxford: Published for the Ecclesiastical History Society by Blackwell Publishers, 1994.

Mauriceau, François. *Des maladies des femmes grosses et accouchées*. Paris: Jean Henault, 1668.

Mauriceau, François. *Observations sur la grossesse et l'accouchement des femmes*. Paris: Chez l'Auteur, 1695.

Mazzio, Carla. "The Senses Divided: Organs, Objects, and Media in Early Modern England." In *The Empire of the Senses: The Sensual Culture Reader*, edited by D. Howes, 85–105. Oxford: Berg, 2005.

McGinn, Bernard. *Antichrist: Two Thousand Years of the Human Fascination with Evil*. New York: Columbia University Press, 2000.

McLaughlin, Terrence. *Coprophilia, or a Peck of Dirt*. London: Cassell, 1971.

McNeil, Alison Kettering. "Terborch's Ladies in Satin." *Art History* 16 (1993): 95–124.

McTavish, Lianne. "Blame and Vindication in the Early Modern Birthing Room." *Medical History* 50, no. 4 (2006): 447–64.

McTavish, Lianne. *Childbirth and the Display of Authority in Early Modern France*. Aldershot, UK: Ashgate, 2005.

Melanchthon, Philip. *Deuttung der zwei greulichen Figuren Papstesels zu Rom und Mönchkalbs zu Freiberg in Meissen gefunden.* Augsburg, Germany: H. Steiner, 1523.

Meller, Peter. "Physiognomical Theory in Renaissance Heroic Portraits." In *The Renaissance and Mannerism: Studies in Western Art. Acts of the XXth International Congress of the History of Art*, 53–69. Princeton, NJ: Princeton University Press, 1963.

Mendelson, Sara H., and Patricia Crawford. *Women in Early Modern England, 1550–1720.* Oxford: Clarendon Press, 1998.

Mennell, Stephen. *All Manners of Food: Eating and Taste in England and France from the Middle Ages to the Present.* Urbana: University of Illinois Press, 1995.

Mercurio, Scipione Girolamo. *La Commare o riccoglitrice.* Venice: C. Cioti, 1596.

Merleau-Ponty, Maurice. *Phénoménologie de la Perception.* Paris: Gallimard, 1945.

Merleau-Ponty, Maurice. *Phenomenology of Perception.* London: Routledge Classics, 2002.

Merrill, Robert V. "The Pléiade and the Androgyne." *Comparative Literature* 1, spring (1949): 97–112.

Michelangelo [Michelangelo Buonarroti]. *The Poems.* Edited and translated by C. Ryan. London: Dent, 1996.

Midelfort, H. C. Erik. *Witch Hunting in Southwestern Germany, 1562–1684: The Social and Intellectual Foundations.* Stanford, CA: Stanford University Press, 1972.

Miller, Naomi J., and Naomi Yavneh, eds. *Maternal Measures: Figuring Caregiving in the Early Modern Period.* Aldershot, UK: Ashgate, 2000.

Miller, William. *The Anatomy of Disgust.* Cambridge, MA: Harvard University Press, 1997.

Miller, William Ian. "Gluttony." *Representations* 60 (1997): 92–122.

Milton, John. *John Milton, A Critical Edition of the Major Works.* Edited by S. Orgel and J. Goldberg. New York: Oxford University Press, 1991.

Milton, John. *Paradise Lost.* London, 1795.

Mirandola, Giovanni Pico della. *De hominis dignitate, Heptaplus, De ente et uno.* Edited by Eugenio Garin. Florence: Vallecchi Editore, 1942.

Mitterauer, Michael. *A History of Youth: Family, Sexuality, and Social Relations in Past Times.* Translated by G. Dunphy. Oxford, Blackwell, 1993.

Montaigne, Michel de. *The Complete Essays of Montaigne.* Translated by Donald M. Frame. Stanford, CA: Stanford University Press, 1958.

Montaigne, Michel de. *Complete Works: Essays, Travel Journal, Letters.* Translated by Donald M. Frame. London: Hamish Hamilton, 1957.

Montaigne, Michel de. "Of the Caniballes." Translated by John Florio, 1603, in *Montaigne's Essays*, 3 vols., vol. 1, Ch. 30, 215–29. Edited by Ernest Rhys. London: Everyman, J. M. Dent & Sons Ltd, 1910.

Montaigne, Michel de. *Travel Journal.* Translated and with an introduction by Donald M. Frame. San Francisco: North Point Press, 1983.

Montrose, Louis. "The Work of Gender in the Discourse of Discovery." In *New World Encounters*, edited by S. Greenblatt, 177–217. Berkeley: University of California Press, 1993.

More, Thomas. *A Dialogue of Comfort against Tribulation*, introduction and notes by Frank Manley. New Haven, CT: Yale University Press, 1977.

Morrall, Andrew. "Defining the Beautiful in Early Renaissance Germany." In *Concepts of Beauty in Renaissance Art*, edited by F. Ames-Lewis and M. Rogers, 80–92. Aldershot, UK: Ashgate, 1998.

Moss, Ann. *Ovid in Renaissance France: A Survey of the Latin Editions of Ovid and Commentaries Printed in France before 1600*. London: Warburg Institute, 1982.

Muchembled, Robert. *A History of the Devil: From the Middle Ages to the Present*. Oxford: Polity, 2003.

Muir, Edward. *Ritual in Early Modern Europe*. Cambridge: Cambridge University Press, 2005.

Munday, Anthony. *Sundry Strange and Inhumaine Murthers*. London, 1591.

Musacchio, Jacqueline Marie. *The Art and Ritual of Childbirth in Renaissance Italy*. New Haven, CT: Yale University Press, 1999.

Nagy, Doreen G. *Popular Medicine in Seventeenth-Century England*. Bowling Green, KY: Bowling Green State University Popular Press, 1988.

Nardi, Patricia. "Mothers at Home: Their Role in Childrearing and Instruction in Early Modern England." Diss., City University of New York, 2007.

Nashe, Thomas. *Christes Teares over Jerusalem*. London, 1594.

Nasus, Johannes. *Ecclesia Militans*. Ingolstadt, Germany: Alexander Weissenhorn, 1569.

Nauert, Charles G. *Agrippa and the Crisis of Renaissance Thought*. Urbana: University of Illinois Press, 1965.

Navarre, Marguerite de. *The Heptameron*. Translated and with an introduction by P. A. Chilton. New York: Penguin, 1984.

Newton, Thomas. *The Touchstone of Complexions*. London, 1576.

Niccholes, Alexander. *A Discourse, of Marriage and Wiving: And of the Greatest Mystery Therein Contained: How to Choose a Good Wife from a Bad*. London: N.O. for Leonard Becket, 1615.

Niccoli, Ottavia. *Prophecy and People in Renaissance Italy*. Translated by Lydia G. Cochrane. Princeton, NJ: Princeton University Press, 1990.

Nichols, Tom. *Tintoretto: Tradition and Identity*. London: Reaktion, 1999.

Nitze, W. A., and T. A. Jenkins, eds. *Le haut livre du graal: Perlesvaus*. Chicago: University of Chicago Press, 1932.

Normand, Lawrence. "The Miraculous Royal Body in James VI and I, Jonson and Shakespeare, 1590–1609." In *The Body in Late Medieval and Early Modern Culture*, edited by D. Grantley and N. Taunton, 143–56. Aldershot, UK: Ashgate, 2000.

Notredame, Michel de (Nostradamus). *Excellent et moult utile opuscule à tous*. Paris: Olivier de Harsy, 1556.

Nussbaum, Felicity. *Torrid Zones: Maternity, Sexuality, and Empire in Eighteenth-Century English Narratives*. Baltimore: Johns Hopkins University Press, 1995.

Nussbaum, Martha C. *Hiding from Humanity: Disgust, Shame, and the Law*. Princeton, NJ: Princeton University Press, 2004.

Nuttall, Paula. *From Flanders to Florence: The Impact of Netherlandish Painting 1400–1500*. New Haven, CT: Yale University Press, 2007.

Nutton, Vivian. "Medicine in Medieval Western Europe, 1000–1500." In *The Western Medical Tradition, 800 BC to AD 1800*, edited by L. I. Conrad, M. Neve, V. Nutton, R. Porter, and A. Wear, 139–206. Cambridge: Cambridge University Press, 1995.

O'Boyle, Cornelius. *The Art of Medicine: Medical Teaching at the University of Paris, 1250–1400*. Leiden, Netherlands: Brill, 1998.

O'Malley, Charles Donald. *The Illustrations from the Works of Andreas Vesalius of Brussels*. New York: Dover, 1973.

Ong, Walter. *Orality and Literacy: The Technologizing of the Word*. London: Routledge, 1982.

Origo, Iris. "The Domestic Enemy: The Eastern Slaves in Tuscany in the Fourteenth and Fifteenth Centuries." *Speculum* 30 (1955): 321–66.

Orme, Nicholas. *English Schools in the Middle Ages*. London: Methuen, 1973.

Orme, Nicholas. *From Childhood to Chivalry: The Education of the English Kings and Aristocracy, 1066–1530*. London: Methuen, 1984.

Otis, Leah Lydia. *Prostitution in Medieval Society: The History of an Urban Institution in Languedoc*. Chicago: University of Chicago Press, 1985.

Ovid (Publius Ovidius Naso). *Metamorphosis cum luculentissimis Raphaelis Regii enarrationibus*. Lyon, France: J. Huguetan, 1518.

Ovid (Publius Ovidius Naso). *Metamorphosis cum luculentissimis Raphaelis Regii enarrationibus*. Lyon, France: J. Mareschal, 1519.

Ovid (Publius Ovidius Naso). *P. Ovidii Nasonis Poetae Sulmonensis Opera Quae Vocantur Amatoria, cum Doctorum Virorum Commentariis partim huscusque etiam alibi editis, partim iam primum adiectis: quorum omnium Catalogum versa pagina reperies. His accesserunt Iacobi Micylli Annotationes*. Basel, Switzerland: Ioannem Hervagium, 1549.

Ozment, Stephen. *Ancestors: The Loving Family in Old Europe*. Cambridge, MA: Harvard University Press, 2001.

Ozment, Steven. *Flesh and Spirit: Private Life in Early Modern Germany*. New York: Viking, 1999.

Pagden, Anthony. *European Encounters with the New World*. New Haven, CT: Yale University Press, 1993.

Pagel, Walter. *Paracelsus: An Introduction to Philosophical Medicine in the Era of the Renaissance*, 2nd rev. ed. Basel, Switzerland: Karger, 1982.

Pagel, Walter. *William Harvey's Biological Ideas*. Basel, Switzerland: Karger, 1967.

Panofsky, Erwin. "The History of the Theory of Human Proportions as a Reflection of the History of Styles." In *Meaning in the Visual Arts*, 55–107. Garden City, NY: Doubleday Anchor, 1955.

Panofsky, Erwin. *Idea: A Concept in Art Theory*. Translated by J. S. Peake. New York: Harper & Row, 1968.

Panofsky, Erwin. *The Life and Art of Albrecht Dürer*. Princeton, NJ: Princeton University Press, 1971.

Panofsky, Erwin. "The Neoplatonic Movement and Michelangelo." In *Studies in Iconology*, 171–230. New York: Harper & Row, 1962.

Panofsky, Erwin. "The Neoplatonic Movement in Florence and North Italy: Bandinelli and Titian." In *Studies in Iconology*, 129–69. New York: Harper & Row, 1962.

Papon, Jean. *Recueil d'arrestz notables des courts souveraines de France*. Paris: Chez J. Macé, 1566.

Paré, Ambroise. *Anatomie universelle du corps humain*. Paris, 1561.

Paré, Ambroise. *Les oeuvres de M. Ambroise Paré*. Paris: Gabriel Buon, 1575.

Park, Katharine. "The Criminal and the Saintly Body: Autopsy and Dissection in Renaissance Italy." *Renaissance Quarterly* 47 (1994): 1–33.

Park, Katharine. *Secrets of Women: Gender, Generation, and the Origins of Human Dissection*. Brooklyn, NY: Zone Books, 2006.

Park, Katharine. "Stones, Bones and Hernias: Surgical Specialists in Fourteenth- and Fifteenth-Century Italy." In *Medicine from the Black Death to the French Disease*, edited by Roger Kenneth French, Jon Arrizabalaga, Andrew Cunningham, and Luis Garcia-Ballester, 110–30. Aldershot, UK: Ashgate, 1998.

Park, Katharine. "Was There a Renaissance Body?" In *The Italian Renaissance in the Twentieth Century*, I Tatti Studies, vol. 19, edited by Walter Kaiser and Michael Rocke, 21–35. Florence: Olschki 2002.

Park, Katharine, and Lorraine Daston. "The Hermaphrodite and the Orders of Nature." *Gay and Lesbian Quarterly* 1 (1995): 419–73.

Park, Katharine, and Lorraine J. Daston. "Unnatural Conception: The Study of Monsters in Sixteenth- and Seventeenth-Century France and England." *Past and Present* 92 (1981): 20–54.

Parker, Charles H. *Social Welfare and Calvinist Charity in Holland, 1572–1620*. Cambridge: Cambridge University Press, 1998.

Paster, Gail Kern. *The Body Embarrassed: Drama and the Disciplines of Shame in Early Modern England*. Ithaca, NY: Cornell University Press, 1993.

Paynell, Thomas. *A Moche Profitable Treatise*. London, 1534.

Paynell, Thomas. *Regimen sanitatis Salerni*. London, 1528.

Pelling, Margaret. "Child Health as a Social Value in Early Modern England." *Social History of Medicine* 1, no. 2 (1988): 135–64.

Pelling, Margaret. *The Common Lot: Sickness, Medical Occupations and the Urban Poor in Early Modern England*. London: Longman, 1998.

Pelling, Margaret. *Medical Conflicts in Early Modern London: Patronage, Physicians, and Irregular Practitioners 1550–1640*. Oxford: Oxford University Press, 2003.

Perkins, Wendy. *Midwifery and Medicine in Early Modern France: Louise Bourgeois*. Exeter, UK: University of Exeter Press, 1996.

Petherbridge, Deanna, and Ludmilla Jordanova. *The Quick and the Dead: Artists and Anatomy*. Berkeley: University of California Press, 1997.

Petrarca, Francesco. *Rime. Trionfi e poesie latine*. Edited by F. Neri, G. Martellotti, E. Bianchi, and N. Sapegno. Milan: Ricciardi, 1951.

Petrarch, Francesco. *On Posterity*. http://petrarch.petersadlon.com/read_letters.html?s=pet01.html.

Petroff, Elizabeth. *Medieval Women's Visionary Literature*. Oxford: Oxford University Press, 1986.

Peu, Philippe. *La pratique des accouchemens*. Paris: Jean Boudot, 1694.

Phillips, Jonathan. *The Fourth Crusade and the Sack of Constantinople*. New York: Viking, 2004.

Pino, Paolo. "Dialogo di Pittura." In *Trattati d'arte del Cinquecento*, edited by P. Barocchi, 1, 95–139. Bari, Italy: Laterza, 1960.

Plato. *The Republic of Plato*. Translated by F. M. Cornford. Oxford: Oxford University Press, 1945.

Plato. *The Symposium*. Translated by Christopher Gill. London: Penguin, 1999.

Pliny. *The Elder Pliny's Chapters on the History of Art*. Edited by E. Sellars and translated by K. Jex-Blake. London: Macmillan, 1896.

Poliziano, Angelo. *Poesie Italiane*. Edited by M. Luzi and S. Orlando. Milan: Rizzoli, 1976.

Pollock, Linda. *With Faith and Physic: The Life of a Tudor Gentlewoman, Lady Grace Mildmay, 1552–1620*. New York: St. Martin's Press, 1993.

Pomata, Gianna. *Contracting a Cure: Patients, Healers, and the Law in Early Modern Bologna*. Baltimore: Johns Hopkins University Press, 1998.

Porta, Giambattista della. *De Humana Physignomonia Ioannis Baptistae Portae Neapolitani*. Urselli, Italy: Cornelio Sutori, 1601.

Porter, Roy. *The Greatest Benefit to All Mankind: A Medical History of Humanity*. New York: W. W. Norton & Company, 1997.

Porter, Roy. *Health for Sale: Quackery in England, 1660–1850*. Manchester, UK: Manchester University Press, 1998.

Porter, Roy. *Medicine: A History of Healing*. New York: Marlowe and Company, 1997.

Porter, Roy, and G. S. Rousseau. *Gout: The Patrician Malady*. New Haven, CT: Yale University Press, 1998.

Prost, Auguste. *Corneille Agrippa: Sa vie et ses œuvres*, 2 vols. Nieuwkoop, Belgium: B. de Graaf, 1965.

Pullan, Brian S. *The Jews of Europe and the Inquisition of Venice, 1550–1670*. Totowa, NJ: Barnes & Noble, 1983.

Purkiss, Diane. "Losing Babies, Losing Women: Attending to Women's Stories in Scottish Witchcraft Trials." In *Culture and Change: Attending to Early Modern Women*, edited by M. Mikesell and A. Seeff, 143–58. Newark: University of Delaware Press, 2003.

Queller, Donald E. *The Fourth Crusade: The Conquest of Constantinople, 1201–1204*. Philadelphia: University of Pennsylvania Press, 1977.

Quintilian. *The Institutio Oratoria of Quntilian*. Translated by H. E. Butler. London: Heinemann, 1921.

Rabisch. *Il grottesco nell'arte del Cinquecento. L'accademia della Val di Blenio, lomazzo e l'ambiente milanese*. Milan: Skira, 1998.

Raleigh, Sir Walter. *The Discoverie of the Large, Rich and Bewtiful Empyre of Guiana*. Translated by N. L. Whitehead. Manchester, UK: Manchester University Press, 1997.

Rambuss, Richard. *Closet Devotions*. Durham, NC: Duke University Press, 1998.

Ravid, David. "The Religious, Economic and Social Background and Context of the Establishment of the Ghetto of Venice." In *Gli Ebrei a Venezia*, edited by G. Cozzi, 211–59. Milan: Edizioni di communità, 1987.

Reulin, Dominique. *Contredicts aux 'Erreurs populaires' de Laurent Joubert*. Montauban, 1580.

Rey, Roselyne. *History of Pain*. Paris: La Découverte, 1993.

Rhodes, Neil, and Jonathan Sawday. *The Renaissance Computer: Knowledge Technology in the First Age of Print*. London: Routledge, 2000.

Riis, Jacob A. *How the Other Half Lives: Studies among the Tenements of New York*. New York: Charles Scribner's Sons, 1890. http://www.yale.edu/amstud/inforev/riis/title.html.

Riolan, Jean. *Discours sur les hermaphrodits*. Paris: Ramier, 1614.

Roberts, K. B. and J.D.W. Tomlinson. *The Fabric of the Body: European Traditions of Anatomical Illustrations*. Oxford: Clarendon, 1992.

Roberts, Penny. "Contesting Sacred Space: Burial Disputes in Sixteenth-Century France." In *The Place of the Dead: Death and Remembrance in Late Medieval and*

Early Modern Europe, edited by B. Gordon and P. Marshall, 131–48. Cambridge: Cambridge University Press, 2000.

Roesslin, Eucharius. *The Birth of Mankinde, otherwise named the Womans Booke. Set forth in English by Thomas Raynalde Phisition, and by him corrected, and augmented.* London: Thomas Adams, 1604.

Rogers, Mary. "The Artist as Beauty." In *Concepts of Beauty in Renaissance Art*, edited by F. Ames-Lewis and M. Rogers, 93–106. Aldershot, UK: Ashgate, 1998.

Rogers, Mary. "The Decorum of Woman's Beauty: Trissino, Firenzuola, Luigini and the Representation of Woman in Sixteenth-Century Painting." *Renaissance Studies* 2 (1988): 47–88.

Rogers, Mary. "Reading the Female Body in Venetian Renaissance Art." In *New Interpretations of Venetian Renaissance Painting*, edited by F. Ames-Lewis, 77–90. London: Birkbeck College, 1994.

Romeo, R. *Le scoperte americane nella coscienza italiana del Cinquecento.* Milan: Ricciardi, 1971.

Root-Bernstein, Robert, and Michelle Root-Bernstein. *Honey, Mud, Maggots, and Other Medical Marvels: The Science behind Folk Remedies and Old Wives' Tales.* New York: Houghton Mifflin, 1997.

Rosenthal, Margaret. *The Honest Courtesan: Veronica Franco, Citizen and Writer in Sixteenth-Century Venice.* Chicago: University of Chicago Press, 1992.

Roskill, Mark W. *Dolce's "Aretino" and Venetian Art Theory of the Cinquecento.* Toronto: University of Toronto Press, 2000.

Ross, J. B. "The Middle-Class Child in Urban Italy, Fourteenth to Early Sixteenth Century." In *The History of Childhood*, edited by L. DeMause, 183–228. New York: Psychohistory Press, 1974.

Rossiaud, Jacques. *Medieval Prostitution: Family, Sexuality, and Social Relations in Past Times.* Translated by L. G. Cochrane. New York: Blackwell, 1988.

Rösslin, Eucharius. *The birth of mankynde, otherwyse named the Woman's Booke. Newly set forth, corrected, and augmented, By Thomas Raynalde Phisition.* (London: Thomas Adams, 1604), 188 (originally published in 1512).

Rösslin, Eucharius. *When Midwifery Became the Male Physician's Province: The Sixteenth Century Handbook, The Rose Garden for Pregnant Women and Midwives, Newly Englished.* Translated by Wendy Arons. Jefferson, NC: McFarland, 1994.

Roth, Cecil. *The Jews in the Renaissance.* Philadelphia: Jewish Publication Society of America, 1959.

Rousset, François. *Traitté nouveau de l'hysterotomotokie, ou enfantement caesarien.* Paris: D. Duval, 1581.

Rubin, Miri. *Gentile Tales: The Narrative Assault on Late Medieval Jews.* New Haven, CT: Yale University Press, 1999.

Rubin, Patricia. "'What Men Saw': Vasari's Life of Leonardo da Vinci and the Image of the Renaissance Artist." *Art History* 13 (1990): 33–45.

Rubin, Patricia Lee. *Giorgio Vasari. Art and History.* New Haven, CT: Yale University Press, 1995.

Rubin, Patricia Lee, and G. Ciappelli, eds. *Art, Memory, and Family in Renaissance Florence.* Cambridge: Cambridge University Press, 2000.

Rublack, Ulinka. "Fluxes: The Early Modern Body and the Emotions." Translated by P. Selwyn. *History Workshop Journal* 53 (2002): 1–16.

Ruderman, David B. *The World of a Renaissance Jew: The Life and Thought of Abraham Ben Mordecai Farissol.* Cincinnati: Hebrew Union College Press, 1981.

Rueff, J. *Ein schön lustig Trostbüchle.* Zurich: C. Froschouer, 1554.

Ruggiero, Guido. *The Boundaries of Eros: Sex Crimes & Sexuality in Renaissance Venice.* Oxford: Oxford University Press, 1985.

Russell, Paul A. "Syphilis, God's Scourge or Nature's Vengeance? The German Printed Response to a Public Problem in the Early Sixteenth Century." *Archiv fur Reformationsgeschichte* 80 (1989): 286–307.

Santore, Cathy. "Julia Lombardo, Somtusoa Meretrize: A Portrait by Property." *Renaissance Quarterly* 41 (1988): 44–83.

Savonarola, Michele. *Il trattato ginecologico-pediatrico in volgare; ad mulieres ferrarienses de regimine pregnantium et noviter natorum usque ad septennium.* Edited by L. Belloni. Milan: Società italiana ostetricia e ginecologia, 1952.

Savonarola, Michele. *Libro della natura et virtu delle cose, che nutriscono, & delle cose non naturali, Con alcune osservationi per conservar la sanità, & alcuni quesiti bellissimi da notare.* Venice: Domenico & Gio. Battista Guerra, 1576.

Sawday, Jonathan. *The Body Emblazoned: Dissection and the Human Body in Renaissance Culture.* London: Routledge, 1996.

Scarry, Elaine. *The Body in Pain.* Oxford: Oxford University Press, 1995.

Schäfer, Daniel. "Medical Representations of Old Age in the Renaissance: The Influence of Non-Medical Texts." In *Growing Old in Early Modern Europe: Cultural Representations,* edited by Erin Campbell, 12–19. Aldershot, UK: Ashgate, 2006.

Schiesari, J. "The Face of Domestication, Physiognomy, Gender Politics, and Humanism's Others." In *Women, "Race" and Writing in the Early Modern Period,* edited by M. Hendricks and P. Parker, 55–70. London: Routledge, 1994.

Schleiner, Winfried. "Infection and Cure through Women: Renaissance Constructions of Syphilis." *Journal of Medieval and Renaissance Studies* 24, no. 3 (1994): 499–517.

Schleiner, Winfried. "Moral Attitudes toward Syphilis and Its Prevention in the Renaissance." *Bulletin of the History of Medicine* 68, no. 3 (1994): 389–410.

Schoeck, R. J. *Erasmus of Europe: The Prince of Humanists 1501–1536.* Edinburgh: Edinburgh University Press, 1993.

Schoenfeldt, Michael C. *Bodies and Selves in Early Modern England: Physiology and Inwardness in Spenser, Shakespeare, Herbert, and Milton.* Cambridge: Cambridge University Press, 1999.

Schoenfeldt, Michael C. "Fables of the Belly in Early Modern England." In *The Body in Parts: Fantasies of Corporeality in Early Modern Europe,* edited by D. Hillman and C. Mazzio, 243–62. New York: Routledge, 1997.

Schofield, Roger. "Did the Mothers Really Die? Three Centuries of Maternal Mortality in the World We Have Lost." In *The World We Have Gained,* edited by L. Bonfield, R. M. Smith, K. Wrightson, and P. Laslett, 231–60. Oxford: Blackwell, 1986.

Schutte, Anne Jacobson. "Suffering from the Stone: The Accounts of Michel de Montaigne and Cecilia Ferrazzi." *Bibliotheque d'Humanisme et Renaissance* 64 (2002): 21–36.

Sennet, Richard. *Flesh and Stone: The Body and the City in Western Civilization.* London: Penguin, 2002.

Seznec, Jean. *La Survivance des dieux antiques,* Studies of the Warburg Institute 11. London: Warburg Institute, 1940.

Shagan, Ethan. "The Anatomy of Opposition in Early Reformation England: The Case of Elizabeth Barton, The Holy Maid of Kent." In *Popular Politics in the English Reformation*, 61–88. Cambridge: Cambridge University Press, 2003.

Shakespeare, *Macbeth*. Edited by R. Gill. Oxford: Oxford University Press, 2004.

Shapin, Steven. *The Scientific Revolution*. Chicago: University of Chicago Press, 1996.

Shea, William R., and Mariano Artigas. *Galileo in Rome: The Rise and Fall of a Troublesome Genius*. Oxford: Oxford University Press, 2003.

Shearman, John. "*Maniera* as an Aesthetic Ideal." In *The Renaissance and Mannerism: Studies in Western Art. Acts of the XXth International Congress of the History of Art*, 200–21. Princeton, NJ: Princeton University Press, 1963.

Shilling, Chris. *The Body and Social Theory*. London: Sage, 2003.

Shulvass, Moses A. *The Jews in the World of the Renaissance*. Leiden, Netherlands: Brill, 1973.

Sigal, Pierre-André. "La Grossesse, l'accouchement et l'attitude envers l'enfant mort-né à la fin du moyen âge d'après les récits de miracles." In *Santé, médecine et assistance au moyen âge*, 23–41. Actes du 110e congrès national des sociétés savantes, Montpellier, 1985.

Siraisi, Nancy G. *Medicine and the Italian Universities, 1250–1600*. Leiden, Netherlands: Brill, 2001.

Siraisi, Nancy G. "Oratory and Rhetoric in Renaissance Medicine." *Journal of the History of Ideas* 65, no. 2 (2004): 191–211.

Slack, Paul. *The Impact of Plague in Tudor and Stuart England*. Oxford: Oxford University Press, 1991.

Slotkin, J. S., ed. *Readings in Early Anthropology*. London: Methuen, 1965.

Smellie, William. *A Treatise on the Theory and Practice of Midwifery*. London: D. Wilson, 1752.

Smith, Bruce R. *The Acoustic World of Early Modern England: Attending to the O-Factor*. Chicago: University of Chicago Press, 1999.

Smith, Bruce R. "Tuning Into London, c. 1600." In *The Auditory Culture Reader*, edited by M. Bull and L. Black, 127–37. Oxford: Berg, 2003.

Smith, Robert. "In Search of Carpaccio's African Gondolier." *Italian Studies* 34 (1979): 45–59.

Spencer, John R. "Ut Rhetorica Pictura. A Study in Quatrrocento Theory of Painting." *Journal of the Warburg and Courtauld Institutes* 20 (1957): 26–44.

Spenser, Edmund. "A Letter to the Right Noble, and Valorous, Sir Walter Raleigh Knight." In *The Faerie Queene*, edited by T. P. Roche. New Haven, CT: Yale University Press.

Spicer, Andrew. "Defyle not Christ's kirk with your carrion: Burial and the Development of Burial Aisles in Post-Reformation Scotland." In *The Place of the Dead: Death and Remembrance in Late Medieval and Early Modern Europe*, edited by B. Gordon and P. Marshall, 149–69. Cambridge: Cambridge University Press, 2000.

Spinks, Jennifer. "Jakob Rueff's 1554 *Trostbüchle*: A Zurich Physician Explains and Interprets Monstrous Births." *Intellectual History Review* 18, no. 1 (2008): 41–59.

Spinks, Jennifer. "Wondrous Monsters: Representing Conjoined Twins in Early Sixteenth-Century German Broadsheets." *Parergon* 22, no. 2 (2005): 77–112.

Solingen, C. *Manuale Operatien Der Chirurgie*. Amsterdam, 1684.

Sontag, Susan. *Illness as Metaphor*. New York: Farrar, Strauss, and Giroux, 1977.

Stallybrass, Peter. "Patriarchal Territories: The Body Enclosed." In *Rewriting the Renaissance: The Discourses of Sexual Difference in Early Modern Europe*, edited by Margaret W. Ferguson, Maureen Quilligan, and Nancy J. Vickers. Chicago: University of Chicago Press, 1986.

Starobinski, J. *Montaigne in Motion*. Translated by A. Goldhammer. Chicago: University of Chicago Press, 1985.

Steinberg, Leo. *The Sexuality of Christ in Renaissance Art and in Modern Oblivion*. 2nd ed. Chicago: University of Chicago Press, 1996.

Stephens, Elizabeth. "Inventing the Bodily Interior: Ecorché Figures in Early Modern Anatomy and von Hagens' Body Worlds." *Social Semiotics* 17, no. 3 (2007): 313–26.

Stephens, Walter. *Demon Lovers: Witchcraft, Sex, and the Crisis of Belief*. Chicago: Chicago University Press, 2003.

Stewart, Alan. *Close Readers: Humanism and Sodomy in Early Modern England*. Princeton, NJ: Princeton University Press, 1997.

Stone, Lawrence J. *The Family, Sex and Marriage in England, 1500–1800*. New York: Harper & Row, 1977.

Stow, John. *The annales of England faithfully collected*. 2nd ed. London: Ralfe Newbery, 1592.

Stow, Kenneth R. *Theater of Acculturation: The Roman Ghetto in the Sixteenth Century*. Seattle: University of Washington Press, 2001.

Stubbs, Philip. *Anatomy of Abuses*. London: John Kingston or Richard Iones, 1583.

Summers, David. "*ARIA II*: The Union of Image and Artist as an Aesthetic Ideal in Renaissance Art." *Artibus et historiae* 20 (1989): 15–31.

Summers, David. "Contrapposto: Style and Meaning in Renaissance Art." *Art Bulletin* 59 (1977): 336–61.

Summers, David. *The Judgement of Sense: Renaissance Naturalism and the Rise of Aesthetics*. Cambridge: Cambridge University Press, 1994.

Summers, David. *Michelangelo and the Language of Art*. Princeton, NJ: Princeton University Press, 1981.

Symonds, Deborah A. *Weep Not for Me: Women, Ballads, and Infanticide in Early Modern Scotland*. University Park: Pennsylvania State University Press, 1997.

Tanner, John. *The Hidden Treasures of the Art of Physick*. London, 1659.

Tanner, John. *The Temperate Man, Or the Right Way of Preserving Health in Three Treastises*. London, 1678.

Tasso, T. *Rime*. Edited by B. Basile. Rome: Salerno Editrice, 1994.

Tatar, Maria. *Off with Their Heads! Fairy Tales and the Culture of Childhood*. Princeton, NJ: Princeton University Press, 1992.

Taylor, B. "The Hand of St James." *Berkshire Archaeological Journal* 75 (1994–1997): 97–102.

Taylor, James Spottiswoode. *Montaigne and Medicine*. New York: Paul B. Hoeber, 1921.

Tebb, William. *Premature Burial and How It May Be Prevented*. London: Swan Sonnenschein, 1905.

Temkin, Owsei. *Galenism: Rise and Fall of a Medical Philosophy*. Ithaca, NY: Cornell University Press, 1973.

Temkin, Owsei. *Hippocrates in an Age of Pagans and Christians*. Baltimore: Johns Hopkins University Press, 1991.

Teresa of Avila, Saint. *The Life of Saint Teresa of Avila by Herself*. Translated by J.M. Cohen. New York: Viking Penguin, 1957.

Terpstra, Nicholas. *Abandoned Children of the Italian Renaissance: Orphan Care in Florence and Bologna*. Baltimore: Johns Hopkins University Press, 2005.

Thomas, Keith. *Religion and the Decline of Magic*. London: Weidenfeld and Nicolson, 1971.

Thompson, Craig R. "Penny Pinching." In *The Colloquies of Erasmus*, translated by Craig R. Thompson. Chicago: University of Chicago Press, 1965.

Thomson, George. *The Direct Method of Curing Chymically*. London, 1675.

Titian [Tiziano Vecellio]. *Le lettere*. Edited by C. Fabbro. Belluno, Italy: Comunità di Cadore, 1997.

Tolnay, Charles de. *The Tomb of Julius II*. Princeton, NJ: Princeton University Press, 1970.

Tomalin, Claire. *Samuel Pepys: The Unequalled Self*. New York: Penguin Viking, 2002.

Trexler, Richard C. *Power and Dependence in Renaissance Florence*. Vol. 1, *The Children of Florence*. Binghamton, NY: Medieval & Renaissance Texts & Studies, 1993.

Turner, Robert. *Botonologia*. London: R. Wood, 1664.

Urquhart, W. P. *Life and Times of Francesco Sforza, Duke of Milan, with a Preliminary Sketch of the History of Italy*. Edinburgh: W. Blackwood and Sons, 1852.

Vaccaro, Mary. "Beauty and Identity in Parmigianino's Portraits." In *Fashioning Identities in Renaissance Art*, edited by M. Rogers, 107–18. Aldershot, UK: Ashgate, 2000.

Vaccaro, Mary. "Resplendent Vessels: Parmigianino at Work in the Steccata." In *Concepts of Beauty in Renaissance Art*, edited by F. Ames-Lewis and M. Rogers, 134–46. Aldershot, UK: Ashgate, 1998.

Vallambert, Simon de. *Cinq Livres, de la maniere de nourrir et gouverner les enfans*. Edited by Colette H. Winn. Geneva, Switzerland: Droz, 2005.

Vasari, Giorgio. "Le vite de' più eccellenti pittori, scultori e architettori." In *Le opere di Giorgio Vasari con nuove annotazioni e commenti di Gaetano Milanesi*, vols. 1–7. Florence: Le Lettere, 1998.

Vaughan, William. *Approved Directions for Health*. London, 1612.

Vecellio, C. *Habiti antichi et moderni di tutto il mondo*. Venice: G. B. Sessa, 1598.

Venier, Lorenzo. *La Puttana Errante*. Milan: Unicopli, 2005.

Vesalius, Andreas. *De corporis humanii fabrica*. Basileae, 1543.

Viardel, Cosme. *Observations sur la pratique des accouchemens naturels, contre nature & monstrueux*. Paris: Edme Couterot, 1671.

Vico, Giambattista. *The New Science of Giambattista Vico*. Edited by T. G. Bergin and M. H. Fisch. Ithaca, NY: Cornell University Press, 1970.

Vigarello, Georges. *Concepts of Cleanliness: Changing Attitudes in France since the Middle Ages*. Translated by J. Birrell. Cambridge: Cambridge University Press, 1988.

Vigarello, Georges. "The Upward Training of the Body from the Age of Chivalry to Courtly Civility." In *Fragments for a History of the Human Body, Part Two*, 149–99. New York: Zone, 1989.

Villon, François. *Selected Poems*. Edited and translated by P. Dale. Harmondsworth, UK: Penguin, 1978.

Vitkus, D. J. "Early Modern Orientalism: Representation of Islam in Sixteenth- and Seventeenth-Century Europe." In *Western Views of Islam in Medieval and Early*

Modern Europe, edited by D.R. Blanks and M. Frassetto, 207–30. New York: St. Martin's Press, 1999.

Vitruvius. *The Ten Books on Architecture*. Edited and translated by M.H. Morgan. New York: Dover, 1960.

Vives, J.L. *The Education of a Christian Woman: A Sixteenth-Century Manual*. Translated and edited by C. Fantazzi. Chicago: University of Chicago Press, 2000.

Vives, Juan Luis. *A Very Frutefull and Pleasant Boke called the Instruction[n] of a Christen Woma[n]*, translated by R. Hyrd. London, 1529.

Vollendorf, Lisa. *The Lives of Women: A New History of Inquisitorial Spain*. Nashville, TN: Vanderbilt University Press, 2005.

Wacquet, Françoise. *Latin or the Empire of a Sign*. Translated by John Howe. London: Verso, 2001.

Warnicke, R.M. *The Rise and Fall of Anne Boleyn: Family Politics at the Court of Henry VIII*. New York: Cambridge University Press, 1989.

Watt, Diane. *Secretaries of God: Women Prophets in Late Medieval and Early Modern England*. Cambridge: D.S. Brewer, 1997.

Wear, Andrew. *Knowledge and Practice in English Medicine, 1550–1680*. Cambridge: Cambridge University Press, 2000.

Wear, Andrew, Roger French, and Iain M. Lonie, eds. *The Medical Renaissance of the Sixteenth Century*. Cambridge: Cambridge University Press, 1985.

Webster, Charles. "Alchemical and Paracelsian Medicine." In *Health, Medicine and Mortality in the Sixteenth Century*, edited by C. Webster. Cambridge: Cambridge University Press, 1979.

Weir, Alison. *The Children of Henry VIII*. New York: Ballantine Books, 1997.

Weiss, R., N. George, and P. O'Reilly, eds. *Comprehensive Urology*. London: Mosby, 2001.

Weyer, Johann. *De praestigiis daemonum*. Frankfurt, Germany: Nicolaus Basseum, 1586.

Whatmore, L.E. "The Sermon against the Holy Maid of Kent and Her Adherents, Delivered at Paul's Cross, November the 23rd, 1533, and at Canterbury, December the 7th." *The English Historical Review* 58, no. 232 (1943): 463–75.

Wilby, Emma. *Cunning Folk and Familiar Spirits: Shamanistic Visionary Traditions in Early Modern British Witchcraft and Magic*. Brighton, UK: Sussex Academic Press, 2006.

Williams, Gerhild Scholz, and Charles D. Gunnoe Jr., eds. *Paracelsan Moments: Science, Medicine, and Astrology in Early Modern Europe*. Kirksville, OW: Truman State University Press, 2002.

Willughby, Percivall. *Observations on Midwifery*. Edited by Henry Blenkinsop. Wakefield: S.R. Publishers, 1972; orig. manuscript 1863.

Wilson, Adrian. "The Ceremony of Childbirth and Its Interpretation." In *Women as Mothers in Pre-Industrial England*, edited by Valerie Fildes, 68–107. London: Routledge, 1990.

Wilson, Adrian. *The Making of Man-Midwifery: Childbirth in England 1660–1770*. London: UCL Press, 1995.

Wilson, Adrian. "The Perils of Early Modern Procreation: Childbirth with or without Fear?" *British Journal for Eighteenth-Century Studies* 16 (Spring 1993): 1–19.

Wilson, Bronwen. *The World in Venice: Print, the City, and Early Modern Identity*. Toronto: Toronto University Press, 2005.

Wilson, Dudley. *Signs and Portents: Monstrous Births from the Middle Ages to the Enlightenment*. London: Routledge, 1993.

Wilson, Fred. *Speak of Me as I Am: The United States Pavilion 50th International Exhibition of Art, The Venice Biennale 2003*. Cambridge, MA: List Visual Arts Center, MIT, 2003.

Wisch, Barbara. "Vested Interest: Redressing Jews on Michelangelo's Sistine Ceiling." *Artibus et Historiae* 24 (2003): 143–72.

Wittkower, Rudolf. *Architectural Principles in the Age of Humanism*. London: Tiranti, 1962.

Wittkower, Rudolf. *Architectural Principles in the Age of Humanism*. London: Academy Editions, 1977.

Woolgar, C. M. *The Senses in Late Medieval England*. New Haven, CT: Yale University Press, 2006.

Wright, P., and A. Treacher, eds. *The Problem of Medical Knowledge*. Edinburgh: Edinburgh University Press, 1982.

Wunderli, R., and G. Broce. "The Final Moment before Death in Early Modern England." *Sixteenth Century Journal* 20, no. 2 (Summer 1989): 259–75.

Yates, Frances A. *Giordano Bruno and the Hermetic Tradition*. Chicago: University of Chicago Press, 1964.

Yates, Frances A. *The Rosicrucian Enlightenment*. London: Routledge and Kegan Paul, 1972.

Young, R.J.C. *Postcolonialism: An Historical Introduction*. Malden, MA: Blackwell, 2001.

CONTRIBUTORS

Patrizia Bettella teaches in the department of Modern Languages and Cultural Studies at the University of Alberta (Canada), where she is the coordinator of the Italian Language Program. She has served as vice president of the Canadian Society for Italian Studies (CSIS) and is currently review editor for *Quaderni d'italianistica,* official journal of CSIS. Her publications include articles on Italian cinema and literature of post-unification Italy. She published various essays on Antipetrarchan poetry and on female literary beauty in treatises of the Italian Renaissance. She is the author of *The Ugly Woman: Transgressive Aesthetic Models in Italian Poetry from the Middle Ages to the Baroque* (Toronto University Press, 2005). Her latest project deals with literary texts against courtesans in Early Modern Italy.

Susan Broomhall is a Winthrop Professor in Early Modern History at The University of Western Australia. Her research focuses on the experiences of women, and ideas of gender and masculinity, in early modern Europe, especially in France, Low Countries, England, and Scotland. Her most recent monographs examine the presentation of early modern women in the Low Countries and the experiences of the poor in sixteenth-century France.

Katherine Crawford is associate professor of history at Vanderbilt University. She is interested in the ways that gender informs sexual practice, ideology, and identity, both in normative and nonnormative formations. Among her ongoing research are projects exploring the presumptions about corporeal color as a product of gender and recuperating the history of pleasure. She has just completed her book *The Sexual Culture of the French Renaissance* (Cambridge University Press, 2010).

Margaret Healy is a reader in English and director of the Centre for Early Modern Studies at the University of Sussex. She has published extensively on the interfaces among early modern literature, art, and medicine and is an editor of the British medical journal, *Medical Humanities*. Healy has coedited a volume of essays titled *Renaissance Transformations: The Making of English Writing 1500–1650* (Edinburgh University Press, 2009) and is the author of *Fictions of Disease in Early Modern England: Bodies, Plagues and Politics* (Palgrave, 2001) and *Richard II* (Northcote House, 1998). She was awarded a Leverhulme Fellowship in 2008 to research her current monograph project, *Shakespeare, Alchemy and the Creative Imagination: The "Sonnets" and "A Lovers Complaint."*

Margaret L. King studies and teaches in the fields of the Italian Renaissance (and of Venice, in particular), humanism, and the classical tradition; the social and cultural history of early modern Europe; women and learning 1300–1800; and the history of childhood from antiquity to the present. Her current projects include continued work as series coeditor of the Other Voice in Early Modern Europe series, and a new editorship of the Oxford University Press online bibliography in Renaissance and Reformation. Above all, she is engaged in writing her book on mothers and sons across time and space, and looks forward to completing it in 2010–2011.

Lianne McTavish is professor in the history of art, design, and visual culture at the University of Alberta, Canada, where she offers courses in early modern visual culture and critical museum theory. Her interdisciplinary research, funded by the Social Sciences and Humanities Research Council of Canada, has centered on early modern French medical imagery, including articles in *Social History of Medicine* (2001), *Medical History* (2006), *Journal of Medical Humanities* (2010), and a monograph, *Childbirth and the Display of Authority in Early Modern France* (2005). Her recent work in this area analyzes representations of cure and convalescence in France, 1600–1800. She has also published widely on the history and theory of museums, and has just completed a book manuscript titled, "Between Museums: Exchanging Objects, Values and Identities, 1842–1950."

Diane Purkiss is fellow in English and director of studies at Keble College Oxford. Her books include *The Witch in History* (Routledge, 1996), *At the Bottom of the Garden* (NYU Press, 2001), and most recently *The English Civil War: Papists, Gentlewomen, Soldiers, and Witchfinders in the Birth of Modern Britain* (Basic Books; Reprint edition 2007). She is now working on a book on reading food in history, and on the dissolution of the monasteries.

Karen Raber is professor of English at the University of Mississippi; she is the author of *Dramatic Difference: Gender, Class and Genre in the Early Modern Closet Drama* (2001); coeditor with Ivo Kamps of *William Shakespeare's* Measure for Measure: *Texts and Contexts* (2004); coeditor with Treva J. Tucker of *The Culture of the Horse: Status, Discipline and Identity in the Early Modern World* (2005); and coeditor with Ivo Kamps and Tom Hallock of *Early Modern Ecostudies: From Shakespeare to the Florentine Codex* (2008). She is also the author of numerous articles on gender, women writers, animals, and ecocriticism, and she is currently working on a monograph on Renaissance animal bodies.

Mary Rogers has been lecturer in the history of art at the Universities of Bristol and Warwick, specializing in the art of Renaissance Europe, especially Venice. Her main research interests have been in the representation of women in Italian Renaissance art in relation to the literature and theoretical writings of the time. She has published in journals such as *Renaissance Studies* and *Word and Image* and in collections of essays, most recently in *Titian: Materiality, Likeness,* Istoria (Brepols, 2008) and *A Cultural History of Woman: The Renaissance* (Berg, forthcoming). She has edited or coedited *Concepts of Beauty in Renaissance Art* (Ashgate, 1998) and *Fashioning Identities in Renaissance Art* (Ashgate, 2000), and is the coauthor of the sourcebooks *Women in Italy, 1350–1650* (Manchester University Press, 2005) and the forthcoming *Women and the Arts in Italy 1350–1650*. She now lives in London as an independent researcher.

INDEX